Springer Series on
Atoms+Plasmas   12

Editor: J. Peter Toennies

Springer Series on
Atoms+Plasmas

Editors: G. Ecker   P. Lambropoulos   I. Sobel'man   H. Walther

Managing Editor: H. K. V. Lotsch

- Volume 1   **Polarized Electrons**   2nd Edition
  By J. Kessler

- Volume 2   **Multiphoton Processes**
  Editors: P. Lambropoulos and S. J. Smith

- Volume 3   **Atomic Many-Body Theory**   2nd Edition
  By I. Lindgren and J. Morrison

- Volume 4   **Elementary Processes in Hydrogen-Helium Plasmas**
  Cross Sections and Reaction Rate Coefficients
  By R. K. Janev, W. D. Langer, K. Evans, Jr. and D. E. Post, Jr.

- Volume 5   **Pulsed Electrical Discharge in Vacuum**
  By G. A. Mesyats and D. I. Proskurovsky

- Volume 6   **Atomic and Molecular Spectroscopy**
  Basic Aspects and Practical Applications
  By S. Svanberg

- Volume 7   **Interference of Atomic States**
  By E. B. Alexandrov, M. P. Chaika and G. I. Khvostenko

- Volume 8   **Plasma Physics**   Basic Theory with Fusion Applications
  By K. Nishikawa and M. Wakatani

- Volume 9   **Plasma Spectroscopy**
  The Influence of Microwave and Laser Fields
  By E. A. Oks

- Volume 10   **Film Deposition by Plasma Techniques**
  By M. Konuma

- Volume 11   **Resonance Phenomena in Electron-Atom Collisions**
  By V. I. Lengyel, V. T. Navrotsky and E. P. Sabad

- Volume 12   **Atomic Spectra and Radiative Transitions**   2nd Edition
  By I. I. Sobelman

Igor I. Sobelman

Atomic Spectra and Radiative Transitions

Second Edition
With 26 Figures

Springer-Verlag
Berlin Heidelberg New York
London Paris Tokyo
Hong Kong Barcelona
Budapest

Professor Dr. Igor I. Sobelman
Academy of Sciences, P.N. Lebedev Physical Institute,
Leninsky Prospect 53, 117924 Moscow, Russia

Guest Editor:
Professor Dr. J. Peter Toennies
Max-Planck-Institut für Strömungsforschung, Bottingerstrasse 6–8,
W-3400 Göttingen, Fed. Rep. of Germany

Series Editors:
Professor Dr. Günter Ecker
Ruhr-Universität Bochum, Institut für Theoretische Physik, Lehrstuhl I, Universitätsstrasse 150,
W-4630 Bochum-Querenburg, Fed. Rep. of Germany

Professor Peter Lambropoulos, Ph. D.
University of Crete, P.O. Box 470, Iraklion, Crete, Greece, and
Department of Physics, University of Southern California, University Park,
Los Angeles, CA 90089-0484, USA

Professor Dr. Igor I. Sobelman
Academy of Sciences, P.N. Lebedev Physical Institute,
Leninsky Prospect 53, 117924 Moscow, Russia

Professor Dr. Herbert Walther
Sektion Physik der Universität München, Am Coulombwall 1,
W-8046 Garching/München, Fed. Rep. of Germany

Managing Editor: Dr. Helmut K.V. Lotsch
Springer-Verlag, Tiergartenstrasse 17, W-6900 Heidelberg, Fed. Rep. of Germany

The first edition appeared as
Springer Series in Chemical Physics, Vol. 1

ISBN 3-540-54518-2 Springer-Verlag Berlin Heidelberg New York
ISBN 0-387-54518-2 Springer-Verlag New York Berlin Heidelberg

Library of Congress Cataloging-in-Publication Data.
Sobelman, I. I. (Igor' Il'ich), 1927– . Atomic spectra and radiative transitions / Igor I. Sobelman. p. cm. – (Springer series on atoms + plasmas ; 12) Includes bibliographical references and index. ISBN 0-387-54518-2 (U.S.) 1. Atomic spectroscopy. 2. Radiative transitions. I. Title. II. Series. QC454.A8S62 1992 539.7'0287–dc20 91-41753

This work is subject to copyright. All rights are reserved, whether the whole or part of the material is concerned, specifically the rights of translation, reprinting, reuse of illustrations, recitation, broadcasting, reproduction on microfilm or in any other way, and storage in data banks. Duplication of this publication or parts thereof is permitted only under the provisions of the German Copyright Law of September 9, 1965, in its current version, and permission for use must always be obtained from Springer-Verlag. Violations are liable for prosecution under the German Copyright Law.

© Springer-Verlag Berlin Heidelberg 1979, 1992
Printed in the United States of America

The use of general descriptive names, registered names, trademarks, etc. in this publication does not imply, even in the absence of a specific statement, that such names are exempt from the relevant protective laws and regulations and therefore free for general use.

54/3140 - 5 4 3 2 1 0 – Printed on acid-free paper

Preface to the Second Edition

Recent years have witnessed increased activity in the investigation of the spectra of multiply charged ions. These ions are of special interest to many modern fields of research, such as X-ray space astronomy and astrophysics, controlled thermonuclear fusion, extreme ultraviolet and soft X-ray lasers. Studies of the structure of highly charged heavy ions are stimulated by the unique possibilities provided by modern accelerators and beam-foil spectroscopic techniques.

This edition of the book includes two new chapters. In Chap. 10, basic information on the Dirac equation for an electron in a Coulomb field is given, together with the treatment of relativistic corrections in the theory of atomic spectra. The reader will gain an understanding of the basic concepts used in the relativistic theory of highly charged ions.

Chapter 11 is devoted to an overview of the spectra of multiply charged ions. No attempt has been made here to reference the vast number of experimental investigations and calculations reported in the literature. Instead, attention is paid to the main features of the spectra of highly charged ions that distinguish them from the spectra of neutral atoms. Among them are the so-called satellite structure and forbidden lines, which appear in the spectra due to violation of the selection rules for radiative transitions by relativistic effects.

The author is very grateful to Dr. Helmut Lotsch. Without his efforts and encouragement this revised edition would not have been realized.

Moscow, April 1991 *I. I. Sobelman*

Preface to the First Edition

My previous book on the theory of atomic spectra was published in Russian about fifteen years ago. Besides the traditional problems usually included in a book on atomic spectroscopy, some other problems arising in various applications of spectroscopic methods were also discussed in the book. These include, for example, continuous spectrum radiation, excitation of atoms, and spectral line broadening. Extensive revisions were made in the English version of the book published by Pergamon Press in 1972, especially in the chapter devoted to the problem of excitation of atoms.

This book is intended as the first part of a two-volume presentation of the theory of atomic spectra, atomic radiative transitions, excitation of atoms, and spectral line broadening. The aim in preparing these new books has been to stress the problems connected with the most interesting applications of atomic spectroscopy to plasma diagnostics, astrophysics, laser physics, and other fields, which have been developed very intensively in recent years.

The content of this first volume, devoted to the systematics of atomic spectra and radiative transitions, is similar to that of Chapters 1 – 6, 8 and 9 of the old book, but considerable revision has been made. Some sections, such as those on the Hartree-Fock method, the Dirac equation, and relativistic corrections, have been deleted. At the same time, more attention is paid to radiative transitions. More extensive tables of oscillator strengths, probabilities, and effective cross sections of radiative transitions in discrete and continuous spectra are given.

The book is based on the courses of lectures on atomic spectroscopy and connected problems which the author and L. A. Vainshtein gave at the Moscow Physics and Technology Institute, and reflects the changes in these courses in recent years. As a rule, references are made only to monographs, reviews, and papers whose results are used in the text.

In conclusion, I wish to express sincere thanks to Dr. V. I. Kogan, who prepared Sect. 9.5, and to Dr. E. A. Yukov, who helped me to prepare Sects. 9.6 and 9.7.

Moscow, November 1978 *I. I. Sobelman*

Contents

Part I
Elementary Information on Atomic Spectra

Chapter 1
The Hydrogen Spectrum

1.1 Schrödinger's Equation for the Hydrogen Atom 3
    1.1.1 Energy Levels 3
    1.1.2 Wave Functions 5
1.2 Series Regularities 7
    1.2.1 Radiative Transition Selection Rules 7
    1.2.2 Spectral Series of the Hydrogen Atom 8
    1.2.3 Hydrogenlike Ions 9
1.3 Fine Structure 10
    1.3.1 Velocity Dependence of Electron Mass 10
    1.3.2 Spin-Orbit Interaction 11
    1.3.3 Fine Structure. Selection Rules 12
    1.3.4 Lamb Shift 15

Chapter 2
Systematics of the Spectra of Multielectron Atoms

2.1 Central Field 16
    2.1.1 Central Field Approximation 16
    2.1.2 Parity of States 18
    2.1.3 Systematics of Electron States in a Central Field 19
2.2 Electrostatic and Spin-Orbit Splitting in the LS Coupling
    Approximation 20
    2.2.1 Spectral Terms. LS Quantum Numbers 20
    2.2.2 Fine Structure of Terms 21
    2.2.3 Finding the Terms of Multielectron Configurations 23
    2.2.4 Radiative Transitions 26
2.3 jj Coupling Approximation 27
    2.3.1 Various Coupling Schemes 27
    2.3.2 Systematics of Electron States with jj Coupling 29

Chapter 3
Spectra of Multielectron Atoms

- 3.1 Periodic System of Elements 32
- 3.2 Spectra of the Alkali Elements 34
  - 3.2.1 Term Scheme 34
  - 3.2.2 Series Regularities 37
  - 3.2.3 Fine Structure 37
  - 3.2.4 Copper, Silver, and Gold Spectra 38
- 3.3 Spectra of the Alkaline Earth Elements 39
  - 3.3.1 He Spectrum 39
  - 3.3.2 Spectra of the Alkaline Earth Elements 40
  - 3.3.3 Zinc, Cadmium, and Mercury Spectra 42
- 3.4 Spectra of Elements with p Valence Electrons 42
  - 3.4.1 One p Electron Outside Filled Shells 42
  - 3.4.2 Configuration p^2 43
  - 3.4.3 Configuration p^3 44
  - 3.4.4 Configuration p^4 45
  - 3.4.5 Configuration p^5 46
  - 3.4.6 Configuration p^6 46
- 3.5 Spectra of Elements with Unfilled d and f Shells 48
  - 3.5.1 Elements with Unfilled d Shells 48
  - 3.5.2 Elements with Unfilled f Shells 49

Part II
Theory of Atomic Spectra

Chapter 4
Angular Momenta

- 4.1 Angular Momentum Operator. Addition of Angular Momenta 53
  - 4.1.1 Angular Momentum Operator 53
  - 4.1.2 Orbital Angular Momentum 54
  - 4.1.3 Electron Spin 54
  - 4.1.4 Addition of Two Angular Momenta 55
  - 4.1.5 Addition of Three or More Angular Momenta 57
- 4.2 Angular Momentum Vector Addition Coefficients 60
  - 4.2.1 Clebsch-Gordan and Associated Coefficients 60
  - 4.2.2 Summary of Formulas for $3j$ Symbols 62
  - 4.2.3 Racah W Coefficients and $6j$ Symbols 66
  - 4.2.4 Summary of Formulas for $6j$ Symbols 70
  - 4.2.5 $9j$ Symbols 72
- 4.3 Irreducible Tensor Operators 74

4.3.1	Spherical Tensors	74
4.3.2	Matrix Elements	76
4.3.3	Some Examples of Calculation of Reduced Matrix Elements	78
4.3.4	Tensor Product of Operators	80
4.3.5	Matrix Elements with Coupled Angular Momenta	84
4.3.6	Direct Product of Operators	86

Chapter 5
Systematics of the States of Multielectron Atoms

5.1	Wave Functions	89
	5.1.1 Central Field Approximation	89
	5.1.2 Two-Electron Wave Functions in LSM_LM_S Representation	90
	5.1.3 Two-Electron Wave Functions in $mm'SM_S$ Representation	93
	5.1.4 Multielectron Wave Functions in a Parentage Scheme Approximation	96
	5.1.5 Fractional Parentage Coefficients	96
	5.1.6 Classification of Identical Terms of l^n Configuration According to Seniority (Seniority Number)	98
5.2	Matrix Elements of Symmetric Operators	106
	5.2.1 Statement of the Problem	106
	5.2.2 F Matrix Elements. Parentage Scheme Approximation	108
	5.2.3 F Matrix Elements. Equivalent Electrons	109
	5.2.4 Q Matrix Elements. Parentage Scheme Approximation	111
	5.2.5 Q Matrix Elements. Equivalent Electrons	113
	5.2.6 Summary of Results	115
5.3	Electrostatic Interaction in LS Coupling. Two-Electron Configuration	115
	5.3.1 Coulomb and Exchange Integrals	115
	5.3.2 Configuration Mixing	118
5.4	Electrostatic Interaction in LS Coupling. Multielectron Configuration	120
	5.4.1 Configurations l^n and $l^n l'$	120
	5.4.2 More Than Half Filled Shells	123
	5.4.3 Filled (Closed) Shells	123
	5.4.4 Applicability of the Single-Configuration Approximation	124
5.5	Multiplet Splitting in LS Coupling	126
	5.5.1 Preliminary Remarks	126
	5.5.2 Landé Interval Rule	126
	5.5.3 One Electron Outside Closed Shells	128
	5.5.4 Configuration l^n	130
	5.5.5 Parentage Scheme Approximation	132
	5.5.6 Fine-Structure Splitting of Levels of He	133
	5.5.7 Spin-Spin and Spin-Other Orbit Interactions	139
5.6	jj Coupling	141
	5.6.1 Wave Functions	141
	5.6.2 Spin-Orbit and Electrostatic Interactions	143

5.7 Intermediate Coupling and Other Types of Coupling 144
    5.7.1 Transformations Between LS and jj Coupling Schemes 144
    5.7.2 Intermediate Coupling 147
    5.7.3 jl Coupling 152
    5.7.4 Experimental Date 153
    5.7.5 Other Types of Coupling 154

Chapter 6
Hyperfine Structure of Spectral Lines

6.1 Nuclear Magnetic Dipole and Electric Quadrupole Moments 156
    6.1.1 Magnetic Moments 156
    6.1.2 Quadrupole Moments 157
6.2 Hyperfine Splitting 159
    6.2.1 General Character of the Splitting 159
    6.2.2 Calculation of the Hyperfine Splitting Constant A 162
    6.2.3 Calculation of the Hyperfine Splitting Constant B 168
    6.2.4 Radiative Transitions Between Hyperfine-Structure Components . 170
    6.2.5 Isotope Shift of the Atomic Levels 170

Chapter 7
The Atom in an External Electric Field

7.1 Quadratic Stark Effect 173
7.2 Hydrogenlike Levels. Linear Stark Effect 177
7.3 Inhomogeneous Field. Quadrupole Splitting 181
7.4 Time-Dependent Field 183
    7.4.1 Amplitude Modulation 183
    7.4.2 The Hydrogen Atom in a Rotating Electric Field 187

Chapter 8
The Atom in an External Magnetic Field

8.1 Zeeman Effect 189
8.2 Paschen-Back Effect 194
    8.2.1 Strong Field 194
    8.2.2 Splitting of Hyperfine Structure Components in a Magnetic Field 198

Chapter 9
Radiative Transitions

9.1 Electromagnetic Radiation 200
    9.1.1 Quantization of the Radiation Field 200
    9.1.2 Radiative Transition Probabilities 201
    9.1.3 Correspondence Principle for Spontaneous Emission 202
    9.1.4 Dipole Radiation 203

9.1.5 Stimulated Emission and Absorption....................... 203
9.1.6 Effective Cross Sections of Absorption and Stimulated Emission 205
9.2 Electric Dipole Radiation....................................... 205
9.2.1 Selection Rules, Polarization and Angular Distribution......... 205
9.2.2 Oscillator Strengths and Line Strengths..................... 208
9.2.3 LS Coupling Approximation. Relative Intensities of Multiplet Components.. 211
9.2.4 One Electron Outside Closed Shell.......................... 213
9.2.5 Multielectron Configurations. Different Coupling Schemes...... 214
9.2.6 Relative Intensities of Zeeman and Stark Components of Lines.. 215
9.3 Multipole Radiation... 216
9.3.1 Fields of Electric and Magnetic Multipole Moments............ 216
9.3.2 Intensity of Multipole Radiation............................. 220
9.3.3 Selection Rules.. 222
9.3.4 Electric Multipole Radiation 223
9.3.5 Magnetic Dipole Radiation.................................. 225
9.3.6 Transitions Between Hyperfine Structure Components. Radio Emission from Hydrogen................................... 227
9.4 Calculation of Radiative Transition Probabilities.................... 229
9.4.1 Approximate Methods...................................... 229
9.4.2 Three Ways of Writing Formulas for Transition Probabilities... 230
9.4.3 Theorems for Sums of Oscillator Strengths 232
9.4.4 Semiempirical Methods of Calculating Oscillator Strengths...... 235
9.4.5 Electric Dipole Transition Probabilities in the Coulomb Approximation... 236
9.4.6 Intercombination Transitions................................ 237
9.5 Continuous Spectrum.. 239
9.5.1 Classification of Processes................................... 239
9.5.2 Photorecombination and Photoionization: General Expressions for Effective Cross Sections................................. 239
9.5.3 Bremsstrahlung: General Expressions for Effective Cross Sections... 245
9.5.4 Radiation and Absorption Coefficients....................... 248
9.5.5 Photorecombination and Photoionization: Hydrogenlike Atoms. 251
9.5.6 Photorecombination and Photoionization: Nonhydrogenlike Atoms.. 255
9.5.7 Bremsstrahlung in a Coulomb Field......................... 257
9.6 Formulas for Q Factors... 274
9.6.1 Symmetry and Sum Rules 274
9.6.2 LS Coupling. Allowed Transitions........................... 275
9.6.3 jl Coupling 279
9.7 Tables of Oscillator Strengths and Radiative Transition Probabilities.. 281
9.7.1 Transition Probabilities for the Hydrogen Atom............... 281
9.7.2 Radiative Transition Probabilities in the Bates-Damgaard Approximation... 283

9.7.3 Oscillator Strengths and Probabilities of Some Selected Transitions 284
9.7.4 Effective Cross Sections and Rates of Photorecombination ... 284

Chapter 10
Relativistic Corrections in the Spectroscopy of Multicharged Ions

10.1 Dirac Equation. Pauli Equation 303
    10.1.1 Dirac Equation 303
    10.1.2 Electron Spin 305
    10.1.3 Non-Relativistic Approximation. Pauli Equation 308
10.2 Central Field 310
    10.2.1 Non-Relativistic Approximation 310
    10.2.2 Second Approximation with Respect to v/c. Fine Splitting.... 313
    10.2.3 Dirac Equation for a Central Field 316
    10.2.4 Coulomb Field. Energy Levels, Fine Splitting 319
    10.2.5 Coulomb Field. Radial Functions 322
10.3 Relativistic Corrections 325
    10.3.1 Calculation of Some Radial Integrals 325
    10.3.2 Hyperfine Splitting Constant A 326
    10.3.3 Hyperfine Splitting Constant B 330
    10.3.4 Nucleus Finite-Size Correction 331
    10.3.5 Radiative Corrections. Lamb Shift 333

Chapter 11
Spectra of Multicharged Ions

11.1 Energy Levels 339
11.2 Forbidden Transitions 341
    11.2.1 H-like Ions 342
    11.2.2 He-like Ions 342
11.3 Satellite Structure 346

References 351

List of Symbols 353

Subject Index 355

Part I

Elementary Information on Atomic Spectra

For the convenience of the reader, the main body of the book, Part II (Chaps. 4–11), is prefaced by a summary of elementary information on atomic spectra in Part I. This section includes the description of atomic spectra in various groups of the periodic table, beginning with hydrogen and hydrogenlike spectra, and a discussion of the main physical principles on which the theory of atomic spectra is based. Experimental data on atomic spectra are discussed for the purpose of illustrating the physical meaning and justifying the approximations used in the theory, such as the concept of the self-consistent field and different coupling schemes for angular momenta.

Chapter 1 The Hydrogen Spectrum

The hydrogen atom and its spectrum treated in this chapter are of special interest in atomic spectroscopy because only for the hydrogen atom can the Schrödinger and Dirac equations be solved analytically. So-called hydrogenlike approximations are widely used in the theory of atomic spectra.

1.1 Schrödinger's Equation for the Hydrogen Atom

1.1.1 Energy Levels

The problem of the relative motion of an electron (mass: m, charge: $-e$) and a nucleus (mass: M, charge: Ze) reduces, as is well known, to the problem of the motion of a particle with an effective mass $\mu = mM/(m+M) \simeq m$ in a Coulomb field of $-Ze^2/r$. The Schrödinger equation for a particle in the field $-Ze^2/r$ has the form

$$\left(\frac{\hbar^2}{2\mu}\Delta + E + \frac{Ze^2}{r}\right)\psi = 0. \tag{1.1}$$

The wave function ψ, which is the solution of this equation, describes the stationary state with a definite value of the energy E. In the case of a centrally symmetric field the angular momentum is conserved. Because of that, we shall consider stationary states which are characterized by definite values of the quantities E, the square of the angular momentum, and the z component of the angular momentum. The wave functions ψ of these stationary states are eigenfunctions of the operators l^2 and l_z and must therefore satisfy the equations

$$l^2\psi = l(l+1)\psi, \tag{1.2}$$

$$l_z\psi = m\psi, \tag{1.3}$$

where $l(l+1)$ and m are eigenvalues of the operators l^2 and l_z. We recall that in quantum mechanics the square of the angular momentum can only take a discrete series of values $\hbar^2 l(l+1)$, where $\hbar = h/2\pi$; h is Planck's constant, and also $l = 0,1,2, \ldots$. In exactly the same way, the z component of the momentum can have the values $\hbar m$, $m = 0, \pm 1, \pm 2, \ldots$ with the additional condition $|m| \leq l$.

For brevity, we shall henceforth speak of the angular momentum l and the z component of the angular momentum m, meaning the angular momentum whose square is equal to $\hbar^2 l(l+1)$ and whose z component equals $\hbar m$.

Let us seek the solution of (1.1) in the form

$$\psi = R(r)\, Y_{lm}(\theta, \varphi), \tag{1.4}$$

where $Y_{lm}(\theta, \varphi)$ is the spherical harmonic. The radial part $R(r)$ satisfies the equation

$$\frac{1}{r^2}\frac{d}{dr}\left(r^2 \frac{dR}{dr}\right) - \frac{l(l+1)}{r^2} R + \frac{2\mu}{\hbar^2}\left(E + \frac{Ze^2}{r}\right) R = 0. \tag{1.5}$$

If $E > 0$, this equation has finite and continuous solutions for any value of E and l. If $E < 0$, such solutions are possible only at certain discrete values of energy

$$E = -\frac{1}{2}\frac{Z^2}{n^2}\frac{\mu e^4}{\hbar^2}, \tag{1.6}$$

where n is an integer, and also $n \geq l + 1$. The number n is called the principal quantum number. For a given value of n, the quantum number l can take the values $0, 1, 2, \ldots n-1$. To each value of l there correspond $(2l+1)$ states, differing by the values of the quantum number m, which is usually called the magnetic quantum number. The energy of an atom in the state nlm is uniquely determined by the principal quantum number and does not depend on l or m. Thus, for a particle in a Coulomb field the energy levels are n^2-fold degenerate. There are $n^2 = 1 + 3 + 5 + \ldots + 2n - 1$ states differing in the quantum numbers l and m. The independence of m for the energy has a simple physical meaning. In a central field, all directions in space are equivalent, and therefore the energy cannot depend on the spatial orientation of the angular momentum. The independence of l is a specific property of the Coulomb field and does not occur in the general case of a centrally symmetric field. The energy level diagram of the hydrogen atom corresponding to (1.6) is shown in Fig. 1.1.

In spectroscopy it is usual to denote states corresponding to the values $l = 0, 1, 2, \ldots$ by letters of the Latin alphabet

$$s, p, d, f, g, h, i, k \ldots$$

Thus the state $n = 1$, $l = 0$ is denoted $1s$, the state $n = 2$, $l = 1$ is denoted $2p$, and so on. So the state $1s$ relates to the level $n = 1$, the states $2s$, $2p$ relate to the level $n = 2$, the states $3s$, $3p$, $3d$ relate to the level $n = 3$, and so on.

If we neglect the difference between the reduced mass $\mu \simeq m(1 - m/M)$ and the electron mass m, which is approximately $m/2000$, we obtain $E_n = -(me^4/\hbar^2) Z^2/2n^2$. The quantity $me^4/\hbar^2 = 4.304 \times 10^{-11}$ ergs ($\simeq 27.07$ eV) is taken as the atomic unit of energy. The Rydberg unit of energy $\text{Ry} = me^4/2\hbar^2$ is also used in spectroscopy; hence $E_n = -\text{Ry}Z^2/n^2$.

For ionization of the hydrogen atom, i.e., for the detachment of an electron from the nucleus, it is necessary to impart to the atom the energy $|E_\infty - E_1| = \mu e^4/2\hbar^2$. This quantity is called the ionization energy (or ionization potential

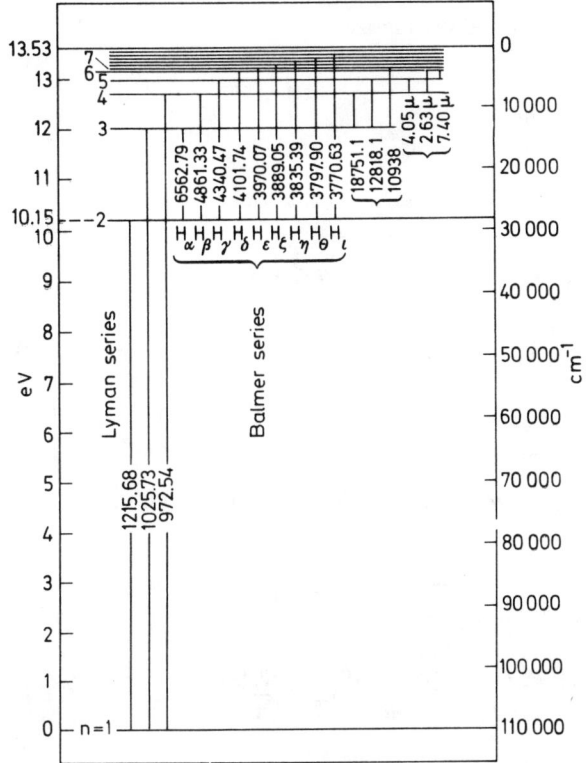

Fig. 1.1. Energy-level diagram for the hydrogen atom

if it is measured in electron volts) and is denoted by E_i. If the difference between μ and m is neglected, $E_i = \text{Ry}$. The level $n = 1$ is called the ground level or ground state. The first excited level, nearest to the ground level, is traditionally called the resonance level. The energy necessary for excitation of the resonance level is called the resonance potential and is denoted E_r. For the hydrogen atom, $E_r = |E_2 - E_1| = 3/4 E_i$. This gives $E_i \simeq 13.53$ eV and $E_r \simeq 10.15$ eV. In atomic spectroscopy, instead of the energy levels E_n, one usually uses the quantities $\sigma_n = |E_n/2\pi\hbar c|$ which are expressed in cm^{-1} as wave numbers. The values of the quantities σ_n for the energy levels of the hydrogen atom are given in Fig. 1.1.

1.1.2 Wave Functions

The angular functions $Y_{lm}(\theta,\varphi)$ can be expressed in terms of the associated Legendre polynomials P_l^m

$$Y_{lm}(\theta, \varphi) = \Theta_{lm}(\theta) \Phi_m(\varphi),$$

$$\Theta_{lm} = (-1)^m \sqrt{\frac{(2l+1)(l-m)!}{2(l+m)!}} P_l^m (\cos \theta), \quad \Phi_m = \frac{e^{im\varphi}}{\sqrt{2\pi}}. \tag{1.7}$$

Here it is assumed that $m \geq 0$. For $m < 0$, $\Theta_{l,-|m|} = (-1)^m \Theta_{l,|m|}$.[1]
The functions Y_{lm} are orthogonal and normalized

$$\int_0^{2\pi} \int_0^{\pi} Y_{l'm'}^* Y_{lm} \sin\theta \, d\theta \, d\varphi = \delta_{ll'} \delta_{mm'}. \tag{1.8}$$

The expressions for the functions Θ_{lm} when $l = 0,1,2$, are

$$\Theta_{00} = \frac{1}{\sqrt{2}}, \quad \Theta_{10} = \sqrt{\frac{3}{2}} \cos\theta, \quad \Theta_{1,\pm 1} = \mp\sqrt{\frac{3}{4}} \sin\theta,$$

$$\Theta_{20} = \sqrt{\frac{5}{2}} \left(\frac{3}{2}\cos^2\theta - \frac{1}{2}\right), \quad \Theta_{2,\pm 1} = \mp\sqrt{\frac{15}{4}} \cos\theta \sin\theta. \tag{1.9}$$

$$\Theta_{2,\pm 2} = \frac{1}{4}\sqrt{15} \sin^2\theta$$

The radial functions for the discrete spectrum are expressed in terms of the generalized Laguerre polynomial

$$L_n^m(x) = (-1)^m \frac{n!}{(n-m)!} e^x x^{-m} \frac{d^{n-m}}{dx^{n-m}} e^{-x} x^n, \tag{1.10}$$

$$R_{nl}(r) = -\sqrt{\frac{(n-l-1)!}{(n+l)!^3 2n}} \left(\frac{2Z}{na_0}\right)^{\frac{3}{2}} e^{-\frac{Zr}{na_0}} \left(\frac{2Zr}{na_0}\right)^l L_{n+l}^{2l+1}\left(\frac{2Zr}{na_0}\right), \tag{1.11}$$

where $a_0 = \hbar^2/me^2 = 0.529 \times 10^{-8}$ cm is the atomic unit of length (Bohr radius). The functions $R_{nl}(r)$ are orthogonal and normalized

$$\int R_{nl}(r) R_{n'l}(r) r^2 dr = \delta_{nn'}. \tag{1.12}$$

For large r, the functions R_{nl} decrease exponentially: $R_{nl} \sim \exp(-Zr/na_0)$. If r is expressed in atomic units a_0 and the energy in Ry, then for $r \to \infty$, $R_{nl} \sim \exp(-\sqrt{|E_n|}r)$.

We shall give explicit expressions for the functions $R_{nl}(r)$ when $n = 1,2,3$, expressing r in units a_0 (for this it is sufficient to make the substitution $r/a_0 \to r$) and omitting the factor $Z^{3/2}a_0^{-3/2}$ common to all the functions,

$$R_{10} = 2e^{-r},$$

$$R_{20} = \frac{1}{\sqrt{2}} e^{-\frac{r}{2}} \left(1 - \frac{1}{2}r\right), \quad R_{21} = \frac{1}{2\sqrt{6}} e^{-\frac{r}{2}} r,$$

$$R_{30} = \frac{2}{3\sqrt{3}} e^{-\frac{r}{3}} \left(1 - \frac{2}{3}r + \frac{2}{27}r^2\right), \tag{1.13}$$

---

[1] Definition of the phases of functions (1.7) corresponds to that adopted in [1].

$$R_{31} = \frac{8}{27\sqrt{6}} e^{-\frac{r}{3}} r\left(1 - \frac{r}{6}\right), \quad R_{32} = \frac{4}{81\sqrt{30}} e^{-\frac{r}{3}} r^2.$$

By using (1.11), one can calculate the mean values of the quantities r^k, which will be necessary later

$$\langle r^k \rangle = \int R_{nl}^2 r^{k+2} dr,$$

$$\langle r \rangle = \frac{1}{2}[3n^2 - l(l+1)] \frac{a_0}{Z},$$

$$\langle r^2 \rangle = \frac{n^2}{2}[5n^2 + 1 - 3l(l+1)] \frac{a_0^2}{Z^2},$$

$$\langle r^3 \rangle = \frac{n^2}{8}[35n^2(n^2 - 1) - 30n^2(l+2)(l-1)$$

$$+ 3(l+2)(l+1)l(l-1)] \frac{a_0^3}{Z^3}, \tag{1.14}$$

$$\langle r^{-1} \rangle = \frac{1}{n^2} \frac{Z}{a_0},$$

$$\langle r^{-2} \rangle = \frac{1}{n^3 \left(l + \frac{1}{2}\right)} \frac{Z^2}{a_0^2},$$

$$\langle r^{-3} \rangle = \frac{1}{n^3(l+1)\left(l + \frac{1}{2}\right)l} \frac{Z^3}{a_0^3}.$$

The radial functions for the continuous spectrum $R_{El}(r)$ can be expressed in terms of confluent hypergeometric functions. Different representations of these functions are given in [2,3].

1.2 Series Regularities

1.2.1 Radiative Transition Selection Rules

Radiative transitions between the states nlm, $n'l'm'$ are possibly only if the quantum numbers l, m change by the quantities

$$\Delta l = l' - l = \pm 1, \quad \Delta m = m' - m = 0, \pm 1. \tag{1.15}$$

There are no limitations on the quantum numbers n, n'.

Relations (1.15) are called the selection rules for dipole radiation. Transitions satisfying conditions (1.15) are called allowed transitions. If conditions (1.15) are not fulfilled, then dipole radiation is forbidden. In this case, quadrupole or magnetic dipole radiation may be possible. The probability of such transitions, however, is approximately 10^5 times less than probability of dipole transitions. They are called forbidden transitions. The probabilities of radiative transitions for the hydrogen atom are given in Sect. 9.7.

1.2.2 Spectral Series of the Hydrogen Atom

The selection rules (1.15) enable one to find out what transitions are responsible for the series of lines observed in the hydrogen spectrum. The hydrogen spectrum consists of clearly defined series of lines with wavelengths λ satisfying the following formulas:

$$\frac{1}{\lambda} = R\left(1 - \frac{1}{n^2}\right), \quad n = 2,3,4, \ldots \quad \text{Lyman series}$$

$$\frac{1}{\lambda} = R\left(\frac{1}{2^2} - \frac{1}{n^2}\right), \quad n = 3,4,5, \ldots \quad \text{Balmer series}$$

$$\frac{1}{\lambda} = R\left(\frac{1}{3^2} - \frac{1}{n^2}\right), \quad n = 4,5,6, \ldots \quad \text{Paschen series}$$

$$\frac{1}{\lambda} = R\left(\frac{1}{4^2} - \frac{1}{n^2}\right), \quad n = 5,6,7, \ldots \quad \text{Brackett series}$$

$$\frac{1}{\lambda} = R\left(\frac{1}{5^2} - \frac{1}{n^2}\right), \quad n = 6,7,8, \ldots \quad \text{Pfund series}$$

Here R is a constant called the Rydberg constant, equal to 109677.581 cm^{-1}.

The longest wavelength lines of these series are, respectively, $\lambda = 1215.68$ Å (vac.), 6562.79 Å, 1.8751 μm, 4.051 μm and 7.456 μm (1 μm = 10^{-4}cm = 10^4 Å). The line $\lambda = 12.37$ μm, corresponding to the sixth series, was observed in absorption. The general form of the series is shown in Fig. 1.2. The distance between the lines decreases as λ decreases. A continuous spectrum adjoins the short-wave edge of the series. The limits of the first three series are located respectively at $\lambda = 912$ Å, 3648 Å, 8208 Å. Thus the Lyman and Balmer series are separated from the others; the other series partially overlap.

It is easy to see that for any two levels n, n' there exist states nl, $n'l'$ between

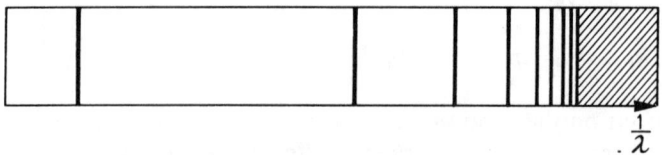

Fig. 1.2. General picture of a series in the hydrogen spectrum

which radiative transitions are allowed. Thus, for $n = 2$, $n' = 1$, transitions are allowed between the states $2p$ and $1s$, for $n = 3$ and $n' = 2$, transitions are allowed between the states $3s$ and $2p$, $3p$ and $2s$, $3d$ and $2p$, and so on.

According to (1.6), in the transition of a one-electron atom from the level n to the level n', a quantum

$$\hbar\omega = E_n - E_{n'} = \frac{\mu e^4 Z^2}{2\hbar^2}\left(\frac{1}{n'^2} - \frac{1}{n^2}\right). \tag{1.16}$$

is emitted. Since the radiation frequency ω is connected with the wavelength λ by the relation $\omega = 2\pi c/\lambda$, where c is the velocity of light, we obtain (for $Z = 1$)

$$\frac{1}{\lambda} = \frac{E_n - E_{n'}}{2\pi\hbar c} = \sigma_n - \sigma_{n'} = \frac{\mu e^4}{4\pi c\hbar^3}\left(\frac{1}{n'^2} - \frac{1}{n^2}\right). \tag{1.17}$$

The quantity $\mu e^4/4\pi c\hbar^3$, within an accuracy determined by the accuracy of the constants m, e, c, $\hbar$, coincides with the experimentally found value of the Rydberg constant R.

For $n' = 1$, (1.17) gives the wavelengths of the lines of the Lyman series (transitions $1s$–np), for $n' = 2$, the wavelengths of the lines of the Balmer series (transitions $2s$–np, $2p$–ns, $2p$–nd), and so on. The continuous background adjoining the limit of the series is due to transitions from the states of the continuous spectrum ($E > 0$) to the states of the discrete spectrum.

Special notations have been adopted for the lines of the hydrogen spectrum. The Lyman series lines, in order of decreasing wavelength, are denoted by L_α, L_β, L_γ, and so on; the Balmer series lines by H_α, H_β, H_γ, and so on. The longest wavelength line of the Lyman series L_α, $\lambda = 1215.68$ Å, is obviously the resonance line of the hydrogen atom, i.e., the line corresponding to a transition from the first excited level to the ground level. This line is located in the vacuum ultraviolet region of the spectrum. The following lines of the Balmer series are the main lines in the visible and near ultraviolet regions of the hydrogen spectrum:

H_α 6562.73 Å, H_ε 3970.07 Å,
H_β 4861.33 Å, H_ζ 3889.06 Å,
H_γ 4340.47 Å, H_η 3835.39 Å,
H_δ 4101.74 Å, H_θ 3797.90 Å.

1.2.3 Hydrogenlike Ions

The level systems of the one-electron ions He$^+$, Li^{++}, Be^{+++}, etc., are similar to the level system for hydrogen. These ions are called hydrogenlike. The constant $R = \mu e^4/4\pi c\hbar^3$ depends on the reduced mass $\mu = mM/(m + M)$ and, consequently, on the nuclear mass M. Since $m \ll M$, the difference between the constants R for two different masses M_1 and M_2 is not great. Thus, for H and He$^+$ spectra, according to (1.17) the ratio $R_H/R_{He} = 0.999596$, which agrees well with experiment. For $M/m \to \infty$, $\mu \to m$. The corresponding value of R is denoted by R_∞. The constant R_∞ is connected with the Rydberg unit of energy, Ry, by the relation $R_\infty = \text{Ry}/2\pi\hbar c$. It is easy to see that for a finite nuclear mass M

$$R_M = R_\infty \left(1 + \frac{m}{M}\right)^{-1}. \tag{1.18}$$

Experimental values of R for hydrogen, deuterium, and a series of ions are given in Table 1.1.

According to (1.6), $E_n \propto Z^2$. Thus, for an ion with a nuclear charge Z the potentials E_i, E_r are Z^2 times greater than for hydrogen, and the wavelength of the resonance transition λ_{res} is Z^2 times less. The values of λ_{res} for a series of hydrogenlike ions are given in Table 1.2. In this table the spectroscopic system of notation is used. The spectra of neutral atoms are denoted by the Roman numeral I following from the symbol of the chemical element, the spectra of singly charged ions by the numeral II, and for doubly charged ions, by the numeral III, and so on.

Table 1.1. Values of the Rydberg constant R for hydrogenlike ions

R	[cm^{-1}]
R_∞	$109,737.311 \pm 0.012$
R_H	$109,677.575 \pm 0.012$
R_D	$109,707.420 \pm 0.012$
R_{He^3}	$109,717.346 \pm 0.012$
R_{He^4}	$109,722.268 \pm 0.012$

Table 1.2. Values of the resonance transition wavelengths λ_{res} for hydrogen and hydrogenlike ions

Z	Spectrum		λ_{res} [Å]
1	H	I	1215.68
2	He	II	303.78
3	Li	III	135.02
4	Be	IV	75.94
5	B	V	48.58
6	C	VI	33.74

1.3 Fine Structure

1.3.1 Velocity Dependence of Electron Mass

For the hydrogen atom and hydrogenlike ions with not very large nuclear charge Z, relativistic effects are not great and can be taken into account within the limits of perturbation theory. The relativistic effects are the velocity dependence of the electron mass and the splitting of the levels connected with electron angular momentum, the spin s. Expanding the relativistic expression for the energy of a particle of mass m in a field $U(r)$, $\mathscr{E} = U + \sqrt{c^2p^2 + m^2c^4}$ in a series in powers of $p^2/m^2c^2 = (v/c)^2$, we obtain $E = \mathscr{E} - mc^2 \simeq p^2/2m + U - p^4/8m^3c^2$. The perturbation $V = -p^4/8m^3c^2 \simeq 1/2mc^2 (E^{(0)} - U)^2$ where $E^{(0)}$ is the nonrelativistic energy $p^2/2m + U$ results in the level shift

$$\Delta E'_{nl} = -(E_n^2 + 2E_n Ze^2 \langle r^{-1}\rangle_{nl} + Z^2e^2\langle r^{-2}\rangle_{nl})/2mc^2.$$

Using (1.6) and (1.14) we obtain

$$\Delta E'_{nl} = -\alpha^2 \left(\frac{1}{l+1/2} - \frac{3}{4n}\right)\frac{Z^4}{n^3}\text{Ry}. \qquad (1.19)$$

Here $\alpha = e^2/\hbar c \simeq 1/137$.

1.3.2 Spin–Orbit Interaction

The existence of a magnetic moment of the electron, connected with electron spin s

$$\mu = -\frac{e\hbar}{mc}s = -2\mu_0 s, \qquad (1.20)$$

where $\mu_0 = e\hbar/2mc$ is the Bohr magneton, leads to an additional interaction $\mu \cdot H$ between the electron and nucleus. H is the magnetic field which is associated with the electron moving in the electric field E. Since $H = -[E, v]/c$, $E = (\partial U/\partial r)r/r$, $m[r, v] = \hbar l$, this additional interaction $V \propto l \cdot s$. Thus this interaction is usually called the spin-orbit interaction. The resulting expression for V is

$$V = \frac{\hbar^2}{2m^2c^2}\frac{\partial U}{\partial r}\frac{1}{r}l \cdot s. \qquad (1.21)$$

Since $s = 1/2$ the eigenvalue of the square of the spin s^2 is

$$s(s+1) = 3/4$$

and the z component of the spin s_z can take two values $\pm 1/2$.

Spin-orbit interaction depends not only on the value of the angular momentum l, but also on the mutual orientation of the angular momenta l and s, i.e., on the value of the total angular momentum of the atom, $j = l + s$. This value is obtained according to the general quantum mechanical rules for the addition of angular momenta.

The eigenvalue of the square of the total angular momentum j^2 equals $j(j+1)$, where for a given value of l, $j = l \pm 1/2$ (for $l = 0$, $j = 1/2$). The z component of the total angular momentum m_j is the sum of the z components of the orbital angular momentum m_l and spin m_s, i.e., $m_j = m_l + m_s$. In the following we shall drop the subscript j on m_j, understanding m to be the z component of the total angular momentum.

For a given value of j, the quantum number m can take $(2j+1)$ different values $j, j-1, \ldots, -j$. Thus, to a level nlj there belong $2j+1$ states differing in the value of the quantum number m. The quantity $2j+1$ is called the statistical weight of the level j. The value of j is usually written as a subscript after the spectroscopic notation of l. Thus, the state $n, l = 1$, $j = 1/2$ is denoted $np_{1/2}$, the state $n = 4$, $l = 2$, $j = 3/2$ is $4d_{3/2}$, and so on. The total angular momentum of any isolated system is conserved; therefore, the state of an atom can be characterized by the value of the total angular momentum j even in the case when the orbital and spin angular momenta are not separately conserved.

Due to the spin-orbit interaction, the energy of an atom in the states $j = l + 1/2$ and $j = l - 1/2$ is different. Thus, the spin-orbit interaction leads to the splitting of the level nl into two components $l + 1/2$ and $l - 1/2$. Before passing on to the calculation of the energy of splitting, we shall express the dependence of the spin-orbit interaction on j in explicit form. Since $j = l + s$

$$j^2 = l^2 + s^2 + 2l \cdot s, \quad l \cdot s = (j^2 - l^2 - s^2)/2 .$$

Remembering also that $U = -Ze^2/r$, we obtain

$$V = \frac{Ze^2 \hbar^2}{2m^2 c^2} \frac{1}{r^3} \frac{1}{2} (j^2 - l^2 - s^2) . \tag{1.22}$$

The mean value of the perturbation (1.22) in the state n, l, j equals obviously

$$\frac{Ze^2 \hbar^2}{2m^2 c^2} \left\langle \frac{1}{r^3} \right\rangle_{nl} \frac{1}{2} [j(j+1) - l(l+1) - s(s+1)] .$$

Therefore, for the correction to the energy due to spin-orbit interaction, we obtain (the value of the matrix element $\langle r^{-3} \rangle_{nl}$ has been given above)

$$\Delta E''_{nlj} = \alpha^2 \frac{j(j+1) - l(l+1) - s(s+1)}{2l(l+1)\left(l+\frac{1}{2}\right)} \frac{Z^4}{n^3} \text{Ry} . \tag{1.23}$$

1.3.3 Fine Structure. Selection Rules

Comparison of (1.19) and (1.23) shows that both effects connected with the electron mass velocity dependence and with electron spin, have the same order of magnitude. It is easy to see that in both possible cases, $j = l+1/2$ and $j = l-1/2$, the total correction to the energy $\Delta E' + \Delta E''$ is given by the same expression

$$\Delta E_{nlj} = \Delta E' + \Delta E'' = \alpha^2 \left(\frac{3}{4n} - \frac{1}{j+1/2}\right)\frac{Z^4}{n^3}\,\text{Ry} \tag{1.24}$$

Thus, owing to relativistic effects, the level nl splits into two components, $j = l+1/2$ and $j = l-1/2$. This splitting is called fine or multiplet splitting. The dimensionless constant $\alpha = e^2/\hbar c \simeq 1/137$ defining the scale of the splitting is called the fine-structure constant. It is significant that whereas each of the corrections $\Delta E'$ and $\Delta E''$ separately depends on l, the total correction ΔE does not depend on l. Thus, for all n, l levels differing only by the value of l, the fine-structure components with one and the same value of j coincide. The fine splitting of levels $n = 1, 2, 3$ is shown in Fig. 1.3. As follows from (1.24), fine splitting decreases with increase of n approximately as $1/n^3$, therefore this splitting is particularly important for lower levels.

According to (1.24), the distance between the levels $j' = l+1/2$ and $j'' = l-1/2$ equals

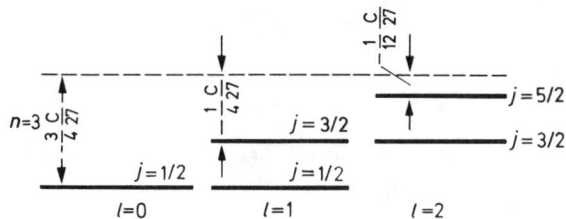

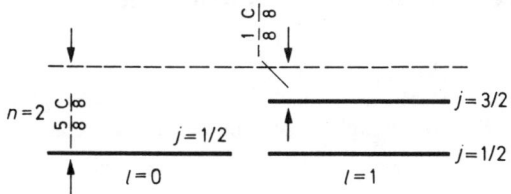

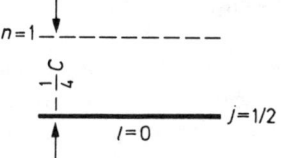

Fig. 1.3. Fine structure of levels $n = 1, 2, 3$. $C = 1/2 a^3 Z^4 \left(\dfrac{me^4}{\hbar^2}\right)$. Dashed lines correspond to the energy given by (1.6)

$$\delta E_{j'j''} = \frac{\alpha^2 Z^4}{n^3 l(l+1)} \text{ Ry}. \tag{1.25}$$

Thus, for the hydrogen atom the splitting of the levels $j = 1/2$ and $j = 3/2$ for $n = 2, 3$ and 4 is, respectively, 0.36, 0.12, and 0.044 cm^{-1}.

The set of lines arising from the transitions between the fine-structure components of the levels nl and $n'l'$ (transitions $nlj \to n'l'j'$) is called a multiplet. The selection rule with respect to the quantum number j is

$$\Delta j = 0, \pm 1. \tag{1.26}$$

Using this rule, it is easy to find how many lines the fine splitting of the hydrogen spectrum results in. For example, the multiplet $n'p - nd$, shown in Fig. 1.4, in accordance with (1.26), consists of three components.

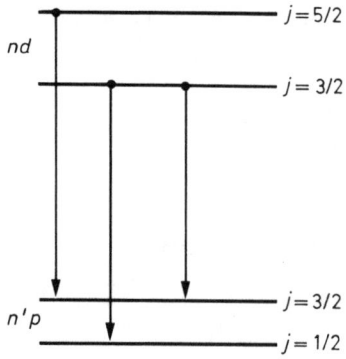

Fig. 1.4. Allowed transitions in the multiplet $nd - n'p$

Further, for the transitions responsible for the Lyman series, both of the following transitions are permitted by the selection rules with respect to j

$$1s_{1/2} - np_{1/2}, \quad 1s_{1/2} - np_{3/2}.$$

In the case of the Balmer series, the following transitions are permitted:

$$2s_{1/2} - np_{1/2}, \quad 2p_{1/2} - ns_{1/2}, \quad 2p_{1/2} - nd_{3/2},$$
$$2p_{3/2} - nd_{3/2},$$
$$2s_{1/2} - np_{3/2}, \quad 2p_{3/2} - ns_{1/2}, \quad 2p_{3/2} - nd_{5/2}.$$

The transition diagram for H_α lines is given in Fig. 1.5. Owing to the fact that the levels $ns_{1/2}$ and $np_{1/2}$, $np_{3/2}$ and $nd_{3/2}$ coincide, each of the Balmer lines must consist of five components. Since, however, the splitting of the lower level considerably exceeds the splitting of the higher levels, the Balmer lines consist of two groups of closely spaced lines. The distance between these two groups equals $0.36\,\text{cm}^{-1}$ and is constant for all lines of the series. The magnitude of splitting within each group falls rapidly in passing from the initial lines of the series to higher ones. For hydrogenlike ions the splitting $\Delta E \propto Z^4$.

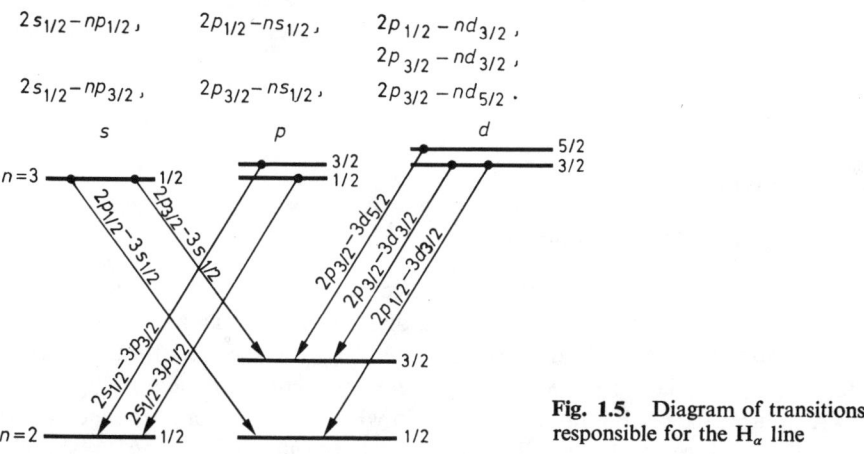

Fig. 1.5. Diagram of transitions responsible for the H_α line

1.3.4 Lamb Shift

Lamb and Retherford discovered in 1947 that the splitting of the hydrogen levels $2s_{1/2}$ and $2p_{1/2}$ equals $0.034\,\text{cm}^{-1}$. Later it was shown that this splitting is caused by the interaction of the electron with the radiation field [2]. The theoretical value of the shift (Table 1.3) and experimental value coincide with great accuracy. For hydrogenlike ions the shift is proportional to Z^4.

Table 1.3. Radiative splitting of the level $n = 2$

Level	Radiative shift [MHz]	Difference [MHz]
$2s_{1/2}$	+1040	1057
$2p_{1/2}$	−17	
$2p_{3/2}$	8	

Chapter 2 Systematics of the Spectra of Multielectron Atoms

Systematics of the spectra of multielectron atoms based on the concept of the self-consistent field are discussed for the two limiting cases of LS and jj coupling of angular momenta.

2.1 Central Field

2.1.1 Central Field Approximation

For atoms containing more than one electron, even for the simplest ones, Schrödinger's equation cannot be solved directly, either analytically, or by numerical methods. For this reason, systematics of the spectra of multielectron atoms must, of necessity, be based on some approximate model.

A suitable schematic treatment is one in which the concept of the individual state of an electron in an atom is accepted, and the state of an atom as a whole is determined by the set of the states of the electrons, taking into account their interaction. In the limit of this approximation, one succeeds in obtaining general information on the system of energy levels possible for a given atom, and on the relative position and grouping of the levels. Also in the limit of this approximation, selection rules for radiative transitions are established, which enable one to predict the structure of the spectrum for each element.

To describe electron states in an atom, one proceeds from the assumption that each electron moves in a certain effective centrally symmetric field created by the nucleus and all the other electrons. This approximation, called the self-consistent field approximation, is taken as the starting point for calculations. For the purpose of systematization of the spectrum, there is no need to know the form of this field. Many results can be obtained on the basis of the general theory of the motion of a particle in a centrally symmetric field.

A more detailed treatment requires a consideration of the noncentral part of the electrostatic interaction between electrons, and also of magnetic interactions, in particular spin-orbit interaction. In the theory of atomic spectra, these interactions are usually considered within perturbation theory as small corrections to the centrally symmetric field. As is known, a perturbation does not alter the number of possible states of a system. The suitability of the above method for the purpose of systematics is determined to a considerable extent by this.

Schrödinger's equation for an electron in an arbitrary centrally symmetric field $U(r)$ has the form

$$\Delta\psi + \frac{2m}{\hbar^2}[E - U(r)]\psi = 0. \tag{2.1}$$

The equation differs from (1.1) for the hydrogen atom only in that the arbitrary potential $U(r)$ appears here instead of the Coulomb potential $-Ze^2/r$. We can therefore use some of the results obtained above. The angular momentum is conserved for motion in an arbitrary centrally symmetric field; therefore each stationary state can be characterized by the assignment of the square of the angular momentum and its z component, i.e., by the assignment of the quantum numbers l and m. The wave functions for stationary states have the form

$$\psi = R(r)\, Y_{lm}(\theta, \varphi), \tag{2.2}$$

where $Y_{lm}(\theta, \varphi)$ are spherical harmonics, defined by (1.7), and the radial part of the function $R(r)$ is defined by the equation

$$\frac{1}{r^2}\frac{d}{dr}\left(r^2 \frac{dR}{dr}\right) - \frac{l(l+1)}{r^2}R + \frac{2m}{\hbar^2}[E - U(r)]R = 0. \tag{2.3}$$

Equation (2.3) has bound solutions only for definite values of E. The set of these values determines the energy spectrum of a particle, i.e., those possible energy values which a particle may have for motion in the given field. The effective potential energy in (2.3),

$$U_l(r) = U(r) + \frac{\hbar^2}{2m}\frac{l(l+1)}{r^2}, \tag{2.4}$$

contains l, but does not depend on the quantum number m. Thus, the energy of a particle also does not depend on m. In other words, the levels are degenerate with respect to m, i.e., with respect to the orientation of the angular momentum; $(2l+1)$ different values of m correspond to the given value of l. Thus, $(2l+1)$ states, differing in angular momentum orientation, correspond to one and the same energy level. The determination of the function $R(r)$, i.e., the solution of (2.3), requires a definition of the form of $U(r)$. As a rule, in this case one has to use different approximate methods. In what follows, we shall only deal with those fields for which $U(r) < 0$ and, in addition,

$$\begin{aligned} U(r) &\to 0, \quad r \to \infty\,; \\ U(r) &\to -\frac{Ze^2}{r}, \quad r \to 0. \end{aligned} \tag{2.5}$$

This enables us to reach a number of general conclusions on the character of the radial motion and on the energy spectrum of a particle. We shall confine ourselves to a statement of results not connected with a special form of $U(r)$.

First of all, it can be shown that the character of the motion of a particle in a centrally symmetric field (2.5) is completely determined by the values of E, l, and m. There do not exist two different wave functions ψ corresponding to one and the same set of numbers E, l, and m. Just as in the case of the Coulomb field, the energy spectrum is discrete for $E < 0$ and continous for $E > 0$. In the general case, the spectrum of E is different for different values of l. It can be shown that the lowest possible value of energy for a given l is lower as l is smaller. This is connected with the increase of the effective potential energy (2.4) with an increase of l, since the centrifugal energy $\hbar^2(l+1)/2mr^2$ is essentially positive. The ground state, i.e., the state with the lowest possible energy value, is always the state with $l = 0$ [see Fig. 2.1, where the typical form of $U(r)$ and $U_l(r)$ curves is shown].

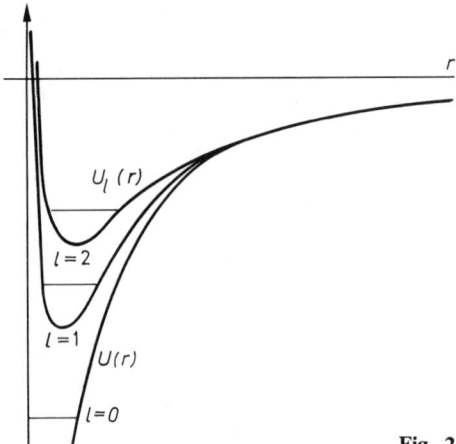

Fig. 2.1. Potentials $U(r)$ and $U_l(r)$

2.1.2 Parity of States

The wave functions $\psi_{Elm} = R_{El}(r)\, Y_{lm}(\theta,\varphi)$, corresponding to different values of the angular momentum of a particle, behave differently upon an inversion transformation ($x \to -x$; $y \to -y$; $z \to -z$). For spherical coordinates, this transformation has the form

$$r \to r, \quad \theta \to \pi - \theta, \quad \varphi \to \varphi + \pi \, .$$

The functions R_{El} do not change in such a transformation. We shall therefore find how the functions $Y_{lm}(\theta,\varphi) \propto P_l^m(\cos\theta)\exp(im\varphi)$ behave. On replacing φ by $\varphi + \pi$, the factor $\exp(im\varphi)$ is multiplied by $(-1)^m$. On replacing θ by $\pi - \theta$, $\cos(\pi - \theta) = -\cos\theta$, $\sin(\pi - \theta) = \sin\theta$, $P_l^m[\cos(\pi - \theta)] = P_l^m(\cos\theta)(-1)^{l-m}$. Consequently $Y_{lm}(\pi - \theta, \varphi + \pi) = Y_{lm}(\theta,\varphi)(-1)^l$.

Thus, the functions ψ_{Elm} corresponding to states with even values of l do not change sign. These states, and also the functions, are called even. For odd l, the functions ψ_{Elm} change sign upon an inversion transformation. In this case a state is called odd. The parity of a state is entirely determined by the value of l and does not depend either on E or m.

The operation of inversion leaves the Hamiltonian of a particle in a centrally symmetric field $H = p^2/2m + U(r)$ unchanged. This means that the parity of the wave function of a stationary state is conserved. Therefore, each state of a particle in a centrally symmetric field is characterized by a definite parity.

The wave function describing the state of a system of noninteracting particles in a centrally symmetric field can be written in the form of a product of functions ψ_{Elm}. Therefore, the parity of this wave function is determined by the factor $(-1)^{l_1}(-1)^{l_2}\ldots(-1)^{l_n}$. Thus the state of a system of particles is even, if the sum of the angular momenta of the particles $\sum_i l_i$ has an even value, and odd for odd values of this sum.

It is important that parity be determined just by the sum of the quantum numbers l_i and not by the vector sum $\sum_i \mathbf{l}_i$. The classification of states with respect to their parity is of great importance when establishing selection rules for radiative transitions. Thus the selection rule $\Delta l = \pm 1$, as will be shown later, is the particular case of a general rule forbidding electric dipole transitions between states of the same parity.

2.1.3 Systematics of Electron States in a Central Field

For a prescribed value l, it is customary to enumerate the states of a particle in ascending order of energy by the principal quantum number n, taking the values $l+1, l+2, \ldots$. It must be noted that the sequence of energy levels in complex atoms is different from that in hydrogen. In hydrogen, E depends only on n and does not depend on l, while $E_{n+1} > E_n$ always. Another sequence of levels frequently occurs with complex atoms: the electron energy in the state $n, l+2$ is greater than in the state $n+1, l$. As a rule, the electron energy is greater, the greater the sum $n + l$.

The distribution of electrons in an atom with respect to states with different values of n and l is spoken of as the electron configuration. Assignment of an electron configuration thus requires the enumeration of the values n and l for all the electrons of an atom. If there are several electrons with the same values of n and l, this is denoted as $(nl)^k$, where k is the number of such electrons; for example $(3s)^2$, $(3p)^3$, and so on, or simply $3s^2$ and $3p^3$.

For a particle with nonzero spin, states with the same values of E, l, and m_l can differ by the values of the z component m_s of the spin. The full characterization of the states of an electron is therefore achieved by the assignment of the four numbers n, l, m_l, and m_s, the energy being defined only by the first two.

For a given l, the number m_l can take $2l+1$ values, while m_s takes only two values $\pm 1/2$. Consequently, there are altogether $2(2l+1)$ states with the same values of n and l, but different values of m_l and m_s. States with the same values of n and l are called equivalent states. According to Pauli's principle there cannot be more than one electron in each n, l, m_l, m_s state. Thus, not more than $2(2l+1)$ electrons can have the same values of n and l in an atom. An assembly of $2(2l+1)$ equivalent electrons is called a closed or filled shell. It is impossible to add another electron with the same values of the quantum numbers n and l to such a shell.

When $l = 0$ s shell $2(2l+1) = 2$
     1 p shell $2(2l+1) = 6$
     2 d shell $2(2l+1) = 10$
     3 f shell $2(2l+1) = 14$

Sometimes a slightly different definition of shells is used: $n = 1$ K shell (states $1s$), $n = 2$ L shell (states $2s, 2p$), $n = 3$ M shell (states $3s, 3p, 3d$). Shells with $n = 4,5,6$ are denoted by the letters N, O, P.

2.2 Electrostatic and Spin-Orbit Splitting in the LS Coupling Approximation

2.2.1 Spectral Terms. LS Quantum Numbers

In the central field approximation the energy of an atom is completely determined by the assignment of the electron configuration, i.e., by the assignment of the values of n and l for all the electrons. To each electron configuration $n_1 l_1$, $n_2 l_2$, $n_3 l_3$, ... there correspond $2(2l_1 + 1)\, 2(2l_2 + 1)\, 2(2l_3 + 1) \ldots$ states, differing by the values of the quantum numbers m_s and m_l or, in other words, by the mutual orientation of the orbital angular momenta and spins of the electrons. Attributing all these states to one and the same energy level of an atom is possible as long as we neglect that part of the electrostatic interaction between electrons which is not taken into account in a centrally symmetric approximation, and also spin-orbit interaction. In reality, both types of interaction always occur, which leads to splitting of the level $n_1 l_1, n_2 l_2, n_3 l_3, \ldots$ into quite a number of sublevels. Joint consideration of both interactions is an extremely complex task. As a rule, therefore, one uses a considerably simpler approach in which one of the interactions is considered small in comparison with the other. Experimental data show that, in quite a number of cases, the electrostatic interaction has a much greater value than the spin-orbit. We shall start with this case.

As will be shown in Sects. 5.3 and 5.4, electrostatic interaction leads to a splitting of the level corresponding to a given electron configuration into quite a number of levels, characterized by different values of the total orbital angular momentum of the electrons L and of the total spin S. The dependence of the energy of splitting on L has a simple physical meaning. To the different values of L there corresponds a different mutual orientation of the orbital angular momenta of the electrons or, roughly speaking, a different orientation of electron orbits. Therefore in states with different values of L, the electrons, on the average, are at different distances from each other, which also leads to a difference in the electrostatic energy of repulsion. The energy dependence on S is not so obvious and becomes apparent indirectly (Sect. 5.3).

The energy of interaction of electrons with the nucleus and the energy of interaction of electrons with each other have different signs. Therefore the electrostatic interaction of electrons with each other leads to a shift of the energy levels upwards (the absolute magnitude of the coupling energy is decreased). It has been established empirically that, for ground configurations and for configurations containing equivalent electrons, electrostatic splitting obeys a definite rule, the so-called Hund's rule. According to this rule, the level with the greatest possible value of S for the given electron configuration and the greatest (possible for this S) value of L has the lowest energy.

Energy levels corresponding to definite values of L and S are called spectral terms or simply terms. Capital letters of the Latin alphabet are usually used for denoting terms, as follows:

$L =$ 0, 1, 2, 3, 4, 5, 6, 7, 8, 9, 10

  S P D F G H I K L M N

2.2.2 Fine Structure of Terms

Just as in the case of the hydrogen atom, relativistic effects, principally spin-orbit interaction, lead to a splitting of the LS term into a number of components, corresponding to different values of the total angular momentum J of the atom. This splitting is called fine or multiplet splitting.

In accordance with the general quantum mechanical rule of addition of angular momenta, the total angular momentum J of an atom can take the values $L+S \geqslant J \geqslant |L-S|$. In the case $L \geqslant S$, $2S+1$ different values of J are possible, i.e., the term splits into $2S+1$ different components. The number $2S+1$, determining in this case the number of components of the term, is called the multiplicity of the term. In the cases $L < S$, the number of components equals $2L+1$; however in this instance, too, the name "multiplicity" is kept for the number $2S+1$. If the multiplicity of a term $2S+1$ equals 1, the term is called singlet; 2: doublet; 3: triplet; 4: quartet; and so on. It is accepted practice to show the multiplicity of a term above and to the left of the term symbol. The value of the number J is shown below and to the right. Thus the full designation of a term has the form $^{2S+1}L_J$. So a term with $L = 0$, $S = 3/2$, and $J = 3/2$ is denoted as $^4S_{3/2}$; the symbols $^2P_{1/2}$, $^2P_{3/2}$ denote the components of the doublet term or simply the doublet $L = 1$, $S = 1/2$ and $J = 1/2, 3/2$, and so on. In cases when it is necessary to show the parity of the states relating to a given term, odd terms are marked with a superscript O(odd), placed to the right of L. For example, $^2P^o_{3/2}$. Absence of the superscript O indicates even parity of the term. To the term LS there belong $(2L + 1)(2S + 1)$ states, differing by values of the z components of the orbital and spin angular momenta M_L and M_S. The spin-orbit interaction does not completely remove this degeneracy. It is obvious that the energy of an isolated atom cannot depend on how the total angular momentum of an atom is oriented in space. Therefore $2J + 1$ states of the atom, corresponding to the different possible values of the z component of the total angular momentum M, pertain to one and the same energy value. In other words, each J component of a term is degenerate with a multiplicity equal to $2J + 1$.

It is easy to verify that

$$\sum_J (2J + 1) = (2L + 1)(2S + 1), \qquad (2.6)$$

i.e., splitting of the term due to spin-orbit interaction does not alter the number of states pertaining to the LS term.

Only if, for any reason, a specific direction in space is preferred, for example, on superimposing a magnetic field, degeneracy with respect to M is removed and each J component in its turn is split into $2J + 1$ components.

Multiplet splitting obeys a rule which is called Landé's interval rule. According to this rule, the splitting of the levels J, $J - 1$ is proportional to J,

$$\Delta E_J - \Delta E_{J-1} = \Delta E_{J, J-1} = A(LS) J. \qquad (2.7)$$

The multiplet splitting constant $A(LS)$ is different for different terms and can be of either sign.

When $A > 0$, the multiplet component with the smallest possible value $J = |L - S|$ has the lowest energy value. These multiplets are called normal.

When $A < 0$, the multiplet component with the greatest possible value $J = L + S$ has the lowest energy value. These multiplets are called inverted.

It has been established empirically that configurations containing n equivalent electrons correspond, when $n < 2l + 1$ (shells which are less than half filled), to normal multiplets and, when $n > 2l + 1$ (shells which are more than half filled), to inverted multiplets. When $n = 2l + 1$, there is no multiplet splitting at all. A grouping of levels, similar to that given in Fig. 2.2, is typical for the case under consideration. The distance between LS terms of configuration is considerably less than that between identical terms of different configurations. Each term, with the exception of singlet terms and S terms, has a fine structure, the distance between the components of this structure being considerably less than the distance between different terms. This grouping of levels is characteristic of the approximation which is called the Russell–Saunders or R-S coupling ap-

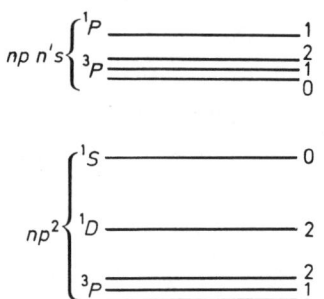

Fig. 2.2. Grouping of levels, typical of LS coupling

proximation. The term LS coupling or normal coupling is also used. The term LS coupling will be used everywhere below.

2.2.3 Finding the Terms of Multielectron Configurations

For configurations of nonequivalent electrons, it is easy to obtain all possible terms on the basis of the general quantum mechanical rule for addition of angular momenta. On adding the angular momenta L_1 and L_2, the absolute value of the resulting angular momentum can take one of the values (Sect. 4.1)

$$L = L_1 + L_2, L_1 + L_2 - 1, \ldots, |L_1 - L_2|.$$

Analogously, on adding the spins

$$S = S_1 + S_2, S_1 + S_2 - 1, \ldots, |S_1 - S_2|.$$

Addition is first carried out for two electrons, then the third is added, then the fourth, and so on.

Let us consider examples:
a) Configuration $np\ n'p$
$L = 0, 1, 2; S = 0, 1$. Therefore the terms $^1S, ^1P, ^1D, ^3S, ^3P, ^3D$ are possible.
b) Configuration $np\ n'p\ n''p$
We shall proceed from the terms of the configuration $np\ n'p$. By combining the ^1S term with $l = 1, S = 1/2$, we obtain the term ^2P. The addition of one p electron to the term ^1P gives the terms $^2S, ^2P, ^2D$; to the term ^1D — the terms $^2P, ^2D, ^2F$; to the term ^3S — the terms ^2P and ^4P; to the term ^3P — the terms $^2S, ^2P, ^2D, ^4S, ^4P, ^4D$; and to the term ^3D — the terms $^2P, ^2D, ^2F, ^4P, ^4D, ^4F$, Thus we obtain altogether: two ^2S terms, six ^2P terms, four ^2D terms, two ^2F terms, one ^4S term, three ^4P terms, two ^4D terms, and one ^4F term;

$np\ n'p\ [^1S]\ p\ ^2P;\ np\ n'p\ [^1P]\ p\ ^2S, ^2P, ^2D;$

np n′p [^1D] *p* $^2P, ^2D, ^2F$; *np n′p* [^3S]*p* $^2P, ^4P$;
np n′p [^3P] *p* $^2S, ^2P, ^2D, ^4S, ^4P, ^4D$
np n′p [^3D] *p* $^2P, ^2D, ^2F, ^4P, ^4D, ^4F$.

In brief form, this is written

$^2S\ P\ D\ F$ $^4S\ P\ D\ F$
  2 6 4 2   3 2

The figure under the term symbol indicates the number of identical terms.

The term of the configuration *np n′p* enclosed in square brackets is called the initial term. The assignment of the initial term is spoken of as the assignment of the term genealogy or origin.

Let us note that the addition of one electron to singlet terms gives doublet terms, to doublet terms—singlet and triplet ones, and to triplet terms—doublet and quartet ones, and so on.

There is a simple method to determine the multiplicity of terms possible for a configuration consisting of nonequivalent electrons, and their relative number. By adding one electron to a term of given multiplicity, we always obtain terms with a multiplicity one more and one less than the initial one, as $S' = S \pm 1/2$ and $2S' + 1 = 2S + 1 \pm 1$. This rule is illustrated in Fig. 2.3. As is evident from this diagram, only singlet and triplet terms are possible for two electrons; for three electrons – doublet and quartet ones, the doublet terms being twice as many as the quartet. For four electrons, singlet, triplet, and quintet terms occur in the ratio 2:3:1, and so on. As is evident from Fig. 2.3, for even *n*, singlet, triplet, and quintet terms are possible ($2S+1$ is odd). Conversely, for odd *n*, doublet and quartet terms are possible ($2S+1$ is even).

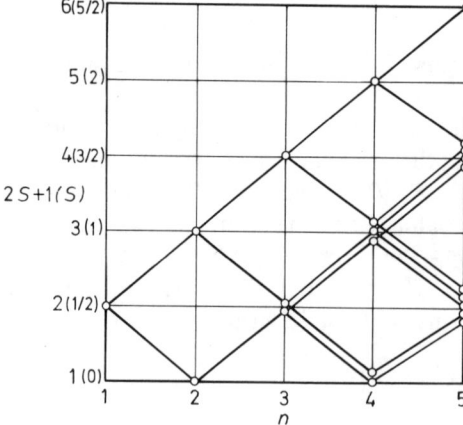

Fig. 2.3. Sequence of even and odd multiplets

2.2 General Picture of Electrostatic and Spin-Orbit Splitting

Thus, even and odd multiplicities alternate for configurations with the number of electrons $n, n+1, n+2, \ldots$. The set of terms of one multiplicity, obtained from the LS term of the initial electron configuration on adding to it one more electron, is called a polyad. Thus, in the example considered above, the terms $npn'p\,[^3P]p$ ^2S, $^2P\,^2D$ and $np'p[^3P]p\,^4S\,^4P\,^4D$ form two different polyads. It is not to easy to find the possible terms for configurations containing equivalent electrons. Among the values of L and S, there can appear those which correspond to states forbidden by the Pauly principle. Thus, for the configuration np^3 there are possible only three terms $^2D\,^2P\,^4S$, in spite of the fact that for configuration $npn'pn''p$ we have 21 terms. Terms of configurations p^n, d^n, f^n are given in Table 2.1.

Table 2.1. Terms of l^n configurations

Configuration		Terms		Statistical weight
s		^2S		2
s^2		^1S		1
p	p^5	$^2P^0$		6
p^2	p^4	^1SD	^3P	15
	p^3	$^2PD^0$	$^4S^0$	20
d	d^9	^2D		10
d^2	d^8	^1SDG	^3PF	45
d^3	d^7	^2PDFGH 2	^4PF	120

d^4	d^6	^1SDFGI
		   2 2 2
			$^3PDFGH\,^5D$
			  4 2
				210

	d^5	$^2SPDFGHI$ 3 2 2	$^4PDFG\,^6S$	252
f	f^{13}	$^2F^0$		14
f^2	f^{12}	^1SDGI	^3PFH	91
f^3	f^{11}	$^2PDFGHIKL^0$ 2 2 2 2	$^4SDFGI^0$	364

f^4	f^{10}	$^1SDFGHIKLN$
		  2 4 4 2 3 2
			$^3PDFGHIKLM\,^5SDFGI$
			  3 2 4 3 4 2 2
				1001

f^5	f^9	$^2PDFGHIKLMNO^0$
		  4 5 7 6 7 5 5 3 2
			$^4SPDFGHIKLM^0\,^6PFH^0$
			  2 3 4 4 3 3 2
				2002

f^6	f^8	$^1SPDFGHIKLMNQ$
		 4 6 4 8 4 7 3 4 2 2
			$^3PDFGHIKLMNO\,^7F$
			  6 5 9 7 9 6 6 3 3
			$^5SPDFGHIKL$
			  3 2 3 2 2
				3003

	f^7	$^2SPDFGHIKLMNOQ^0$
		  2 5 7 10 10 9 9 7 5 4 2
			$^4SPDFGHIKLMN^0\,^8S^0$
			  2 2 6 5 7 5 5 3 3
			$^6PDFGHI^0$
				3432

The statistical weight of the configuration (total number of states pertaining to the given configuration) is shown in the last column of this table. For configurations not containing equivalent electrons, the statistical weight equals $2(2l_1 + 1)\,2(2l_2 + 1)\ldots$. For the configuration l^n, the statistical weight is determined by the number of possible combinations which can be formed from the quantum numbers m_l, m_s, taking into account the Pauli principle. The number of such configurations, as is easy to show, is equal to $N_l!/n!(N_l - n)!$; $N_l = 2(2l + 1)$. The statistical weight of a configuration can also be calculated in another way. The statistical weight of each J level equals $(2J + 1)$, and the statistical weight of the term LS equals $(2L + 1)(2S + 1)$, whereupon $\Sigma (2J + 1) = (2L + 1)(2S + 1)$. Therefore, the sum $\Sigma (2J + 1)$ with respect to all J levels of a given configuration, and also the sum $\Sigma (2L + 1)(2S + 1)$ extended to all terms of a given configuration, gives the statistical weight of this configuration. Thus the statistical weight 45 is given in Table 2.1 for the configuration d^2. By summing the statistical weights of the terms ^1SDG, ^3PF (1, 5, 9, 9, 21), we obtain the same number.

For configurations of the greatest possible number of equivalent electrons, i.e., for a filled shell, only one term is possible, namely the ^1S term. In fact, in this case M_L is simply the sum of all possible values $m_l = 0, \pm 1, \pm 2 \ldots$ which obviously is equal to zero. Similarly, only one value, $M_S = 0$, is possible for M_S. One and the same term corresponds to the configurations l^k and $l^{2(2l+1)-k}$, i.e., to configurations mutually completing each other to form a filled shell.

In the case of an electron configuration containing more than one group of equivalent electrons, it is necessary first of all to find the possible terms for the individual groups of equivalent electrons separately, and then, by using the rule for addition of angular momenta, to find the terms of the given configuration. Let us consider, for example, the configuration $p^4 d$. In accordance with Table 2.1, we have the terms ^1S, ^1D, and ^3P for the configuration p^4. Combining them with $l = 2$, $s = 1/2$ we obtain: from the term ^1S the term ^2D; from the term ^1D the terms ^2G, ^2F, ^2D, ^2P, ^2S; from the term ^3P the terms ^2F, ^2D, ^2P, ^4F, ^4D, ^4P. Thus the terms ^2SPDFG, ^4PDF correspond to the configuration $p^4 d$.
$\phantom{\text{terms}^2S}$2 3 2

2.2.4 Radiative Transitions[1]

The selection rule (1.26) generalizes in the following way for the case of a multi-electron atom. Electric dipole radiative transitions $LSJM$–$L'S'J'M'$ are allowed, provided

$$\Delta J = 0, \pm 1; \; J + J' \geqslant 1 \tag{2.8}$$

odd term $\rightleftarrows$ even term. $\tag{2.9}$

---

[1] For a detailed discussion of the problems connected with radiative transitions see Chap. 9. All formulas necessary for calculating transition probabilities are given there.

The selection rules (2.8) and (2.9) are not associated with any approximation. According to (2.9), transitions are possible only between terms of different parity. The probability of an electric dipole transition is determined by the matrix element of the electric dipole moment, which is not dependent on the spin coordinates of the electrons. If the spin-orbit interaction is small, as is assumed in the case of LS coupling, the spin of an atom does not change in an electric dipole transition. Therefore

$$\Delta S = 0 \tag{2.10}$$

$$\Delta L = 0, \pm 1; \; L + L' \geqslant 1. \tag{2.11}$$

In accordance with (2.10), transitions are possible only between terms of one multiplicity. Transitions between terms of different multiplicities, the so-called intercombination transitions, are forbidden. This selection rule is valid as long as the spin-orbit interaction is small; it is violated in some cases. Fulfillment of (2.10) is evidence in favor of the applicability of the LS coupling approximation.

The relative intensities of the components of a multiplet obey the following sum rule. The sum of intensities of all the components of a multiplet $LSJ \rightarrow LSJ'$ having the same initial level J is proportional to the statistical weight of this level $(2J+1)$. The sum of the intensities of all the components of a multiplet having the same final level J' is proportional to the statistical weight of this level $(2J'+1)$. There are additional sum rules determining the relative intensity of the components of a supermultiplet and of a set of transitions (Sect. 9.2). By a "supermultiplet" is understood all transitions between two polyads, and by a "set of transitions", all transitions between the terms of two electron configurations.

2.3 jj Coupling Approximation

2.3.1 Various Coupling Schemes

Analysis of experimental data shows that the range of applicability of the LS coupling approximation is limited. The system of levels of many atoms differs substantially from that to which the LS coupling approximation corresponds. It is therefore of interest to consider another limiting case, when the spin-orbit interaction considerably exceeds the electrostatic interaction. This case is called jj type coupling or simply jj coupling. If the spin-orbit interaction is large, the concepts of separate orbital angular moment and spin angular momentum of an electron lose meaning. One can only speak of the total angular momentum of an electron j, as only this angular momentum is conserved. jj coupling is rarely found in pure form in atomic spectra; however, the structure of the spectra of the heavy elements very closely approaches the structure characteristic of jj coupling. Generally speaking, in passing from the light to heavy elements, a more or less continuous transition from LS coupling to jj coupling occurs, i.e., there is an intermediate type of coupling. jj coupling is of particular interest for multiply charged ions. The electrostatic interaction $\langle e^2/|r_1-r_2|\rangle$ between electrons which are in the field of a nuclear charge Ze is approximately proportional to Z. We recall that the radius of the first Bohr orbit for a hydrogenlike ion with a charge Ze is proportional to $1/Z$. But the energy of the spin-orbit interaction is proportional to Z^4 (Sect. 1.3). Thus the role of spin-orbit interaction rapidly increases with increase of Z.

jj coupling is also of interest for nuclear theory, as precisely this type of coupling frequently occurs in nuclear shells. The choice between the different types of couplings, i.e., the answer to the question as to which interaction, electrostatic or spin-orbit, is greater, is frequently different for different levels of one and the same atom. As a rule, the levels of atoms of the beginning and middle of the periodic systems of elements, which correspond to lower excited states, are well described in the LS coupling approximation. This approximation, however, is not applicable to highly excited levels of atoms. States in which one of the electrons is on the average at a great distance from the nucleus and from the remaining electrons of the atom correspond to these levels. The electrostatic interaction of the electrons of an atomic core with the outer electron is small in comparison with their spin-orbit interaction. In this case the value of the electrostatic interaction is determined by the mutual orientation of the total angular momentum of the atomic core J' and of the orbital angular momentum of the outer electron l.

It is significant that, with few exceptions, all real spectra can be systematized with respect to LS or jj coupling schemes, even if neither of these limiting cases is, strictly speaking, applicable. Comparing the systems of terms for the two limiting cases of LS and jj coupling, one can obtain an idea of the system of levels in the case of an intermediate type of coupling. As a rule, such a qualitative treatment proves to be sufficient for purposes of systematization of spectra.

Speaking of the different types of coupling, we mean only the fact that one of the interactions, spin-orbit or electrostatic, is small in comparison with the other. This terminology is associated with one's being able to interpret electrostatic and spin-orbit interactions as couplings of different types between the vectors l and s. In the LS coupling approximation, the electrostatic interaction can be treated as a coupling of the vectors l_i, l_j and s_i, s_j. For all states pertaining to a given LS term, the condition $\sum_i l_i = L$ and $\sum_i s_i = S$ is imposed on the vectors l_i and s_i. The energy depends on how the angular momenta l_i sum up into the total angular momentum L and the spins s_i into the total spin S. The spin-orbit interaction and splitting with respect to J, associated with this interaction, can be considered as the consequence of the coupling between the angular momenta L and S. The energy depends on how the vectors L and S sum up into the vector of the total angular momentum $J = L + S$. Bearing this interpretation in mind, the Russell–Saunders approximation is spoken of as coupling of type LS.

In the event of the decisive role being played by the spin-orbit interaction, the energy depends in the first place on how the orbital and spin angular momenta of each electron l_i and s_i sum up into the total angular momentum j_i of the electron. Consequently, one speaks of a breakdown of coupling between the vectors l_i, l_j and s_i, s_j and of the appearance of a coupling between the vectors l_i, s_i, and l_j, s_j. Electrostatic interaction now leads to splitting, depending on how the vectors j_i sum up into the total angular momentum J. Hence the term jj coupling.

2.3.2 Systematics of Electron States with jj Coupling

In the jj coupling scheme, the state of each electron is described by the four quantum numbers $nljm$. For a given value of j, $l = j \pm 1/2$. One of these values is even and the other odd; therefore, the assignment of j and of the parity of the state uniquely determines l. The value of j is usually shown on the right and below the value of l, for example, $p_{1/2}$, $d_{5/2}$, and so on.

Obviously, the following states are possible:

$$s_{1/2}, p_{1/2}, p_{3/2}, d_{3/2}, d_{5/2}, f_{5/2}, f_{7/2}, g_{7/2}, g_{9/2}, h_{9/2}, h_{11/2},$$

the states $s, d, g, \ldots$ being even, and the states $p, f \ldots$ odd.

The states $j = l + 1/2$ and $j = l - 1/2$, owing to the spin-orbit interaction, correspond to different energy levels. If electrostatic interaction between the electrons is completely neglected, the energy of each electron does not depend on the orientation of its total angular momentum j in space, i.e., it is entirely determined by the assignment of the three quantum numbers nlj. Each j state in this case is $(2j + 1)$-fold degenerate. When $j = l + 1/2$, $2j + 1 = 2l + 2$; when $j = l - 1/2$, $2j + 1 = 2l$. Thus, $2l + 2$ states with different values of m pertain to the level $j = l + 1/2$, and $2l$ states to the level $j = l - 1/2$. On taking into account electrostatic interaction, a level described by a set of quantum numbers $n_i l_i j_i$, assigned to each electron, splits into a number of levels characterized by definite values of the total angular momentum J. Finding the possible values of J is carried out in exactly the same way as finding the possible terms in LS coupling. In the case of nonequivalent electrons, it is easy to find the allowed values of J by means of the general rule for addition of quantum mechanical angular momenta. Let us consider, for example, the configuration $np\,nd$. For a p electron, $j = 1/2, 3/2$; for a d electron, $j = 3/2, 5/2$. Possible values of the total angular momentum are given in Table 2.2. States with given values of $j_1, j_2,$ and J are denoted by means of $(j_1 j_2)_J$. Thus, the states $j_1 = 1/2, j = 3/2$ and $J = 1, 2$ are the states $(1/2\ 3/2)_1$ and $(1/2\ 3/2)_2$. The appropriate notations are given in the last column of Table 2.2. The total number of levels with a given value J for a specific electron configuration must be one and the same, both in the case of LS and in the case of jj coupling. In the case of equivalent electrons, just as in LS coupling, it is necessary to take into account the Pauli principle.[2]

Table 2.2. Terms of the configuration $np\ nd$ in the jj coupling approximation

j_1	j_2	$\|j_1 - j_2\| \leqslant J \leqslant j_1 + j_2$	Terms
$1/2$	$3/2$	1 2	$(1/2\ 3/2)_{1,2}$
$1/2$	$5/2$	2 3	$(1/2\ 5/2)_{2,3}$
$3/2$	$3/2$	0 1 2 3	$(3/2\ 3/2)_{0,1,2,3}$
$3/2$	$5/2$	1 2 3 4	$(3/2\ 5/2)_{1,2,3,4}$

Allowed levels for configurations j^n are given in Table 2.3. In the event of a given level occurring several times, the corresponding number is shown below.

Table 2.3. j^n configuration terms

Configuration	J	g
$(1/2)$	$1/2$	2
$(1/2)^2$	0	1
$(3/2)^1\ (3/2)^3$	$3/2$	4
$(3/2)^2$	0 2	6
$(5/2)^1\ (5/2)^5$	$5/2$	6
$(5/2)^2\ (5/2)^4$	0 2 4	15
$(5/2)^3$	$3/2\ 5/2\ 9/2$	20
$(7/2)^1\ (7/2)^7$	$7/2$	8
$(7/2)^2\ (7/2)^6$	0 2 4 6	28
$(7/2)^3\ (7/2)^5$	$3/2\ 5/2\ 7/2\ 9/2\ 11/2\ 15/2$	56
$(7/2)^4$	0 2 4 5 6 8	70
	2 2	

In conclusion, let us note one important fact. If spin-orbit splitting is completely neglected in the case of LS coupling, and electrostatic in the case of jj coupling, then we obtain a different number of levels. For example, in the case of LS coupling, for a two-electron configuration the number of terms equals $2(2l_{min} + 1)$, where l_{min} is the minimum of the numbers $l_1\ l_2$; when $l_{min} = 1,2,3,4$ we obtain 6, 10, 14, 18 … terms. But in the case of jj coupling only four different combinations of the numbers $j_1 j_2$ are possible, as $j_1 = l_1 \pm 1/2$ and $j_2 = l_2 \pm 1/2$. Thus, if a spectrum is being investigated by means of an apparatus which cannot resolve small splitting, then in the case of jj coupling a spectrum

---

[2] Let us note that it is nonequivalent electrons that are of greatest interest for jj coupling. Electrostatic interaction is always large for equivalent electrons.

will have considerably fewer lines than in the case of LS coupling. The same will also occur if broadening of spectral lines makes the resolution of closely spaced lines impossible.

Chapter 3 Spectra of Multielectron Atoms

This chapter contains a brief discussion of the specific features of the spectra of multielectron atoms belonging to various groups of the periodic table according to successive filling of electron shells.

3.1 Periodic System of Elements

The electrons of an atom in the ground state occupy those levels allowed by the Pauli principle with the lowest energy. The number of electrons of an atom increases by one in passing from an atom with atomic number Z to an atom with atomic number $Z + 1$. The added electron occupies the lowest of the states not occupied by other electrons. This process of successive filling of electron shells is illustrated by Table 3.1. The electron configurations of the ground states of atoms (inner filled shells have been omitted) and also the ground state term and ionization potentials are given in this table. Knowing the electron configuration, the ground state term can be determined by Hund's rule.

The table begins with hydrogen, the ground state of which is the state $1s$. The next element, He, corresponds to the configuration $1s^2$. The third element, Li, has the ground configuration $1s^2 2s$. In accordance with the Pauli principle, there cannot be more than two electrons in the state $1s$; therefore the third electron of the Li atom occupies the lowest free state $2s$. Filling of the states $n = 2$ begins from the Li atom. Then comes Be with the configuration $1s^2 2s^2$. The states $2p$ are filled beginning with B right up to Ne. The states with the quantum number $n = 3$, first the $3s$ and then the $3p$ states, are successively filled beginning with Na. This continues up to Ar, which corresponds to the configuration $1s^2\,2s^2\,2p^6\,3s^2\,3p^6$. Then the process of filling the states with $n = 3$ is temporarily interrupted. The added electrons in the K and Ca atoms do not occupy $3d$ states but the states $4s$ and $4s^2$, which are found to be energetically more favorable. Filling of the first principal groups of the periodic system ends with the Ca atom. Elements not containing d or f electrons at all, or containing filled d or f shells, belong to the principal groups. Filling of the $3d$ states begins in the elements of the first intermediate (transition) group, the so-called iron group, Sc, Ti, and so on. This process is not so regular as the filling of the s and p states in the elements of the principal groups. From Sc to V, the added electrons successively occupy the states $3d4s^2$, $3d^2 4s^2$ and $3d^3 4s^2$. In the next element, Cr, the state $3d^5 4s$ is energetically more favorable and not $3d^4 4s^2$ as might have been expected. In Mn atom the added electron occupies the $4s$ state to yield the configura-

Table 3.1. Electron configuration of atoms

Element	Electron configuration	Ground state term	E_i [eV]	Element	Electron configuration	Ground state term	E_i [eV]
1 H	$1s$	$^2S_{1/2}$	13.598	51 Sb	$5s^2\ 5p^3$	$^4S_{3/2}$	8.641
2 He	$1s^2$	$^1S_0$	24.587	52 Te	$5s^2\ 5p^4$	$^3P_2$	9.009
3 Li	$2s$	$^2S_{1/2}$	5.392	53 I	$5s^2\ 5p^5$	$^2P_{3/2}$	10.451
4 Be	$2s^2$	$^1S_0$	9.322	54 Xe	$5s^2\ 5p^6$	$^1S_0$	12.130
5 B	$2s^2\ 2p$	$^2P_{1/2}$	8.298	55 Cs	$6s$	$^2S_{1/2}$	3.894
6 C	$2s^2\ 2p^2$	$^3P_0$	11.260	56 Ba	$6s^2$	$^1S_0$	5.212
7 N	$2s^2\ 2p^3$	$^4S_{3/2}$	14.534	57 La	$5d\ 6s^2$	$^2D_{3/2}$	5.577
8 O	$2s^2\ 2p^4$	$^3P_2$	13.618	58 Ce	$4f\ 5d\ 6s^2$	$^1G_4$	5.47
9 F	$2s^2\ 2p^5$	$^2P_{3/2}$	17.422	59 Pr	$4f^3\ 6s^2$	$^4I_{9/2}$	5.42
10 Ne	$2s^2\ 2p^6$	$^1S_0$	21.564	60 Nd	$4f^4\ 6s^2$	$^5I_4$	5.49
11 Na	$3s$	$^2S_{1/2}$	5.139	61 Pm	$4f^5\ 6s^2$	$^6H_{5/2}$	5.55
12 Mg	$3s^2$	$^1S_0$	7.646	62 Sm	$4f^6\ 6s^2$	$^7F_0$	5.63
13 Al	$3s^2\ 3p$	$^2P_{1/2}$	5.986	63 Eu	$4f^7\ 6s^2$	$^8S_{7/2}$	5.67
14 Si	$3s^2\ 3p^2$	$^3P_0$	8.151	64 Gd	$4f^7\ 5d\ 6s^2$	$^9D_2$	6.14
15 P	$3s^2\ 3p^3$	$^4S_{3/2}$	10.486	65 Tb	$4f^9\ 6s^2$	$^6H_{15/2}$	5.85
16 S	$3s^2\ 3p^4$	$^3P_2$	10.360	66 Dy	$4f^{10}\ 6s^2$	$^5I_8$?	5.93
17 Cl	$3s^2\ 3p^5$	$^2P_{3/2}$	12.967	67 Ho	$4f^{11}\ 6s^2$	$^4I_{15/2}$	6.02
18 Ar	$3s^2\ 3p^6$	$^1S_0$	15.759	68 Er	$4f^{12}\ 6s^2$	$^3H_6$	6.10
19 K	$4s$	$^2S_{1/2}$	4.341	69 Tm	$4f^{13}\ 6s^2$	$^2F_{7/2}$	6.18
20 Ca	$4s^2$	$^1S_0$	6.113	70 Yb	$4f^{14}\ 6s^2$	$^1S_0$	6.254
21 Sc	$3d\ 4s^2$	$^2D_{3/2}$	6.54	71 Lu	$5d\ 6s^2$	$^2D_{3/2}$	5.426
22 Ti	$3d^2\ 4s^2$	$^3F_2$	6.82	72 Hf	$5d^2\ 6s^2$	$^3F_2$	7.0
23 V	$3d^3\ 4s^2$	$^4F_{3/2}$	6.74	73 Ta	$5d^3\ 6s^2$	$^4F_{3/2}$	7.89
24 Cr	$3d^5\ 4s$	$^7S_3$	6.766	74 W	$5d^4\ 6s^2$	$^5D_0$	7.98
25 Mn	$3d^5\ 4s^2$	$^6S_{5/2}$	7.435	75 Re	$5d^5\ 6s^2$	$^6S_{5/2}$	7.88
26 Fe	$3d^6\ 4s^2$	$^5D_4$	7.870	76 Os	$5d^6\ 6s^2$	$^5D_4$	8.7
27 Co	$3d^7\ 4s^2$	$^4F_{9/2}$	7.86	77 Ir	$5d^7\ 6s^2$	$^4F_{9/2}$?	9.1
28 Ni	$3d^8\ 4s^2$	$^3F_4$	7.635	78 Pt	$5d^9\ 6s$	$^3D_3$	9.0
29 Cu	$4s$	$^2S_{1/2}$	7.726	79 Au	$6s$	$^2S_{1/2}$	9.225
30 Zn	$4s^2$	$^1S_0$	9.394	80 Hg	$6s^2$	$^1S_0$	10.437
31 Ga	$4s^2\ 4p$	$^2P_{1/2}$	5.999	81 Tl	$6s^2\ 6p$	$^2P_{1/2}$	6.108
32 Ge	$4s^2\ 4p^2$	$^3P_0$	7.899	82 Pb	$6s^2\ 6p^2$	$^3P_0$	7.416
33 As	$4s^2\ 4p^3$	$^4S_{3/2}$	9.81	83 Bi	$6s^2\ 6p^3$	$^4S_{3/2}$	7.289
34 Se	$4s^2\ 4p^4$	$^3P_2$	9.752	84 Po	$6s^2\ 6p^4$	$^3P_2$	8.42
35 Br	$4s^2\ 4p^5$	$^2P_{3/2}$	11.814	85 At	$6s^2\ 6p^5$	$^2P_{3/2}$	9.5
36 Kr	$4s^2\ 4p^6$	$^1S_0$	13.999	86 Rn	$6s^2\ 6p^6$	$^1S_0$	10.748
37 Rb	$5s$	$^2S_{1/2}$	4.177	87 Fr	$7s$	$^2S_{1/2}$	4.0
38 Sr	$5s^2$	$^1S_0$	5.695	88 Ra	$7s^2$	$^1S_0$	5.279
39 Y	$4d\ 5s^2$	$^2D_{3/2}$	6.38	89 Ac	$6d\ 7s^2$	$^2D_{3/2}$	5.17
40 Zr	$4d^2\ 5s^2$	$^3F_2$	6.84	90 Th	$6d^2\ 7s^2$	$^3F_2$	6.08
41 Nb	$4d^4\ 5s$	$^6D_{1/2}$	6.88	91 Pa	$5f^2\ 6d\ 7s^2$	$^4K_{11/2}$	5.9
42 Mo	$4d^5\ 5s$	$^7S_3$	7.099	92 U	$5f^3\ 6d\ 7s^2$	$^5L_6$	6.05
43 Tc	$4d^5\ 5s^2$	$^6S_{5/2}$	7.28	93 Np	$5f^4\ 6d\ 7s^2$	$^6L_{11/2}$	6.2
44 Ru	$4d^7\ 5s$	$^5F_5$	7.37	94 Pu	$5f^6\ 7s^2$	$^7F_0$	6.06
45 Rh	$4d^8\ 5s$	$^4F_{9/2}$	7.46	95 Am	$5f^7\ 7s^2$	$^8S_{7/2}$	6.0
46 Pd	$4d^{10}$	$^1S_0$	8.34	96 Cm	$5f^7\ 6d\ 7s^2$	$^9D_2$	6.02
47 Ag	$5s$	$^2S_{1/2}$	7.576	97 Bk	$5f^9\ 7s^2$	$^6H_{15/2}$	6.23
48 Cd	$5s^2$	$^1S_0$	8.993	98 Cf	$5f^{10}\ 7s^2$	$^5I_8$	6.30
49 In	$5s^2\ 5p$	$^2P_{1/2}$	5.786	99 Es	$5f^{11}\ 7s^2$	$^4I_{15/2}$	6.4
50 Sn	$5s^2\ 5p^2$	$^3P_0$	7.344	100 Fm	$5f^{12}\ 7s^2$	$^4I_{15/2}$	6.5

tion $3d^54s^2$. Then come Fe with the configuration $3d^64s^2$, Co with the configuration $3d^74s^2$, and Ni with the configuration $3d^84s^2$. The regularity of filling of shells is again broken in the next element Cu; the configuration $3d^{10}4s$ occurs instead of the configuration $3d^94s^2$. Thus, Cu contains a completely filled $3d$ shell and therefore belongs to elements of the principal groups. The $4s$, $4p$, and $5s$ states are successively filled in the next elements. After this the $4d$ shell is filled in the elements of the second transition group, the palladium group. Here again there occurs a peculiar competition between $4d$ and $5s$ states. As a result, after Zr with configuration $4d^25s^2$, there follows Nb with configuration $4d^45s$, and after Rh with configuration $4d^85s$ there is Pd with configuration $4d^{10}$. Irregularities of this type are also met in the filling of the shells of the elements of the platinum group. The f shells are filled even more irregularly. The $4f$ states begin to be filled in the rare-earth elements later than the $5p$ and $6s$ states, a competition also occurring between the $4f$, $5d$ and $6s$ states. The rare-earth elements, as a rule, hardly differ from each other as regards their chemical properties. This is because in the $4f$ state the electron is on the average considerably closer to the nucleus than, for example, in the $5p$ or $6s$ states. The chemical properties are determined basically by the peripheral electrons, in this case the s and p electrons of the earlier filled shells.

If the anomalies mentioned above are not taken into account, the sequence of filling of states is determined in general by the value of $n + l$. States are filled in the following order: $1s$ — 2 electrons, $2s2p$ — 8 electrons, $3s3p$ — 8 electrons, $4s3d4p$ — 18 electrons, $5s4d5p$ — 18 electrons, $6s4f5d6p$ — 32 electrons, and so on.

The principal regularities of the structure of electron shells discussed above are reflected in the periodic system of elements of Mendeleev. The whole set of elements was subdivided by Mendeleev according to their physicochemical properties into seven periods; this subdivision is still retained now and includes a number of elements discovered later. Each of the periods begins with an alkali element and ends with an atom of a noble gas (with the exception of the last incomplete period). Thus the beginning of a period coincides with the beginning of the filling of a new shell. The ionization potential, which is determined by the binding energy of the electron in an atom, increases on the whole, although not monotonically, in proportion to the filling of the shells. The greatest ionization potential value is reached in the atoms of the noble gases, which correspond to completely filled shells. The ionization potential drops sharply in passing to the alkali elements (Table 3.1).

3.2 Spectra of the Alkali Elements

3.2.1 Term Scheme

The electron shells of the atoms of the alkali elements Li, Na, K, Rb, and Fr have the same structure; there is one electron in the state ns outside the filled shells. The term $^2S_{1/2}$ is the ground state term. The filled shells are very stable, as their

3.2 Spectra of the Alkali Elements

structure is the same as that of atoms of the noble gases. For this reason, the spectra of alkali metal atoms are determined solely by the transitions of the outer and most weakly bound electron. The effective field in which this electron moves is centrally symmetric, since the filled shells always have a total orbital angular momentum and total spin equal to zero. At great distances the effective field coincides with the Coulomb field of charge e, because the electrons of the closed shells screen the nuclear field. At short distances (near the nucleus), screening does not occur and the role of the filled shells reduces to the creation of a certain constant potential. Thus

$$U(r) \to -e^2/r, \quad r \to \infty, \quad U(r) \to -Ze^2/r, \quad r \to 0. \tag{3.1}$$

Since the curve $U(r)$ lies below the Coulomb potential $-e^2/r$ at all distances, the level n,l lies below the corresponding level of the hydrogen atom

$$E_{nl} < -\text{Ry}/n^2. \tag{3.2}$$

The further the electron is from the filles shell, the more hydrogenlike is the field; therefore it can be expected that for large n,l the system of levels is close to that of hydrogen.

These general considerations are confirmed by experimental data. The term schemes of Li, Na, K, Rb, and Cs are given in Fig. 3.1. The corresponding

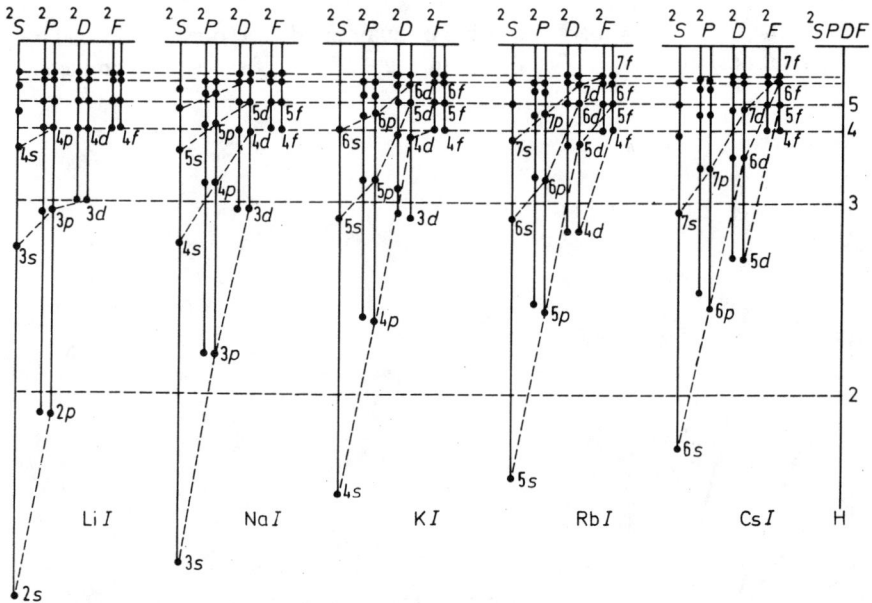

Fig. 3.1. Term scheme of alkali atoms

hydrogen terms are plotted with a broken line. The term scheme of Li for small n and l is essentially different from that of hydrogen. The degeneracy with respect to l, typical for hydrogen, does not occur. With the increase of n and l, the terms coincide more and more with the hydrogen ones. The distance between the levels E_{nl} and $E_{nl'}$ decreases with increasing n and l. For a given n, the levels are more hydrogenlike the greater l is. This relationship has a simple physical meaning. On the average, the optical electron in the state n,l spends more time at large distances from the nucleus, where the field is close to the Coulomb one, the larger l is.

With Na, the difference of the field from the Coulomb one appears even more strongly than with Li. The arrangement of the lower levels differs even more from what is characteristic of hydrogen. Thus the $4s$ level lies lower than the $3d$.

A similar picture also occurs with Rb. The $5s$ and $5p$ levels lie considerably lower than the $4d$ and $4f$ levels. Just as in the case of Li, hydrogenlikeness is restored for large n,l.

It is customary to describe the terms of alkali metal atoms by analogy with hydrogen by the formula

$$E_{nl} = -\mathrm{Ry}/n_*^2 , \qquad (3.3)$$

where n_* is the effective principal quantum number, which is selected so as to satisfy experimental data. Comparison of (3.3) with experiment shows that n_* can with good accuracy be represented in the form of the difference

$$n_* = n - \Delta_l , \qquad (3.4)$$

where Δ_l, the so-called Rydberg correction or quantum defect, does not depend on n. The relationship of Δ_l to l is shown in Fig. 3.2. The f states are completely

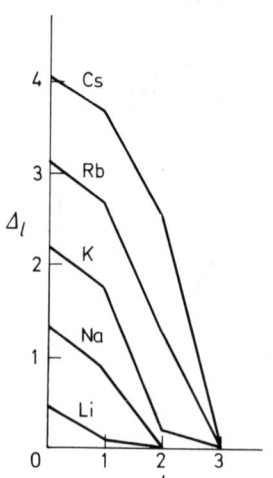

Fig. 3.2. Magnitude of the quantum defect for the series of alkali atoms

hydrogenlike in all cases – even for Cs, which corresponds to the greatest values of Δ_l, when $l = 3$, $\Delta_l = 0$.

It is significant that the lowest of the values of n_* is always greater than one. For example, $(n_*)_{min} = 1.627$ for Na, $(n_*)_{min} = 1.8$ for Rb, and so on. Thus the ionization potential E_i and resonance potentials E_r of the alkali metals are considerably less than for hydrogen.

3.2.2 Series Regularities

In the spectra of alkali elements one can distinguish a number of series of the same type as in hydrogen. A number of the series are overlapped in the visible region of the spectrum. The following four series are the main ones:

$ns\ S - n'p\ P$ Principal series

$np\ P - n's\ S$ Sharp series

$np\ P - n'd\ D$ Diffuse series

$nd\ D - n'f\ F$ Fundamental series

The transitions $S \rightleftarrows P$, $P \rightleftarrows D$, $D \rightleftarrows F$ ... are the only transition allowed by the selection rule $\Delta L = 0, \pm 1$; even term $\rightleftarrows$ odd term. The sharp series is also called the first subordinate, the diffuse series – the second subordinate, and the fundamental series – the Bergmann series. The ionization potentials and resonance potentials of the atoms of alkali metals are small – E_r is of the order of $1.5 - 2$ eV. Therefore, alkali metal atoms are easily excited even in comparatively low temperature sources. The main spectral series are located in the visible and infrared regions of the spectrum. The resonance lines are located in the visible part of the spectrum.

3.2.3 Fine Structure

All the terms of the alkali atoms, with the exception of ^2S terms, are doublet. As a rule, the level $j = 1/2$ lies below the level $j = 3/2$. We have, therefore, the following terms:

$^2S_{1/2}$, $^2P_{1/2\ 3/2}$, $^2D_{3/2\ 5/2}$, $^2F_{5/2\ 7/2}$.

The value of the fine splitting increases rapidly with increase of atomic number Z and decreases with increase of n.

The selection rule with respect to j allows the transitions $\Delta j = 0, \pm 1$. Taking this into account, we obtain: principal series – doublets $^2S_{1/2} - {}^2P_{1/2\ 3/2}$; sharp series – doublets $^2P_{1/2}, {}^2P_{2/3} - {}^2S_{1/2}$; diffuse series – triplets $^2P_{1/2} - {}^2D_{3/2}, {}^2P_{3/2} - {}^2D_{3/2}, {}^2P_{3/2} - {}^2D_{5/2}$; fundamental series – triplets $^2D_{3/2} - {}^2F_{5/2}, {}^2D_{5/2} - {}^2F_{5/2}, {}^2D_{5/2} - {}^2D_{7/2}$..

Doublet splitting of the lines of a principal series is determined by the fine structure of the terms ^2P. Thus doublet splitting is particularly large for the leading lines of a principal series. The splitting decreases rapidly with increase in n'.

Doublet splitting of the lines of a sharp series is entirely determined by the fine structure of the lower term ^2P. Therefore all lines of a sharp series have the same doublet splitting on a frequency scale.

The splitting of the two triplet components $^2P_{1/2} - {}^2D_{3/2}$ and $^2P_{3/2} - {}^2D_{3/2}$ is constant for all lines of a diffuse series. But the distance between the components $^2P_{3/2} - {}^2D_{3/2}$ and $^2P_{3/2} - {}^2D_{5/2}$ is considerably less and drops rapidly for the higher members of a series. The structure of lines of the fundamental series can easily be established in a similar way.

According to the rule formulated above, the ratio of the intensities of doublet components originating from the levels j_1 and j_2 equals $(2j_1 + 1) : (2j_2 + 1)$. This ratio is 1:2 for the principal series.

3.2.4 Copper, Silver, and Gold Spectra

The atoms of Cu, Ag, and Au also have one ns electron outside filled shells in the ground state. The Ag atom is preceded in the periodic system by the Pd atom, the $4d$ shell of which is completely filled. Therefore in the case of Ag, only the outer $5s$ electron is comparatively easily excited and the spectrum completely resembles the spectra of alkali elements. The situation is somewhat different for Cu and Au. The Cu atom is preceded by Ni with the configuration $3d^8\,4s^2$ and not $3d^{10}$. This is due to the above-mentioned competition of the s and d states. Similarly, Au is preceded by Pt with the configuration $5d^9 6s$. This indicates that the binding energies of the s and d electrons are approximately the same in the case of Cu and Au; therefore, besides excitation of the s electron, excitation of the d electron is possible. The excited states of the s electron of Cu and Au correspond to systems of terms of the same type as in the case of atoms of the alkali metals. New states are also possible upon the excitation of the d electron. Thus for Cu, such states are $3d^9 4s^2$, $3d^9 4sns$, $3d^9 4snp$, $3d^9 4snd$, etc., and, in the general case, $3d^9 4snl$.

The ionization limits of the alkalilike systems of terms of Cu and Au are determined by the energy of the ground states of the Cu$^+$ ions $3d^{10}\,^1S_0$ and of the Au$^+$ ions $5d^{10}\,^1S_0$. But if ionization occurs because one of the d electrons is removed, the Cu$^+$ ion is in one of the states $3d^9 4s\,^1D_2, {}^3D_{1,2,3}$. Therefore, the terms associated with the d electron excitation converge to the ionization limits $3d^9 4s\,^1D_2$ and $3d^9 4s\,^3D_{1,2,3}$. New ionization limits $5d^9 4s\,^1D_2, {}^3D_{1,2,3}$ also appear in the case of Au. The existence of additional systems of terms leads to the spectra of Cu and Au being considerably more complex than the spectra of the alkali elements.

3.3 Spectra of the Alkaline Earth Elements

3.3.1 He Spectrum

The atoms of He, Be, Mg, Ca, Sr, Ba, Ra, Hg, Zn and Cd have two s electrons outside filled shells. The ground state of He is the state $1s^2\,{}^1S_0$. Two systems of terms are possible upon the excitation of one of the s electrons – the singlet system, $S = 0$, $2S + 1 = 1$, and the triplet $S = 1$, $2S + 1 = 3$. The closed shell $1s^2$ is extremely stable, and thus the He ground state term lies very deep, considerably deeper than in the case of hydrogen. The ionization potential of helium is greater than that of any other element, $E_i = 24.5$ eV. The binding energy of the electron in the excited state is considerably less than in the ground state because the second electron, remaining in the $1s$ state, screens the nuclear charge. The first excited level is therefore located very high above the ground state: $E_r \simeq 20$ eV ($\lambda_t \simeq 600$ Å). Transitions between triplet and singlet terms are forbidden in the LS coupling approximation.

Thus two independent systems of lines must be observed in the spectrum. This is just what happens in the case of He. Intercombination lines, corresponding to transitions between triplet and singlet terms, are absent. In view of this, two species of helium, orthohelium and parahelium, have been spoken of for a long time. This terminology is retained even now. The singlet system of terms is the parahelium system and the triplet system, the orthohelium system of terms.

Transition of the following types are allowed by the selection rules within each of the systems of terms:

$1s^2\,{}^1S_0 \;-\; 1snp\,{}^1P_1$, $\qquad\qquad$ $1s2s\,{}^3S_1 \;-\; 1snp\,{}^3P_{0,1,2}$,

$1s2p\,{}^1P_1 \;-\; 1sns\,{}^1S_0$, $\qquad\qquad$ $1s2p\,{}^3P_{0,1,2} \;-\; 1sns\,{}^3S_1$,

$1s2p\,{}^1P_1 \;-\; 1snd\,{}^1D_2$, $\qquad\qquad$ $1s2p\,{}^3P_{0,1,2} \;-\; 1snd\,{}^3D_{1,2,3}$,

$1s3d\,{}^1D_2 \;-\; 1snf\,{}^1F_3$, $\qquad\qquad$ $1s3d\,{}^3D_{1,2,3} \;-\; 1snf\,{}^3F_{2,3,4}$,

and so on. Just as in the case of alkali spectra, these series are often called principal, sharp, diffuse, and fundamental. The lowest triplet state of He is the term $1s2s\,{}^3S_1$. As the transition $1s2s\,{}^3S_1 - 1s^2\,{}^1S_0$ is forbidden, this state is metastable.

A sharp deviation from Landé's interval rule is noticeable when analyzing the multiplet splitting of the He triplet terms. The splitting has an inverted order. The interval ratio approximately equals 1:14 instead of 2:1 by Landé's rule. It is impossible to attribute the observed splitting to a deviation from the LS coupling approximation, since intercombination transitions are not observed in the He spectrum. This is characteristic of the LS coupling, as already noted above. It will be shown in Sect. 5.5 that this divergence is in fact due to other reasons.

Only the spectral lines due to transitions between triplet terms, obviously, have a multiplet structure. We shall examine, by way of example, the transitions $1s2s\,^3S_1$—$1snp\,^3P_{0,1,2}$ $1s2p\,^3P_{0,1,2}$—$1sns\,^3S_1$, and $1s2p\,^3P_{0,1,2}$—$1snd\,^3D_{1,2,3}$. In the first case, all splitting is determined by the fine structure of the upper level. This splitting falls rapidly with increasing n. The corresponding lines are triplets, but the triplet structure can be resolved only for small values of n. On the other hand, in the case of the transition $1snp\,^3P_{0,1,2}$—$1sns\,^3S_1$ the splitting is determined by the lower level; therefore the triplet structure is not dependent on n and is the same for all lines of this series. As has just been remarked, the splitting of the levels $^3P_0$, $^3P_1$ is 14 times the splitting of the levels $^3P_1$, $^3P_2$. If this last splitting is not resolved by the apparatus, the lines will have the appearance of doublets.

Six transitions $0 \rightarrow 1$; $1 \rightarrow 1,2$; $2 \rightarrow 1,2,3$ are allowed by the selection rules with respect to J for the lines of the series $1s2p\,^3P_{0,1,2}$—$1snd\,^3D_{1,2,3}$. The lines of this series are therefore sextets. The splitting of the upper level is much less than that of the lower and, in addition, rapidly drops with increasing n. The sextet structure is thus difficult to resolve. The majority of the lines of this series have the appearance of triplets under ordinary conditions. The relative strengths of the components of the multiplets under consideration can be calculated on the basis of the sum rule.

The resonance line He $\lambda_r = 600$ Å lies in the vacuum ultraviolet region of the spectrum. Only the lines corresponding to transitions between excited levels can be observed with the aid of ordinary spectral apparatus. A number of very strong lines of He are located in the infrared region of the spectrum. All these lines require 21–24 eV for their excitation; thus the He spectrum is excited only in high-temperature sources. The He ion is absolutely hydrogenlike and thus does not need special discussion.

3.3.2 Spectra of the Alkaline Earth Elements

The atoms of Be, Mg, Ca, Sr, Ba, and Ra have two s electrons outside filled shells in the ground state. The term $^1S_0$ is the ground term. The nuclear charge is screened by the electrons of the filled shells; therefore the effective charge of the atomic core is approximately equal to two. In the present instance, however, the electrons are at a considerably greater distance from the nucleus than in the case of He. As a result of this, the atoms of alkaline earth elements are characterized by considerably lower excitation and ionization energies than the He atom. Just as in the case of He, the systems of terms, singlet and triplet, occur on the excitation of one of the s electrons. The lowest term of the triplet system $ns\,np$ $^3P_{0,1,2}$ is metastable. In the case of alkaline earth elements, however, the selection rule $\Delta S = 0$ is not so strictly fulfilled as in the case of He. Intercombination lines are observed in the spectra of all these elements, corresponding to transitions from the levels $^3P_1$ to the ground level $ns^2\,^1S_0$. The strength of these lines increases with increasing Z.

3.3 Spectra of the Alkaline Earth Elements

As can be seen from Fig. 3.3, where the term schemes of Be and Mg are given, the term $ns\,np\,^3P$ lies below the first excited singlet term $nsnp\,^1P$ in the case of all alkaline earth atoms. Nevertheless, it is usual to consider the transition $ns^2\,^1S_0$—$ns\,np\,^1P_1$ as the resonance transition in the case of alkaline earth elements, since the corresponding line is considerably stronger than the intercombination one. The term $ns\,np\,^3P$ is called metastable for the same reason.

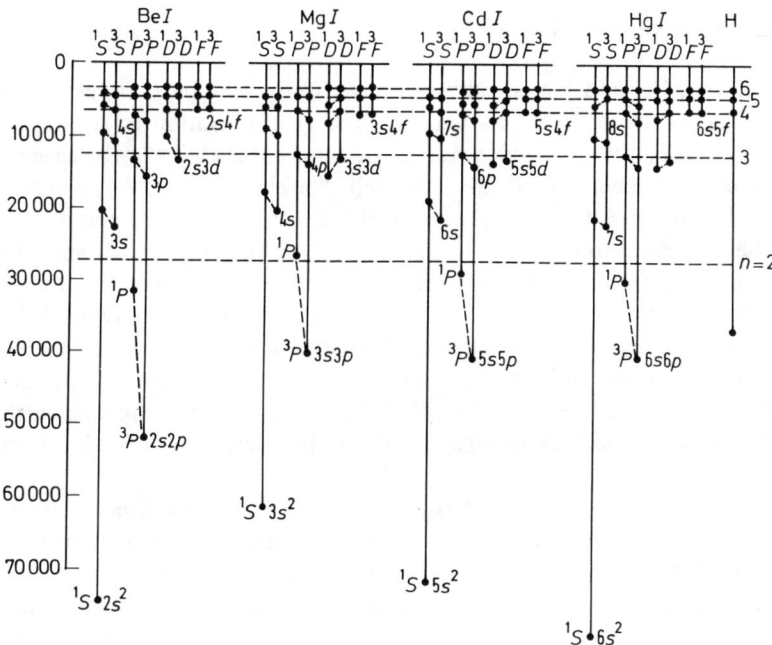

Fig. 3.3. Term scheme of alkaline earth atoms

Just as in the case of alkali spectra, one can distinguish the principal, sharp, diffuse, and fundamental series in the spectra of alkaline earth elements. The lines associated with the transitions between the terms of the triplet system are triplets (principal and sharp series) and sextets (diffuse and fundamental series); here both a normal and an inverted order of splitting are encountered. Alkaline earth atoms are characterized by comparatively small excitation energies. Besides the resonance lines, the leading lines of the sharp and diffuse series in both the singlet and the triplet system of terms are strong in the spectra of the elements under consideration.

The low ionization potential values of the elements being examined determine their easy ionization. The spectra of alkaline earth element ions are fully analogous to the spectra of alkali metals. The excitation energy of these ions is relatively low; for this reason, the lines of alkaline earth element ions are extremely strong even in such sources as an arc source. All alkaline earth elements have a so-called displaced system of terms, due to the simultaneous excitation of the two electrons. For Ca, these terms correspond to the electron configurations $3dns$, $3dnp$, $3dnd$, ..., $4pnp$, and so on. The probabilities of two-electron radiative transitions are negligibly small in comparison with single-electron transitions; the displaced terms do not, therefore, combine with the terms of the main system.

3.3.3 Zinc, Cadmium, and Mercury Spectra

The elements Zn, Cd, and Hg occupy in relation to the alkaline earth elements the same place as do the elements Cu, Ag, and Au in relation to the alkali elements. Two s electrons are added not to the filled np^6 shell, as in the case of the alkaline earth elements, but to the nd^{10} shell. The elements Cu, Ag, and Au, which stand before Zn, Cd, and Hg, respectively, in Table 3.1, have a completely filled nd shell. The binding energy of the nd electron in the Zn, Cd, and Hg atoms considerably exceeds the binding energy of the $(n+1)$ s electrons; therefore only the s electron is excited. Thus the spectra of Zn, Cd, and Hg are fully analogous to the spectra of the alkaline earth elements. The term scheme of Hg is shown by way of example in Fig. 3.3. The intercombination lines in the spectra of these elements are even stronger than in the spectra of alkaline earth elements. Thus, in the mercury spectrum, some of the intercombination lines are very strong.

The spectra of the Zn^+, Cd^+, and Hg^+ ions are similar to the spectra of the ions of the alkaline earth elements and of the neutral atoms of the alkali metals. A competition between the s and d electrons appears, however, in the spectra of these ions. Excitation of the s electron and also excitation of the d electron are possible.

3.4 Spectra of Elements with p Valence Electrons

3.4.1 One p Electron Outside Filled Shells

A p electron first appears in the B atom with the configuration $1s^2 2s^2 2p$. The atoms of Al, Ga, In, and Tl also have ground configurations of the same type, i.e., one p electron outside filled shells. The ground term of all these atoms is the doublet term $^2P_{1/2\,3/2}$, the level $^2P_{1/2}$ being located below the level $^2P_{3/2}$. Dipole radiative transitions between the levels $^2P_{1/2}$ and $^2P_{3/2}$ are forbidden, because both these levels belong to one electron configuration and thus have the same parity. The level $^2P_{3/2}$ is therefore metastable.

The distance between the levels $^2P_{1/2}$ and $^2P_{3/2}$ rapidly increases with increase of atomic number. In the case of B it is only 16 cm^{-1} and in the case of Tl it is 7793 cm^{-1}. The level $2s\ ^2S_{1/2}$ is the resonance level of B and therefore the resonance line is a doublet with splitting of 16 cm^{-1} (transitions $2p\ ^2P_{1/2}$—$3s\ ^2S_{1/2}$ and $2p\ ^2P_{3/2}$—$3s\ ^2S_{1/2}$). Since this splitting is determined by the lower level, the other lines corresponding to the transitions $2p\ ^2P_{1/2,\ 3/2}$—$ns\ ^2S_{1/2}$ also have the same structure.

Also allowed by the dipole selection rules are the transitions $2p\ ^2P_{1/2,\ 3/2}$—$nd\ ^2D_{3/2,\ 5/2}$ to which corresponds the series of triplet lines: $^2P_{1/2}$—$^2D_{3/2}$, $^2P_{3/2}$—$^2D_{3/2}$, and $^2P_{3/2}$—$^2D_{5/2}$. The transition $2p\ ^2P_{1/2,\ 3/2}$—$3d\ ^2D_{3/2,\ 5/2}$ gives the longest wavelength line of this series. We recall that the state $2d$ is impossible because we must have $n \geqslant l+1$.

For the other atoms of the isoelectronic sequence under consideration, Al, Ga,..., the states $nd\ ^2D_{3/2,\ 5/2}$ and $(n+1)s\ ^2S_{1/2}$ will be the closest to the ground state $np\ ^2P_{1/2,\ 3/2}$ ($n \geqslant 3$). The level $(n+1)s\ ^2S_{1/2}$, which is the resonance one, lies lower in all cases. The distance between the ground and resonance levels rapidly decreases with increasing n; the resonance lines therefore displace into the long-wave region of the spectrum with increase of atomic number. Splitting of the resonance line increases at the same time. As already noted above, in the case of Tl one component of the resonance line is located in the visible region and the other in the ultraviolet region. Deviation from LS coupling becomes substantial with such large splitting.

In addition to the terms considered, a number of others are possible corresponding to excitation of one of the s electrons, e.g., belonging to configurations of the type $ns\ np\ n'l$, for example $ns\ np^2$, $ns\ np\ n's$, $ns\ np\ n'd$, etc. The total spin S can have the two values $1/2$ and $3/2$ for three electrons. Accordingly, doublet and quartet terms are possible. These additional terms converge to a limit, which is determined by the energy of the corresponding ion in the excited state $ns\ np$.

The configuration $ns\ np\ n'l$ can be obtained from the ground configuration $ns^2\ np$ by exciting two electrons

$$ns^2\ np \to ns^2\ n'l \to ns\ np\ n'l.$$

Accordingly, it can be assumed that the energies of the state $ns^2 n'l$ and $ns\ np\ n'l$ differ approximately by the excitation energy $E' = E(ns\ np) - E(ns^2)$. It follows from this that the terms of the configuration $ns\ np\ n'l$ are displaced upwards relative to the terms of the configuration $ns^2\ n'l$ approximately by the quantity E'. As noted above, these terms are called displaced terms.

The ground configuration of the ions B^+, Al^+, ... is a configuration of the same type as in the case of the alkaline earth elements, i.e., the configuration ns^2. The spectra of these ions are therefore similar to the spectra of alkaline earth elements.

3.4.2 Configuration p^2

Two equivalent p electrons outside filled shells appear in the ground configurations of the elements C, Si, Ge, Sn, and Pb. The configuration np^2 gives the three terms: $^1S_0$, $^1D_2$, and $^3P_{0,1,2}$ (Table 2.1). In accordance with Hund's rule, the ground state term is the term of maximum multiplicity, i.e., the term ^3P. As the p shell is less than half filled in this instance, the levels $J = 0, 1, 2$ are arranged in normal order, i.e., the level $J = 0$ lies lowest. Electric dipole transitions between the terms $^1S_0$, $^1D_2$ and the ground state term are forbidden by the selection rules with respect to parity. The terms $np^2\,^1S_0$ and $np^2\,^1D_2$ are therefore metastable. Excitation either of one of the p electrons or of one of the s electrons is possible. In the first case, we obtain electron configurations of the type $ns^2npn'l$ (singlet and triplet terms) and, in the other, configurations of the type $ns\,np^2n'l$ (singlet, triplet, and quintet terms). Let us consider, as an example, the terms scheme of carbon. The state $2s^22p^2\,^3P_0$ is the ground state of the C atom. The metastable terms $^1S_0$ and $^1D_2$ also belong to this configuration.

The levels $2s^22p3s\,^1P_1$ and $^3P_{0,1,2}$ are the resonance levels of the carbon atom. The term $^1P_1$ can combine with the terms $^1S_0$ and $^1D_0$ of the ground configuration, and the term ^3P with the term ^3P. Let us note that in this instance the resonance levels are not the lowest excited levels. The level $2s\,2p^3\,^5S_2$ is located somewhat lower. Transitions from this level to the ground level are forbidden by the selection rule $\Delta S = 0$ in the LS coupling approximation. Lines of this type have in fact been detected in the carbon spectrum. The intensity of these lines is very small, and therefore, it is usual to consider the levels of the configuration $2s^22p3s$ as the resonance levels.

Excitation of the resonance levels requires comparatively large energies ($E_r \sim 7.5$ eV); for this reason the carbon spectrum belongs among those relatively difficult to excite.

The term schemes of Si, Ge, Sn and Pb are basically of similar form. The excitation energy of the resonance terms of these atoms is somewhat lower than in the case of carbon. Thus E_r is approximately 5 eV for Si; therefore the resonance lines of Si lie in an accessible part of the ultraviolet region of the spectrum.

A noticeable divergence from LS coupling and a transition to jj coupling is observed for the heavy atoms of the isoelectronic sequence being examined. The C^+, Si^+, ... ions have a ground configuration of the same type as B, Al, ... , i.e., ns^2np. Accordingly, the spectra of these ions are analogous to the spectra of the elements B, Al, ...

3.4.3 Configuration p^3

The atoms of N, P, As, Sb, and Bi have a p^3 configuration in the ground state. The terms ^2P, ^2D, and ^4S correspond to this configuration. In accordance with Hund's rule, the ^4S state is the ground state term. The levels ^2P and ^2D are metastable. Among the terms of the excited configurations $np^2\,n'l$, only the even terms can combine with the terms of the ground configuration. Such terms are given, for example, by the configurations $np^2\,n's$ and $np^2\,n'd$. Also possible is the even configuration $ns\,np^4$, corresponding to the excitation of one of the electrons of the group ns^2.

We shall consider, by way of example, the term scheme of N. The term $2p^3\,^4S_{3/2}$ is the ground state term of N and the term $2p^2[^3P]3s\,^4P$ is the resonance state. The remaining terms of the configuration $2p^2 3s$, namely $2p^2[^1S]3s\,^2S$, $2p^2[^1D]3s\,^2D$, and $2p^2[^3P]3s\,^2P$, cannot combine with the ground state term due to the selection rule $\Delta S = 0$. These terms can combine only with the metastables $2p^3\,^2P$ and $2p^3\,^2D$. The selection rule $\Delta S = 0$ in the N spectrum is violated and some of the intercombination lines are observed.

The resonance potential of nitrogen is comparatively large and is approximately 10 eV; transitions between the terms of the ground and first excited configurations give lines in the vacuum ultraviolet region of the spectrum.

The other excited levels of N lie in a comparatively narrow region of energy. Lines lying in the visible and infrared regions of the spectrum correspond to transitions between these levels. The terms of nitrogen converge to three ionization limits, which correspond to the three possible terms of the ground configuration of the N⁺ ion—$2p^2\,^1S_0$, $2p^2\,^1D_2$, and $2p^2\,^3P_{0,1,2}$. The difference in the energies of the states $2p^2\,^3P_0$, $2p^2\,^3P_1$, and $2p^2\,^3P_2$ is small and can be disregarded. Ionization is also possible on account of one of the s electrons. The systems of terms of the other elements having ground configuration p^3 have a similar structure. Here the values of E_r and E_i rapidly decrease with increase of atomic number. Thus, in the case of P, a majority of the lines corresponding to transitions between the levels of the ground configuration $3p^3$ and of the first excited configuration $3p^2 4s$ are located in an accessible part of the ultraviolet region of the spectrum.

3.4.4 Configuration p^4

The elements O, S, Se, Te, and Po have a p^4 configuration in the ground state. The configuration p^4 gives the same terms as the configuration p^2. The difference lies only in the inverted order of the multiplet structure. For this reason, the ground state term, just as in the case of the configuration p^2, is ^3P. However, the ground level is not $^3P_0$ but the level $^3P_2$.

The excited states correspond to the configurations $np^3\,n's, np^3\,n'p, np^3\,n'd,$ …. In the case of oxygen, the term $2s2p^5\,^3P$, corresponding to the excitation of one of the $2s$ electrons, is also well known. The excitation energy of the lowest excited states of oxygen is about 9 eV. The corresponding lines lie in the vaccum ultraviolet region. The lines associated with the transitions between the excited states lie in the visible region of the spectrum. The oxygen ion in the ground state has the same electron configuration as the C atom. Accordingly, a number of systems can be distinguished in the terms scheme of oxygen, converging to different ionization limits $2p^3\,^4S$, $2p^3\,^2D$, and $2p^3\,^2P$ (13.55, 16.86, and 18.54 eV). The systems of terms of S, Se, Te and Po have approximately the same form as in the case of oxygen. The values of E_r and E_i, just as in the case of the nitrogen and other atoms with ground configurations p^3, decrease with increasing atomic number. Thus $E_r = 6.6$ eV for S. This regularity has a simple phydical meaning. Approximately the same effective nuclear charge corresponds to all elements of the series being considered. At the same time, the electron is on average further from the nucleus in the elements with higher atomic number.

3.4.5 Configuration p^5

The halogens F, Cl, Br, I, and At have a p^5 configuration. The configuration np^5 gives only one term $^2P_{3/2\,1/2}$. Again, the difference from the configuration np lies in the inversion of the order of the multiplet splitting. Just as in the previous cases, several ionization limits are possible on excitation. The values of E_r and E_i are very large for the halogens, because the remaining np electrons do not screen the charge of the atomic core very well and $Z_{\text{eff}} \simeq 4$. Thus, for F, $E_r = 12.9$ eV and $E_i = 17.42$ eV and for Cl, $E_r = 9.16$ eV and $E_i = 13.01$ eV. The resonance lines lie in the vacuum ultraviolet region of the spectrum. Transitions between excited states gives lines in the visible and infrared regions of the spectrum.

3.4.6 Configuration p^6

The last group of elements having p optical (lowest ionization potential) electrons is formed by the noble gases Ne, Ar, Kr, Xe, and Rn. Six p electrons form a completely filled shell; the state $^1S_0$ is thus the ground state. The binding energy of the p electrons in atoms of the noble gases is greater than in atoms of the halogens; $Z_{\text{eff}} \simeq 5$. As a result, the ionization potentials and the resonance potentials are very large, in fact the largest in the whole periodic system of elements. The excited levels, just as in the case of halogens, lie in a comparatively narrow energy region. For this reason, the main lines of the spectra of these elements lie in the vacuum ultraviolet region (transitions to the ground level) and in the visible and infrared regions (transitions between excited levels). A rather peculiar type of coupling is observed for the excited states of atoms of the noble gases. The excited states are obtained by excitation of one of the np electrons into the states $n's$, $n'p$, $n'd$, $\ldots$. The binding energy of the electron $n'l$ is much less than the binding energy of the p electrons ($Z_{\text{eff}} \simeq 1$ for the electron $n'l$ and $Z_{\text{eff}} \simeq 5$ for the p electrons), and on an average this electron is at a comparatively great distance from the other electrons of the atomic core, including the p shell electrons. Thus the spin-orbit interaction of the electrons of the atomic core is greater than the electrostatic interaction of these electrons with the excited electron. Accordingly, the levels of the noble gas atoms are conveniently classified by the following scheme.

The atomic core is characterized by the quantum numbers L, S, and j, where L is the orbital angular momentum of the atomic core, S is the spin of the atomic core, and j is the total angular momentum of the atomic core. Due to the electrostatic interaction of the excited electron with the electrons of the atomic core, the state $LSjl$ gives a series of levels, each of which is described by the quantum number K corresponding to the angular momentum $\boldsymbol{K} = \boldsymbol{j} + \boldsymbol{l}$. Finally the spin-orbit interaction of the excited electron leads to the splitting of each level of $LSjlK$ into the two J components. The total angular momentum of the atom is denoted as before by J, where $J = K \pm 1/2$.

The level is described by the set of quantum numbers $LSjlKJ$ when classified by this scheme. The following notation is usually used:

$$^{2S+1}L_j nl[K]_J.$$

Let us consider as an example, the configurations $np^5\, n's$ and $np^5\, n'p$. In the first case we have four levels

$$np^5\, {}^2P_{3/2}\, n's\,[3/2]_{2,1};\ np^5\, {}^2P_{1/2}\, n's\,[1/2]_{1,0}.$$

The term $^2P_{3/2}$ of the atomic core gives one pair of levels $J = 2,1$ and the term $^2P_{1/2}$ gives another pair. In the second case the initial terms are also the terms $^2P_{1/2}$ and $^2P_{3/2}$. Now, however,

$$K = j+l, j+l-1, \ldots\ |j-l|$$

can take the following values:

when $j = 1/2$, $K = 1/2, 3/2$
when $j = 3/2$, $K = 1/2, 3/2, 5/2$.

We thus have the following levels:

$$np^5\, {}^2P_{3/2}\, n'p\,[1/2]_{0,1} \quad np^5\, {}^2P_{3/2}\, n'p\,[3/2]_{1,2},$$
$$np^5\, {}^2P_{3/2}\, n'p\,[5/2]_{2,3} \quad np^5\, {}^2P_{1/2}\, n'p\,[3/2]_{1,2},$$
$$np^5\, {}^2P_{1/2}\, n'p\,[1/2]_{0,1}.$$

In this instance there are obviously two ionization limits, which can be denoted $(^2P_{3/2})$ and $(^2P_{1/2})$.

The type of coupling described above is called a jl coupling. The following groupings of levels are characteristic of this type of coupling. The distance between the levels $LSjK$ and $LSjK'$ is considerably less than the distance between the levels $LSjK$ and $L'S'j'K$ belonging to different states of the atomic core. Splitting of the levels $LSjK$ with respect to J is small in comparison with the distance between the levels $LSjK$ and $LSjK'$. Since each level $LSjK$ splits into the two components $J = K \pm 1/2$ as a result of spin-orbit interaction, the system of terms recalls in its structure the system of doublet terms of the alkali elements. The difference lies only in the fact that K can now take half-integer values and J integer ones. In the case of LS coupling, singlets and triplets correspond to the configuration $p^5 l$.

jl coupling also appears in the spectra of some other atoms, for highly excited states when one of the electrons is on average at a large distance from the atomic core. The spectrum of Cu II is an example of this type.

3.5 Spectra of Elements with Unfilled d and f Shells

3.5.1 Elements with Unfilled d Shells

The shells $3d$, $4d$, and $5d$ are filled, respectively, in the elements of the iron group

Sc, Ti, V, Cr, Mn, Fe, Co, Ni;

of the palladium group

Y, Zr, Nb, Mo, Tc, Ru, Rh, Pd;

and of the platinum group

Lu, Hf, Ta, W, Re, Os, Ir, Pt.

As already noted, a distinctive competition occurs between the d and s states in the filling of the d shells. As a result, in the case of some of the elements enumerated, the ground state configuration is the configuration $nd^{k+1}(n+1)s$ (e.g., $Cr-3d^54s$; $Mo-4d^55s$) or even nd^{k+2} (e.g., $Pd-4d^{10}$) instead of $nd^k(n+1)s^2$.

For the majority of the atoms of the groups being considered, the electron configurations $nd^k(n+1)s^2$, $nd^{k+1}(n+1)s$ and nd^{k+2} correspond to comparatively closely spaced energy levels, the order in which the levels are arranged being different for the different atoms.

A large number of terms, some of which have a high multiplicity, correspond to electron configurations containing several d electrons. For example, we have 16 terms ^1PDFGH, ^3PDFGH, ^3PF and ^5PF and 38 levels for the configuration $3d^34s$. As a result of this, the spectra of the elements we are considering are characterized by an extraordinary large number of lines.

Since the levels of the first excited configurations and of the ground configuration are comparatively close, there are a large number of lines in the visible and ultraviolet regions of the spectra of the elements with d optical electrons. A typical feature of the spectra of these elements is also the absence of very intense lines like those in the spectra of the alkali and alkali earth elements. This feature is, obviously, due to the fact that a large number of levels belong to each electron configuration and the transitions between the levels of the two configurations give a very large number of spectral lines. A comparatively large group of lines, as a rule, plays the role of resonance lines for each element. The closely spaced levels of the configurations $nd^k(n+1)s^2$, $nd^{k+1}(n+1)s$, and nd^{k+2} have the same parity; electric dipole transitions between these levels are thus not possible. The nearest odd configuration, as a rule, is the configuration obtained by the excitation of one of the nd or $(n+1)s$ electrons into the state $(n+1)p$.

Let us consider, as an example, the spectrum of iron. The ground state configuration of the Fe atom is $3d^6 4s^2$. The terms $^1 S D F G I\atop{22}$ and $^3 P D F G H\,^5D\atop{22}$ correspond to this configuration. In accordance with Hund's rule, the ground state term is the term $^5D_{4,3,2,1,0}$. Because the number of d electrons in the present case is more than half the maximum possible, the multiplet splitting has an inverted order, and the lowest level is $^5D_4$. The lowest excited terms belong to the configuration $3d^7 4s$;

$$3d^7\,[^4F]4s\;^5F_{5,4,3,2,1};\;3d^7\,[^4F]4s\;^3F_{4,3,2};3d^7\,[^4P]4s^5P_{3,2,1},\text{ and so on.}$$

Sixteen terms altogether belong to the configuration $3d^7 4s$. All these terms are even and thus metastable. The lowest odd term is $3d^6 4s[^6D]4p^7 D^0_{5,4,3,2,1}$. This term, however, has the multiplicity 7, whereas the multiplicity of the ground state term equals 5. Therefore, the resonance transition is the transition

$$3d^6 4s^2\;{}^5D_{4,3,2,1,0} - 3d^6 4s\,[^6D]\,4p\;^5D^\circ_{4,3,2,1,0}\,.$$

The resonance term $3d^6 4s\,[^6D]\,4p\;^5D^\circ$ can combine with the lowest excited term $3d^7\,[^4F]\,4s\;^5F$. The corresponding lines can also be called resonance lines. The other lowest odd terms of multiplicity 5 are the terms $3d^6 4s\,[^6D]\,4p\;^5F^\circ$, and $3d^7\,[^4F]\,4p\;^5F^\circ$. As a result of the nonregularity of filling of the d shell for the elements with d optical electrons, there is not such a strict correspondence between the spectra of elements occupying the same places in different periods as there is for elements with s optical electrons.

3.5.2 Elements with Unfilled f Shells

In the sixth period, the lanthanides Ce, Pr, Nd, Pm, Sm, Eu, Gd, Tb, Dy, Ho, Er, Tm, and Yb, and in the seventh period the actinides Ac, Th, Pa, U, Np, Pu, Am, Cm, Bk, and Cf have ground state configurations containing f optical electrons. Although the ground state configurations of lanthanum ($5d6s^2$) and actinium ($6d7s^2$) do not contain f electrons, it is usual to consider these elements with the other rare-earth elements.

The spectra of elements with f optical electrons are even more complex and richer in lines than the spectra with d optical electrons. This is due to the fact that the electron configurations containing f electrons give an extremely large number of terms and levels. Thus, for example, the configuration f^7 gives 119 terms of multiplicity 2,4,6,8, and 327 levels. For configurations containing a group f^k and also s,p, and d electrons, the number of terms can be increased up

3. Spectra of Multielectron Atoms

to several thousand and the number of levels can exceed 10^4. As a result, the spectra of some of the lanthanide and actinide atoms (for example, U and Th) are a continuous network of lines of comparable intensity. (Detailed discussion of the spectra of the rare-earth elements is given in [4, 5].)

Part II

Theory of Atomic Spectra

This part of the book presents a systematic treatment of the theory of atomic spectra. A brief outline is given of the Racah techniques in the theory of angular momenta, and of the method of fractional parentage coefficients. There are numerous examples and tables throughout the text which enable one to use these very effective mathematical tools in different problems of atomic spectroscopy.

Radiation phenomena are dealt with very comprehensively, including multipole radiation, bremsstrahlung, photorecombination, and photoionization. Special attention is paid to approximation methods for calculating radiative transition probabilities and cross sections. Tables containing the results of approximate calculations of electric dipole and quadrupole oscillator strengths and photorecombination cross sections are given. The choice of material has been made from the standpoint of those interested in various applications of atomic spectroscopy to other branches of physics.

Chapter 4 Angular Momenta

This chapter consists of a brief outline of the Racah techniques for calculating the matrix elements of spherical tensor operators encountered in different problems of atomic physics, together with a summary of the necessary formulas and tables. It is difficult to overestimate the significance of Racah methods for the theory of atomic spectra. Many calculations which previously required tiresome and laborious calculations are carried out almost instantaneously by means of the Racah technique, the results being expressed in terms of tabulated coefficients [6–10].

4.1 Angular Momentum Operator. Addition of Angular Momenta

4.1.1 Angular Momentum Operator

The angular momentum operator $\boldsymbol{J}$ can be defined by requiring its components J_x, J_y, J_z to satisfy commutation relations

$$[J_x, J_y] = iJ_z, \quad [J_y, J_z] = iJ_x, \quad [J_z, J_x] = iJ_y. \tag{4.1}$$

Each of the angular momentum components J_x, J_y, J_z commutes with the square of the momentum operator $\boldsymbol{J}^2$, i.e., can have a definite value simultaneously with $\boldsymbol{J}^2$. States are usually considered in which the square of the angular momentum and its z component have been determined.

Using (4.1), we find that the eigenvalues of the operators $\boldsymbol{J}^2$ and J_z are $J(J+1)$ and M respectively, where

$$J = 0, 1/2, 1, 3/2, 2, \ldots, \quad M = J, J-1, J-2, \ldots \tag{4.2}$$

Thus J can assume both integer and half-integer values.

Let us denote Ψ_{JM} the eigenfunctions of the operators $\boldsymbol{J}^2, J_z$. Then

$$\begin{aligned}
\boldsymbol{J}^2 \Psi_{JM} &= J(J+1)\, \Psi_{JM}, \\
J_z \Psi_{JM} &= M \Psi_{JM}, \\
(J_x + iJ_y)\, \Psi_{JM} &= \sqrt{(J-M)(J+M+1)}\, \Psi_{JM+1}, \\
(J_x - iJ_y)\, \Psi_{JM} &= \sqrt{(J+M)(J-M+1)}\, \Psi_{JM-1}.
\end{aligned} \tag{4.3}$$

4.1.2 Orbital Angular Momentum

The orbital angular momentum operator of a particle

$$L = -i[r, \nabla] \tag{4.4}$$

satisfies the general commutation relations (4.1) and also (4.3). The eigenfunctions of the operators L^2, L_z are the spherical harmonics $Y_{lm}(\theta, \varphi)$ determined above by formulas (1.7, 9)

$$L^2 Y_{lm} = l(l+1) Y_{lm}, \quad L_z Y_{lm} = m Y_{lm}, \tag{4.5}$$

where

$$l = 0, 1, 2, \ldots, \quad m = 0, \pm 1, \pm 2, \ldots, \pm l. \tag{4.6}$$

In many cases it is helpful to introduce the functions

$$C_m^l(\theta, \varphi) = \sqrt{\frac{4\pi}{2l+1}} Y_{lm}(\theta, \varphi). \tag{4.7}$$

The advantage of the functions (4.7) is that the well-known theorem of addition of spherical harmonics

$$P_l(\cos \omega) = \frac{4\pi}{2l+1} \sum_{m=-l}^{l} Y_{lm}^*(\theta_1, \varphi_1) Y_{lm}(\theta_2, \varphi_2), \tag{4.8}$$

where $P_l(\cos \omega)$ is the Legendre polynomial and ω is the angle between the directions θ_1, φ_1 and θ_2, φ_2, acquires for the functions C_m^l the particularly simple form

$$P_l(\cos \omega) = \sum_m C_m^{l*}(\theta_1, \varphi_1) C_m^l(\theta_2, \varphi_2)$$

$$= \sum_m (-1)^m C_m^l(\theta_1, \varphi_1) C_{-m}^l(\theta_2, \varphi_2). \tag{4.9}$$

4.1.3 Electron Spin

In the general case the eigenfunctions of the operators J^2 and J_z are neither spherical harmonics (the latter are defined only for integer values of J) nor generally functions of the variables θ, φ. Eigenfunctions of the electron spin operator are precisely functions of this type.

The z component of the spin momentum of an electron can assume only two values $\pm 1/2$. Hence, it follows that $s = 1/2$ and $s(s+1) = 3/4$. In (4.3), setting $J = s = 1/2$, $M = \mu$, we obtain

$$\langle\tfrac{1}{2}\mu|s_x|\tfrac{1}{2}\mu'\rangle = \begin{pmatrix} 0 & 1/2 \\ 1/2 & 0 \end{pmatrix} = \tfrac{1}{2}\sigma_x,$$

$$\langle\tfrac{1}{2}\mu|s_y|\tfrac{1}{2}\mu'\rangle = \begin{pmatrix} 0 & -i/2 \\ i/2 & 0 \end{pmatrix} = \tfrac{1}{2}\sigma_y, \quad (4.10)$$

$$\langle\tfrac{1}{2}\mu|s_z|\tfrac{1}{2}\mu'\rangle = \begin{pmatrix} 1/2 & 0 \\ 0 & -1/2 \end{pmatrix} = \tfrac{1}{2}\sigma_z.$$

Here σ_x, σ_y, σ_z are the Pauli spin matrices.

In nonrelativistic theory the existence of a spin angular momentum can be described by the introduction of the additional spin variable λ, taking discrete values $1/2$, $-1/2$.

In the state with given value of μ

$$\psi_{a\mu} = \psi_a(r)\, q_\mu(\lambda),$$
$$q_{1/2}(\lambda) = \delta_{\lambda,1/2}, \quad q_{-1/2}(\lambda) = \delta_{\lambda,-1/2}. \quad (4.11)$$

In the following, the set of the three coordinates of r together with the spin variable λ will be denoted by the shorthand ξ.

$$\int d\xi = \sum_{\lambda=-1/2}^{1/2} \int dr, \quad (4.12)$$

$$\int \Psi^* \Phi \, d\xi = \sum_\lambda \int \Psi^*(r,\lambda)\, \Phi(r,\lambda)\, dr. \quad (4.13)$$

4.1.4 Addition of Two Angular Momenta

The problem of adding the angular momenta J_1 and J_2 of two noninteracting systems consists in finding the eigenvalues of the operators

$$J^2 = (J_1 + J_2)^2, \quad J_z = J_{1z} + J_{2z} \quad (4.14)$$

and their eigenfunctions Ψ_{JM}, if the eigenvalues of the operators J_1^2, J_2^2, J_{1z}, J_{2z} and the functions $\Psi_{J_1 M_1}$ and $\Psi_{J_2 M_2}$ are known. For the values of J and M, we have

$$J = J_1 + J_2, \ J_1 + J_2 - 1, \ \ldots, \ |J_1 - J_2|, \quad (4.15)$$

$$M = M_1 + M_2. \quad (4.16)$$

Let us represent the eigenfunctions Ψ_{JM} of the operators J^2, J_z in the form of an expansion in terms of the functions

$$\Psi_{M_1M_2} = \Psi_{J_1M_1} \Psi_{J_2M_2}. \tag{4.17}$$

According to (4.16) only the functions $\Psi_{M_1M_2}$ with $M_1 + M_2 = M$ can arise in the expansion; therefore,

$$\Psi_{JM} = \sum_{M=M_1+M_2} C^J_{M_1M_2} \cdot \Psi_{M_1M_2}. \tag{4.18}$$

The coefficients of the expansion $C^J_{M_1M_2}$, for which we shall also use the notation

$$C^J_{M_1M_2} = (J_1J_2M_1M_2|J_1J_2JM), \tag{4.19}$$

are called Clebsch-Gordan coefficients. The main properties of these coefficients are discussed in Sect. 4.2.

Since the functions Ψ_{JM} and $\Psi_{M_1M_2}$ are orthogonal and normalized, the transformation inverse to (4.18) has the form

$$\Psi_{M_1M_2} = \sum_{J \geqslant M} C^{J*}_{M_1M_2} \cdot \Psi_{JM}. \tag{4.20}$$

Summation in (4.20) is carried out with respect to all values of J consistent with (4.15) and satisfying the condition $J \geqslant M = M_1 + M_2$. In the general case, many different values of J appear on the right-hand side of (4.20). The probability of one or another value of J in the state $J_1J_2M_1M_2$ equals $|C^J_{M_1M_2}|^2$. Conversely, if a set of numbers J_1, J_2, J, M is given, i.e., states of a system are considered in which the total angular momentum and its z component also have definite values together with the angular momenta of each of the systems, then M_1 and M_2 are not defined. The probability of definite values of M_1 and M_2 when there are given values of J and M is determined by the square of the modulus of the corresponding coefficient in the expansion of the wave function $\Psi_{J_1J_2JM}$ in terms of the functions $\Psi_{J_1J_2M_1M_2}$.

Let us consider as an example the addition of the orbital angular momentum and the spin. According to (4.15) the total angular momentum of the electron

$$j = l + s \tag{4.21}$$

can have two values

$$j = l \pm 1/2.$$

Therefore

$$\Psi_{jm_j} = \sum_{m+\mu=m_j} C^j_{m\mu} \Psi_{m\mu} = \sum_{\mu} C^j_{m_j-\mu,\,\mu} \Psi_{m_j-\mu,\,\mu}$$
$$= C^j_{m_j-1/2,\,1/2} \psi_{l,m_j-1/2} \cdot q_{s,1/2} + C^j_{m_j+1/2,\,-1/2} \psi_{l,m_j+1/2} \cdot q_{s,-1/2}.$$

4.1 Angular Momentum Operator. Addition of Angular Momenta

Values of the coefficients $C'_{m\mu} = (l\frac{1}{2}m\mu | l\frac{1}{2}jm_j)$ are given in Sect. 4.2. Finally

$$j = l + 1/2 \quad \Psi_{jm_j} = \sqrt{\frac{l + m_j + 1/2}{2l + 1}} \psi_{l, m_j - 1/2} \cdot q_{s, 1/2}$$

$$- \sqrt{\frac{l - m_j + 1/2}{2l + 1}} \psi_{l, m_j + 1/2} q_{s, -1/2},$$

$$j = l - 1/2 \quad \Psi_{jm_j} = \sqrt{\frac{l - m_j + 1/2}{2l + 1}} \psi_{l, m_j - 1/2} \cdot q_{s, 1/2}$$

$$+ \sqrt{\frac{l + m_j + 1/2}{2l + 1}} \psi_{l, m_j + 1/2} q_{s, -1/2}.$$

(4.22)

The expressions (4.22) enable one to find the probability of definite values $m\mu$ for given values jm_j. For example, when $l = 1$, $j = 3/2$, $m_j = 1/2$, the probability of the values $m = 0$, $\mu = 1/2$ and $m = 1$, $\mu = -1/2$, respectively, equals

$$\frac{l + m_j + 1/2}{2l + 1} = \frac{2}{3}, \quad \frac{l - m_j + 1/2}{2l + 1} = \frac{1}{3}.$$

When $l = 0$, the total angular momentum is entirely determined by the spin $j = s = 1/2$. In this case from (4.22) the obvious result follows: when $m_j = 1/2$, the probability of the values $\mu = 1/2, -1/2$ equals 1, 0, respectively. Conversely, when $m_j = -1/2$, only one value is possible: $\mu = -1/2$.

In the following we shall speak of the description of a system with the aid of the wave functions $\Psi_{J_1 J_2 JM}$ and $\Psi_{J_1 J_2 M_1 M_2}$ as of the different representations of the state of the system, or simply as the JM representation and the $M_1 M_2$ representation. Different representations of the states of an arbitrary system can be described in a similar way. In the general case we shall understand by the γ representation the description of a system by the wave function Ψ_γ where γ is the total set of quantum numbers describing a definite state of a system. According to this terminology, the matrix of the operator F calculated with the aid of functions Ψ_γ will be spoken of as the γ representation of the operator and the functions Ψ_γ as the basis of the representation.

4.1.5 Addition of Three or More Angular Momenta

In adding the two angular momenta J_1 and J_2, the values of J and M entirely determine the state of a system. This is because the total number of quantum numbers describing the state of a system remains the same. The values $J_1 J_2 JM$ in the same way as $J_1 J_2 M_1 M_2$ make up a complete set. This no longer occurs when adding several angular momenta. Different states of a system may correspond to the same values JM. It is therefore necessary to define more specifically the method of addition of the angular momenta. We shall show this in

4. Angular Momenta

the example of adding the three angular momenta J_1, J_2, and J_3. Let us carry out the addition of the angular momenta in two different ways. In the first case we shall add first J_1 and J_2 and then add J_3. In accordance with (4.15), addition of J_1 and J_2 gives

$$J' = J_1 + J_2, J_1 + J_2 - 1, \ldots |J_1 - J_2|; \quad M' = M_1 + M_2.$$

Then by adding to each of these values of J' the angular momentum J_3, we obtain

$$J = J' + J_3, J' + J_3 - 1, \ldots |J' - J_3|; \quad M = M' + M_3 = M_1 + M_2 + M_3.$$

In the second case we shall add first J_2 and J_3:

$$J'' = J_2 + J_3, J_2 + J_3 - 1, \ldots |J_2 - J_3|; \quad M'' = M_2 + M_3;$$

and then J_1 and J'':

$$J = J_1 + J'', J_1 + J'' - 1, \ldots |J'' - J_1|;$$
$$M = M'' + M_1 = M_1 + M_2 + M_3.$$

We shall denote the wave functions of the states obtained in the first and second cases $\Psi_{JM}(J_1 J_2 [J'] J_3)$ and $\Psi_{JM}(J_1; J_2 J_3 [J''])$. It is obvious that in the general case

$$\Psi_{JM}(J_1 J_2 [J'] J_3) \neq \Psi_{JM}(J_1; J_2 J_3 [J'']).$$

We shall obtain one further scheme of addition of angular momenta if we simultaneously alter both the sequence and the order of the angular momenta,

$$\Psi_{JM}(J_1 J_2 [J'] J_3) \neq \Psi_{JM}(J_1 J_3 [J''] J_2).$$

The transition from one scheme of addition of angular momenta to the other

$$\Psi_{JM}(J_1; J_2 J_3 [J'']) = \sum_{J'} (J_1 J_2 [J'] J_3 | J_1, J_2 J_3 [J'']) \, \Psi_{JM}(J_1 J_2 [J'] J_3),$$

$$\Psi_{JM}(J_1 J_3 [J''] J_2) = \sum_{J'} (J_1 J_2 [J'] J_3 | J_1 J_3 [J''] J_2) \, \Psi_{JM}(J_1 J_2 [J'] J_3)$$

is determined by the so-called Racah W coefficients:

$$(J_1 J_2 [J'] J_3 | J_1, J_2 J_3 [J'']) = \sqrt{(2J' + 1)(2J'' + 1)} \, W(J_1 J_2 J J_3; J' J''). \quad (4.23)$$

$$(J_1 J_2 [J'] J_3 | J_1 J_3 [J''] J_2) = \sqrt{(2J' + 1)(2J'' + 1)} \, W(J' J_3 J_2 J''; J J_1). \quad (4.24)$$

The Racah W coefficients, which are functions of six arguments, play a very important role in the theory of complex spectra. As will be seen below, one has to deal with these coefficients in solving very different problems. A discussion of the properties of these coefficients and also the formulas necessary to calculate them are given in Sect. 4.2. In the case of addition of three angular momenta considered above, the two numbers J and M are not sufficient for a complete description of the states of a system. It is necessary to give the value of the sum of any two angular momenta, for example, J' or J''. The set of quantum numbers $J_1 J_2 [J'] J_3 JM$ or $J_1; J_2 J_3 [J''] JM$ will form the complete set in this case.

Similarly, in adding a larger number of angular momenta it is necessary for the complete description of a state to give, as well as JM, the values of the angular momenta of subsystems of any two particles, three particles, and so on. For example, in the case of four particles a state can be described by the set of quantum numbers $J_1 J_2 [J'] J_3 [J''] J_4 JM$. Other schemes of addition of angular momenta are, of course, possible; for example,

$$J_1 J_2 [J']; J_3 J_4 [J''] JM, J_1; J_2 J_3 [J'] J'' J_4 JM,$$

and so on.

Two schemes of additon of orbital angular momenta and electron spins are of greatest interest: the LS coupling scheme

$$l_1 l_2 [L], s_1 s_2 [S] JM; \tag{4.25}$$

and the jj coupling scheme

$$l_1 s_1 [j_1] l_2 s_2 [j_2] JM. \tag{4.26}$$

In the case of (4.25) we have

$$L = l_1 + l_2, \; l_1 + l_2 - 1, \ldots, |l_1 - l_2|;$$
$$S = 0, 1;$$
$$J = L + S, \; L + S - 1, \ldots, |L - S|;$$

$$\left. \begin{aligned} \Psi_{LM_L} &= \sum_{M_1 + M_2 = M_L} C^L_{m_1 m_2} \Psi_{m_1} \Psi_{m_2}, \\ \Psi_{SM_S} &= \sum_{\mu_1 + \mu_2 = M_S} C^S_{\mu_1 \mu_2} q_{\mu_1} q_{\mu_2}, \\ \Psi_{JM} &= \sum_{M_L + M_S = M} C^J_{M_L M_S} \Psi_{LM_L} \Psi_{SM_S}, \end{aligned} \right\} \tag{4.27}$$

and in the case of (4.26)

$$j_1 = l_1 + 1/2, \; l_1 - 1/2;$$
$$j_2 = l_2 + 1/2, \; l_2 - 1/2;$$
$$J = j_1 + j_2, \; j_1 + j_2 - 1, \ldots, |j_1 - j_2|;$$

$$\left.\begin{aligned}\Psi_{j_1 m_{j1}} &= \sum_{m_1+\mu_1=m_{j1}} C^{j_1}_{m_1\mu_1} \Psi_{m_1} q_{\mu_1}, \\ \Psi_{j_2 m_{j2}} &= \sum_{m_2+\mu_2=m_{j2}} C^{j_2}_{m_2\mu_2} \Psi_{m_2} q_{\mu_2}, \\ \Psi_{JM} &= \sum_{m_{j1}+m_{j2}=M} C^{J}_{M_{j1} M_{j2}} \Psi_{j_1 M_{j1}} \Psi_{j_2 M_{j2}}.\end{aligned}\right\} \quad (4.28)$$

4.2 Angular Momentum Vector Addition Coefficients

4.2.1 Clebsch–Gordan and Associated Coefficients

In this section we shall consider the main properties of Clebsch–Gordan coefficients

$$C^{j}_{m_1 m_2} = (j_1 j_2 m_1 m_2 | j_1 j_2 j m) \tag{4.29}$$

and coefficients associated with them – Racah V coefficients

$$V(j_1 j_2 j;\, m_1 m_2 m) \tag{4.30}$$

and Wigner $3j$ symbols

$$\begin{pmatrix} j_1 & j_2 & j \\ m_1 & m_2 & m \end{pmatrix}. \tag{4.31}$$

As will be seen later, these coefficients appear in many problems and they play an important role in the theory of atomic spectra.

The Clebsch–Gordan coefficients determine the expansion of the eigenfunctions of the operators $j_1^2 j_2^2 j^2 j_z\, (j = j_1 + j_2)$ in eigenfunctions of the operators $j_1^2 j_{z1} j_2^2 j_{z2}$

$$\Psi_{j_1 j_2 j m} = \sum_{m_1 m_2} (j_1 j_2 m_1 m_2 | j_1 j_2 j m)\, \Psi_{j_1 m_1 j_2 m_2}. \tag{4.32}$$

These coefficients are determined for integer and half-integer values of the arguments and are nonzero if the following two conditions are fulfilled

$$m_1 + m_2 = m, \tag{4.33}$$

$$j = j_1 + j_2,\ j_1 + j_2 - 1,\ \ldots,\ |j_1 - j_2|. \tag{4.34}$$

The differences $j_1 - m_1$, $j_2 - m_2$, $j - m$, and also the sum $j_1 + j_2 + j$ are integers. Condition (4.34) is often called the triangle condition and is denoted by $\Delta(j_1 j_2 j)$. According to this condition, any of the numbers $j_1 j_2 j$ is greater than or equal to the difference of the other two and less than or equal to the sum of the other two.

4.2 Angular Momentum Vector Addition Coefficients

The Racah V coefficients and the $3j$ symbols are connected with the Clebsch–Gordan coefficients by the following relation

$$(j_1 j_2 m_1 m_2 | j_1 j_2 jm) = (-1)^{-j_1+j_2-m} \sqrt{2j+1} \begin{pmatrix} j_1 & j_2 & j \\ m_1 & m_2 & -m \end{pmatrix}, \tag{4.35}$$

$$(j_1 j_2 m_1 m_2 | j_1 j_2 jm) = (-1)^{j+m} \sqrt{2j+1} \, V(j_1 j_2 j; m_1 m_2 -m), \tag{4.36}$$

$$V(j_1 j_2 j; m_1 m_2 m) = (-1)^{-j_1+j_2+j} \begin{pmatrix} j_1 & j_2 & j \\ m_1 & m_2 & m \end{pmatrix}. \tag{4.37}$$

It follows from (4.35, 36) that

$$V(j_1 j_2 j; m_1 m_2 -m) = (-1)^{-j_1+j_2+j} \begin{pmatrix} j_1 & j_2 & j \\ m_1 & m_2 & -m \end{pmatrix}.$$

Since $j - m$ is an integer and $2j - 2m$ is even, (4.35, 36) are equivalent to (4.37).

According to (4.35, 36) the coefficients (4.30, 31) are nonzero if the condition (4.34) and the slightly modified condition (4.33)

$$m_1 + m_2 + m = 0$$

are fulfilled.

The advantage of the V coefficients and especially of the $3j$ symbols is that they have a considerably higher symmetry than Clebsch–Gordan coefficients. The $3j$ symbols have the following symmetries

$$\begin{pmatrix} j_1 & j_2 & j \\ m_1 & m_2 & m \end{pmatrix} = \begin{pmatrix} j_2 & j & j_1 \\ m_2 & m & m_1 \end{pmatrix} = \begin{pmatrix} j & j_1 & j_2 \\ m & m_1 & m_2 \end{pmatrix} = (-1)^{j_1+j_2+j} \begin{pmatrix} j_2 & j_1 & j \\ m_2 & m_1 & m \end{pmatrix}$$

$$= (-1)^{j_1+j_2+j} \begin{pmatrix} j_1 & j & j_2 \\ m_1 & m & m_2 \end{pmatrix} = (-1)^{j_1+j_2+j} \begin{pmatrix} j & j_2 & j_1 \\ m & m_2 & m_1 \end{pmatrix}. \tag{4.38}$$

Thus an even transposition of the columns of a $3j$ symbol does not alter its value; an odd transposition multiplies the initial value by $(-1)^{j_1+j_2+j}$. Moreover,

$$\begin{pmatrix} j_1 & j_2 & j \\ m_1 & m_2 & m \end{pmatrix} = (-1)^{j_1+j_2+j} \begin{pmatrix} j_1 & j_2 & j \\ -m_1 & -m_2 & -m \end{pmatrix}. \tag{4.39}$$

Using (4.35, 37) it is not difficult to obtain similar relations for the coefficients (4.29) and (4.30). In particular, it follows from (4.35, 38) that

$$(j_1 j_2 m_1 m_2 | j_1 j_2 jm) = (-1)^{j_1+j_2-j} (j_2 j_1 m_2 m_1 | j_2 j_1 jm). \tag{4.40}$$

The 3j symbols obey the following conditions of orthogonality

$$\sum_{j,m} (2j+1) \begin{pmatrix} j_1 & j_2 & j \\ m_1 & m_2 & m \end{pmatrix} \begin{pmatrix} j_1 & j_2 & j \\ m'_1 & m'_2 & m \end{pmatrix} = \delta_{m_1 m'_1} \delta_{m_2 m'_2}, \qquad (4.41)$$

$$\sum_{m_1 m_2} \begin{pmatrix} j_1 & j_2 & j \\ m_1 & m_2 & m \end{pmatrix} \begin{pmatrix} j_1 & j_2 & j' \\ m_1 & m_2 & m' \end{pmatrix} = \frac{1}{2j+1} \delta_{jj'} \delta_{mm'}. \qquad (4.42)$$

The coefficients (4.29, 30) also satisfy similar relations in accordance with (4.35–37). Thus

$$\sum_{j,m} (j_1 j_2 m_1 m_2 | j_1 j_2 jm)(j_1 j_2 jm | j_1 j_2 m'_1 m'_2) = \delta_{m_1 m'_1} \delta_{m_2 m'_2}, \qquad (4.43)$$

$$\sum_{m_1 m_2} (j_1 j_2 m_1 m_2 | j_1 j_2 jm)(j_1 j_2 j'm' | j_1 j_2 m_1 m_2) = \delta_{jj'} \delta_{mm'}. \qquad (4.44)$$

The phases of the Clebsch–Gordan coefficients can be chosen in different ways. In all further formulas, the phases are chosen so that the Clebsch–Gordan coefficients are real. This convention agrees with that of [1].

When $j_2 = 0$, it follows from the definition of the Clebsch–Gordan coefficients (4.32) that

$$(j_1 0 m_1 0 | j_1 0 jm) = \delta_{jj_1} \delta_{mm_1}, \qquad (4.45)$$

$$V(j_1 0 j; m_1 0 m) = (-1)^{-j+m} (2j+1)^{-1/2} \delta_{jj_1} \delta_{-m_1 m}, \qquad (4.46)$$

$$\begin{pmatrix} j_1 & 0 & j \\ m_1 & 0 & m \end{pmatrix} = (-1)^{j_1 - m} (2j+1)^{-1/2} \delta_{jj_1} \delta_{-m_1 m}. \qquad (4.47)$$

General formulas defining numerical values of the coefficients of vector addition of angular momenta are very complicated. In the case when one of the arguments $j_1 j_2 j$ equals 1/2, 1, 3/2, 2, the formulas given below can be used.[1]

Let us note that in carrying out calculations it is convenient to use 3j symbols. A summary of formulas for 3j symbols is given below. The corresponding expressions for the Clebsch–Gordan coefficients and the V coefficients can be found using (4.35–37). Formulas for Clebsch–Gordan coefficients are given only for $j_2 = 1/2$.

4.2.2 Summary of Formulas for 3j Symbols

The general formula for 3j symbols takes a comparatively simple form in the following cases

---

[1] Numerical values of Clebsch–Gordan coefficients for $j \leq 9/2, j_1 \leq j_2 \leq j$ can be found in [11].

4.2 Angular Momentum Vector Addition Coefficients

$j = j_1 + j_2$

$$\begin{pmatrix} j_1 & j_2 & j_1 + j_2 \\ m_1 & m_2 & -m_1 - m_2 \end{pmatrix} = (-1)^{j_1 - j_2 + m_1 + m_2}$$
$$\times \sqrt{\frac{(2j_1)!\,(2j_2)!\,(j_1 + j_2 + m_1 + m_2)!\,(j_1 + j_2 - m_1 - m_2)!}{(2j_1 + 2j_2 + 1)!\,(j_1 + m_1)!\,(j_1 - m_1)!\,(j_2 + m_2)!\,(j_2 - m_2)!}},$$
(4.48)

$m_1 = j_1$

$$\begin{pmatrix} j_1 & j_2 & j \\ j_1 & -j_1 - m & m \end{pmatrix} = (-1)^{-j_1 + j_2 + m}$$
$$\times \sqrt{\frac{(2j_1)!\,(-j_1 + j_2 + j)!\,(j_1 + j_2 + m)!\,(j - m)!}{(j_1 + j_2 + j + 1)!\,(j_1 - j_2 + j)!\,(j_1 + j_2 - j)!\,(-j_1 + j_2 - m)!\,(j + m)!}},$$
(4.49)

$m_1 = m_2 = m = 0$
$j_1 + j_2 + j = 2g$

$$\begin{pmatrix} j_1 & j_2 & j \\ 0 & 0 & 0 \end{pmatrix} = (-1)^g \sqrt{\frac{(2g - 2j_1)!\,(2g - 2j_2)!\,(2g - 2j)!}{(2g + 1)!}}$$
$$\times \frac{g!}{(g - j_1)!\,(g - j_2)!\,(g - j)!} \quad (4.50)$$

$m_1 = m_2 = m = 0$
$j_1 + j_2 + j = 2g + 1$

$$\begin{pmatrix} j_1 & j_2 & j \\ 0 & 0 & 0 \end{pmatrix} = 0 \quad (4.51)$$

where g is an integer.

For the values $j = 0, 1/2, 1, 3/2, 2$, the general formula gives

$j = 0$
$$\begin{pmatrix} j_1 & j_2 & 0 \\ m_1 & m_2 & 0 \end{pmatrix} = (-1)^{-j_2 - m_2} \frac{\delta_{j_1 j_2} \delta_{m_1, -m_2}}{\sqrt{(2j_1 + 1)}}, \quad (4.52)$$

$j = 1/2$
$$\begin{pmatrix} j + 1/2 & j & 1/2 \\ m & -m - 1/2 & 1/2 \end{pmatrix} = (-1)^{j - m - 1/2} \left[\frac{j - m + 1/2}{(2j + 2)(2j + 1)} \right]^{1/2}.$$
(4.53)

From this formula it follows for $(j_1\ 1/2\ m_1\ m_2|j_1\ 1/2\ jm)$

m_2	$1/2$	$-1/2$
j		
$j_1 + 1/2$	$\sqrt{\dfrac{j_1+m+1/2}{2j_1+1}}$	$\sqrt{\dfrac{j_1-m+1/2}{2j_1+1}}$
$j_1 - 1/2$	$-\sqrt{\dfrac{j_1-m+1/2}{2j_1+1}}$	$\sqrt{\dfrac{j_1+m+1/2}{2j_1+1}}$

(4.54)

$$j = 1$$

$j+1, j, 1$

$$\begin{pmatrix} j+1 & j & 1 \\ m & -m-1 & 1 \end{pmatrix} = (-1)^{-J-m-1} \left[\frac{(j-m)(j-m+1)}{(2j+3)(2j+2)(2j+1)}\right]^{1/2},$$

$$\begin{pmatrix} j+1 & j & 1 \\ m & -m & 0 \end{pmatrix} = (-1)^{J-m-1} \left[\frac{(j+m+1)(j-m+1)}{(2j+3)(j+1)(2j+1)}\right]^{1/2},$$

$j, j, 1$

$$\begin{pmatrix} j & j & 1 \\ m & -m-1 & 1 \end{pmatrix} = (-1)^{J-m} \left[\frac{(j-m)(j+m+1)}{(j+1)(2j+1)2j}\right]^{1/2}, \quad (4.55)$$

$$\begin{pmatrix} j & j & 1 \\ m & -m & 0 \end{pmatrix} = (-1)^{J-m} \frac{m}{[(2j+1)(j+1)j]^{1/2}},$$

$$j = 3/2$$

$j + 3/2, j, 3/2$

$$\begin{pmatrix} j+3/2 & j & 3/2 \\ m & -m-3/2 & 2/2 \end{pmatrix}$$

$$= (-1)^{J-m+1/2} \left[\frac{(j-m-1/2)(j-m+1/2)(j-m+3/2)}{(2j+4)(2j+3)(2j+2)(2j+1)}\right]^{1/2}, \quad (4.56)$$

$$\begin{pmatrix} j+3/2 & j & 3/2 \\ m & -m-1/2 & 1/2 \end{pmatrix}$$

$$= (-1)^{J-m+1/2} \left[\frac{3(j-m+1/2)(j-m+3/2)(j+m+3/2)}{(2j+4)(2j+3)(2j+2)(2j+1)}\right]^{1/2},$$

$j + 1/2, j, 3/2$

$$\begin{pmatrix} j+1/2 & j & 1/2 \\ m & -m-3/2 & 3/2 \end{pmatrix}$$

4.2 Angular Momentum Vector Addition Coefficients

$$= (-1)^{j-m-1/2} \left[\frac{3(j-m-1/2)(j-m+1/2)(j+m+3/2)}{(2j+3)(2j+2)(2j+1)2j} \right]^{1/2}, \quad (4.57)$$

$$\begin{pmatrix} j+1/2 & j & 3/2 \\ m & -m-1/2 & 1/2 \end{pmatrix}$$

$$= (-1)^{j-m-1/2} (j+3m+3/2) \left[\frac{j-m+1/2}{(2j+3)(2j+2)(2j+1)2j} \right]^{1/2},$$

$$j = 2$$

$j+2, j, 2$

$$\begin{pmatrix} j+2 & j & 2 \\ m & -m-2 & 2 \end{pmatrix} = (-1)^{j-m} \left[\frac{(j-m-1)(j-m)(j-m+1)(j-m+2)}{(2j+5)(2j+4)(2j+3)(2j+2)(2j+1)} \right]^{1/2},$$

$$\begin{pmatrix} j+2 & j & 2 \\ m & -m-1 & 1 \end{pmatrix} = 2(-1)^{j-m} \left[\frac{(j+m+2)(j-m+2)(j-m+1)(j-m)}{(2j+5)(2j+4)(2j+3)(2j+2)(2j+1)} \right]^{1/2},$$

(4.58)

$$\begin{pmatrix} j+2 & j & 2 \\ m & -m & 0 \end{pmatrix} = (-1)^{j-m} \left[\frac{6(j+m+2)(j+m+1)(j-m+2)(j-m+1)}{(2j+5)(2j+4)(2j+3)(2j+2)(2j+1)} \right]^{1/2},$$

$j+1, j, 2$

$$\begin{pmatrix} j+1 & j & 2 \\ m & -m-2 & 2 \end{pmatrix} = 2(-1)^{j-m+1} \left[\frac{(j-m-1)(j-m)(j-m+1)(j+m+2)}{(2j+4)(2j+3)(2j+2)(2j+1)2j} \right]^{1/2},$$

$$\begin{pmatrix} j+1 & j & 2 \\ m & -m-1 & 1 \end{pmatrix}$$

$$= (-1)^{j-m+1} 2(j+2m+2) \left[\frac{(j-m+1)(j-m)}{(2j+4)(2j+3)(2j+2)(2j+1)2j} \right]^{1/2}, \quad (4.59)$$

$$\begin{pmatrix} j+1 & j & 2 \\ m & -m & 0 \end{pmatrix} = (-1)^{j-m+1} 2m \left[\frac{6(j+m+1)(j-m+1)}{(2j+4)(2j+3)(2j+2)(2j+1)2j} \right]^{1/2},$$

$j, j, 2$

$$\begin{pmatrix} j & j & 2 \\ m & -m-2 & 2 \end{pmatrix} = (-1)^{j-m} \left[\frac{6(j-m-1)(j-m)(j+m+1)(j+m+2)}{(2j+3)(2j+2)(2j+1)2j(2j-1)} \right]^{1/2},$$

$$\begin{pmatrix} j & j & 2 \\ m & -m-1 & 1 \end{pmatrix}$$

$$= (-1)^{j-m}(1+2m) \left[\frac{6(j+m+1)(j-m)}{(2j+3)(2j+2)(2j+1)2j(2j-1)} \right]^{1/2}, \quad (4.60)$$

$$\begin{pmatrix} j & j & 2 \\ m & -m & 0 \end{pmatrix} = (-1)^{j-m} \left[\frac{[3m^2 - j(j+1)]}{[(2j+3)(j+1)(2j+1)j(2j-1)]^{1/2}} \right].$$

The following formula containing 3j symbols is also important for various applications:

$$Y_{l_1m_1}(\theta, \varphi) Y_{l_2m_2}(\theta, \varphi) = \sum_{l,m} \sqrt{\frac{(2l_1 + 1)(2l_2 + 1)(2l + 1)}{4\pi}}$$

$$\times \begin{pmatrix} l_1 & l_2 & l \\ m_1 & m_2 & m \end{pmatrix} Y_{lm}^*(\theta, \varphi) \begin{pmatrix} l_1 & l_2 & l \\ 0 & 0 & 0 \end{pmatrix}. \quad (4.61)$$

Let us multiply (4.61) by $Y_{l_3m_3}(\theta, \varphi)$ and integrate over all angles. For the values of l_3 obeying the triangle condition $\Delta(l_1l_2l_3)$, it follows from (4.61) that

$$\int Y_{l_1m_1}(\theta, \varphi) Y_{l_2m_2}(\theta, \varphi) Y_{l_3m_3}(\theta, \varphi) \sin\theta \, d\theta \, d\varphi$$

$$= \sqrt{\frac{(2l_1 + 1)(2l_2 + 1)(2l_3 + 1)}{4\pi}} \begin{pmatrix} l_1 & l_2 & l_3 \\ 0 & 0 & 0 \end{pmatrix} \begin{pmatrix} l_1 & l_2 & l_3 \\ m_1 & m_2 & m_3 \end{pmatrix} \quad (4.62)$$

and

$$\frac{1}{2} \int P_{l_1}(\cos\theta) P_{l_2}(\cos\theta) P_{l_3}(\cos\theta) \sin\theta \, d\theta = \begin{pmatrix} l_1 & l_2 & l_3 \\ 0 & 0 & 0 \end{pmatrix}^2, \quad (4.63)$$

The integrals of the three Legendre polynomials in (4.63) are often denoted by $C_{l_1l_2l_3}$. According to (4.63)

$$C_{l_1l_2l_3} = 2 \begin{pmatrix} l_1 & l_2 & l_3 \\ 0 & 0 & 0 \end{pmatrix}^2. \quad (4.64)$$

4.2.3 Racah W Coefficients and 6j Symbols

Let us consider the two schemes of addition of the angular momenta j_1, j_2 and j_3

$$j_1 + j_2 = J', \quad J' + j_3 = J, \quad (4.65)$$

$$j_2 + j_3 = J'', \quad j_1 + J'' = J. \quad (4.66)$$

In the first case

$$\Psi_{JM}(j_1j_2[J']j_3) = \sum_{m_3M'} (J'j_3M'm_3 | J'j_3JM) \Psi_{J'M'}\Psi_{j_3m_3}$$

$$= \sum_{m_1m_2m_3M'} (j_1j_2m_1m_2 | j_1j_2J'M')(J'j_3M'm_3 | J'j_3JM) \Psi_{j_1m_1}\Psi_{j_2m_2}\Psi_{j_3m_3}. \quad (4.67)$$

In the second case

$$\Psi_{JM}(j_1, j_2 j_3[J'']) = \sum_{m_1 M''} (j_1 J'' m_1 M'' | j_1 J'' JM) \Psi_{j_1 m_1} \Psi_{J'' M''}$$
$$= \sum_{m_1 m_2 m_3 M''} (j_2 j_3 m_2 m_3 | j_2 j_3 J'' M'') (j_1 J'' m_1 M'' | j_1 J'' JM) \Psi_{j_1 m_1} \Psi_{j_2 m_2} \Psi_{j_3 m_3}. \tag{4.68}$$

The functions $\Psi_{JM}(j_1, j_2 j_3[J''])$ can be expanded in terms of the functions $\Psi_{JM}(j_1 j_2[J']j_3)$,

$$\Psi_{JM}(j_1, j_2 j_3[J'']) = \sum_{J'} (j_1 j_2[J']j_3 J | j_1, j_2 j_3[J''] J) \Psi_{JM}(j_1 j_2[J']j_3). \tag{4.69}$$

With the help of the expressions given above for the functions $\Psi_{JM}(j_1 j_2[J']j_3)$ and $\Psi_{JM}(j_1, j_2 j_3[J''])$ the coefficients $(j_1 j_2[J']j_3 J | j_1, j_2 j_3[J''] J)$ can be expressed in terms of Clebsch–Gordan coefficients

$$(j_1 j_2[J']j_3 J | j_1, j_2 j_3[J''] J) = \sum_{m_1 m_2 m_3 M' M''} (J' j_3 JM | J' j_3 M' m_3)$$
$$\times (j_1 j_2 J' M' | j_1 j_2 m_1 m_2)(j_2 j_3 m_2 m_3 | j_2 j_3 J'' M'')(j_1 J'' m_1 M'' | j_1 J'' JM). \tag{4.70}$$

The sum on the right-hand side is independent of $m_1 m_2 m_3 M' M''$ with respect to which summing is carried out, and is a function of the six arguments $j_1 j_2 j_3 J' J'' J$. Expression (4.70) can thus be rewritten in the following form

$$(j_1 j_2[J']j_3 J | j_1, j_2 j_3[J''] J) = \sqrt{(2J'+1)(2J''+1)}\, W(j_1 j_2 J j_3; J' J''). \tag{4.71}$$

The function W on the right-hand side of (4.71) is the Racah W coefficient.

If we interchange the vectors j_2 and j_3, we obtain the following scheme of addition of angular momenta:

$$j_1 + j_3 = J'', \quad J'' + j_2 = J. \tag{4.72}$$

In this case

$$(j_1 j_2[J']j_3 J | j_1 j_3[J''] j_2 J) = \sqrt{(2J'+1)(2J''+1)}\, W(J' j_3 j_2 J''; J j_1). \tag{4.73}$$

Formulas (4.71, 73) generalize naturally to the case when the order of addition of orbital angular momenta and spins of three electrons changes simultaneously. For example, for the transition from the scheme

$$l_1 + l_2 = L', \quad s_1 + s_2 = S', \quad L' + l_3 = L, \quad S' + s_3 = S \tag{4.74}$$

to the scheme

$$l_2 + l_3 = L'', \quad s_2 + s_3 = S'', \quad l_1 + L'' = L, \quad s_1 + S'' = S \tag{4.75}$$

we have

$$\begin{aligned}(l_1s_1,\, l_2s_2\,[L'S']\,l_3s_3LS\,|\,l_1s_1;\, l_2s_2,\, l_3s_3\,[L''S'']\,LS) \\ = \sqrt{(2L'+1)\,(2L''+1)\,(2S'+1)\,(2S''+1)} \\ \times W(l_1l_2Ll_3;\, L'L'')\, W(s_1s_2Ss_3;\, S'S'')\,.\end{aligned} \tag{4.76}$$

It follows from (4.70) that $W(abcd, ef)$ is nonzero if the following triangle conditions are fulfilled

$$\Delta(abe),\ \Delta(cde),\ \Delta(acf),\ \Delta(bdf)\,. \tag{4.77}$$

W coefficients satisfy a series of symmetry relations. It is convenient to write these relations expressing W in terms of more symmetrical coefficients, the so-called $6j$ symbols

$$\begin{Bmatrix} j_1 & j_2 & j_3 \\ l_1 & l_2 & l_3 \end{Bmatrix}, \tag{4.78}$$

$$W(j_1 j_2 l_2 l_1;\, j_3 l_3) = (-1)^{-j_1-j_2-l_1-l_2} \begin{Bmatrix} j_1 & j_2 & j_3 \\ l_1 & l_2 & l_3 \end{Bmatrix}. \tag{4.79}$$

The $6j$ symbol remains invariant under any transposition of its column and also under transposition of the lower and upper arguments in each of any two columns.

From (4.79) it is easy to obtain the symmetry relations for W coefficients

$$\begin{aligned}W(abcd;\, ef) = W(badc;\, ef) = W(cdab;\, ef) = W(acbd;\, fe) \\ = (-1)^{e+f-a-d}\, W(ebcf;\, ad) = (-1)^{e+f-b-c}\, W(aefd;\, bc)\,.\end{aligned} \tag{4.80}$$

When $e = 0$,

$$W(abcd;\, 0f) = (-1)^{b+c-f}\, \delta_{ab}\delta_{cd}\, [(2b+1)(2c+1)]^{-1/2}\,. \tag{4.81}$$

It follows from (4.80, 81)

$$W(abcd;\, e0) = (-1)^{c+b-e}\, \delta_{ac}\delta_{bd}\, [(2c+1)(2b+1)]^{-1/2},$$
$$W(0bcd;\, ef) = \delta_{eb}\delta_{cf}\, [(2e+1)(2f+1)]^{-1/2},$$

$$W(a0cd; ef) = \delta_{ae}\delta_{fd}[(2e+1)(2f+1)]^{-1/2}, \tag{4.82}$$
$$W(ab0d; ef) = \delta_{de}\delta_{af}[(2e+1)(2f+1)]^{-1/2},$$
$$W(abc0; ef) = \delta_{ec}\delta_{fb}[(2e+1)(2f+1)]^{-1/2}.$$

$6j$ symbols obey the following sum rule

$$\sum_j (2j+1)(2j''+1) \begin{Bmatrix} j_1 & j_2 & j' \\ j_3 & j_4 & j \end{Bmatrix} \begin{Bmatrix} j_3 & j_2 & j \\ j_1 & j_4 & j'' \end{Bmatrix} = \delta_{j'j''}, \tag{4.83}$$

$$\sum_j (-1)^{j+j'+j''}(2j+1) \begin{Bmatrix} j_1 & j_2 & j' \\ j_3 & j_4 & j \end{Bmatrix} \begin{Bmatrix} j_2 & j_3 & j \\ j_1 & j_4 & j'' \end{Bmatrix} = \begin{Bmatrix} j_2 & j_1 & j' \\ j_3 & j_4 & j'' \end{Bmatrix}, \tag{4.84}$$

and also

$$\sum_x (-1)^{l_1+l_2+l_3+l_1'+l_2'+l_3'+l_1'+l_2'+l_3'+x}(2x+1) \begin{Bmatrix} l_1 & x & l_1' \\ l_3' & j_2 & l_3 \end{Bmatrix}$$
$$\times \begin{Bmatrix} l_2 & x & l_2' \\ l_1' & j_3 & l_1 \end{Bmatrix} \begin{Bmatrix} l_3 & x & l_3' \\ l_2' & j_1 & l_2 \end{Bmatrix} = \begin{Bmatrix} j_1 & j_2 & j_3 \\ l_1 & l_2 & l_3 \end{Bmatrix} \begin{Bmatrix} j_1 & j_2 & j_3 \\ l_1' & l_2' & l_3' \end{Bmatrix}. \tag{4.85}$$

Using (4.79), it is easy to obtain similar relations also for W coefficients. For example,

$$\sum_e (2e+1) W(acfd; be) W(acgd; be) = \frac{1}{2f+1} \cdot \delta_{fg}, \tag{4.86}$$

$$\sum_e (-1)^{a+b+c+d+e+f+g}(2e+1) W(acbd; fe) W(abdc; eg) = W(acdb; fg). \tag{4.87}$$

To conclude this section, we give the formula for the sum of the products of three $3j$ symbols:

$$\sum_{\mu_1\mu_2\mu_3} (-1)^{l_1+l_2+l_3+\mu_1+\mu_2+\mu_3} \begin{pmatrix} j_1 & l_2 & l_3 \\ m_1 & \mu_2 & -\mu_3 \end{pmatrix} \begin{pmatrix} l_1 & j_2 & l_3 \\ -\mu_1 & m_2 & \mu_3 \end{pmatrix}$$
$$\times \begin{pmatrix} l_1 & l_2 & j_3 \\ \mu_1 & -\mu_2 & m_3 \end{pmatrix} = \begin{pmatrix} j_1 & j_2 & j_3 \\ m_1 & m_2 & m_3 \end{pmatrix} \begin{Bmatrix} j_1 & j_2 & j_3 \\ l_1 & l_2 & l_3 \end{Bmatrix}, \tag{4.88}$$

and one important asymptotic expression for the $6j$ symbol; when $j_1, j_2 \gg x$

$$\begin{Bmatrix} j & j_2 & j_1 \\ x & j_1 & j_2 \end{Bmatrix} \to (-1)^{j_1+j_2+j} \frac{P_x(\cos(\mathbf{j}_1 \cdot \mathbf{j}_2))}{\sqrt{(2j_1+1)(2j_2+1)}}. \tag{4.89}$$

4.2.4 Summary of Formulas for 6j Symbols [8]

$$\begin{Bmatrix} a & b & c \\ 0 & c & b \end{Bmatrix} = (-1)^{a+b+c} [(2b+1)(2c+1)]^{-1/2}, \tag{4.90}$$

$$\begin{Bmatrix} a & b & c \\ \tfrac{1}{2} & c-\tfrac{1}{2} & b+\tfrac{1}{2} \end{Bmatrix} = (-1)^{a+b+c} \left[\frac{(a+c-b)(a+b-c+1)}{(2b+1)(2b+2)\,2c\,(2c+1)} \right]^{1/2},$$

$$\begin{Bmatrix} a & b & c \\ \tfrac{1}{2} & c-\tfrac{1}{2} & b-\tfrac{1}{2} \end{Bmatrix} = (-1)^{a+b+c} \left[\frac{(a+b+c+1)(b+c-a)}{2b(2b+1)\,2c\,(2c+1)} \right]^{1/2}, \tag{4.91}$$

$$\begin{Bmatrix} a & b & c \\ 1 & c-1 & b-1 \end{Bmatrix} = (-1)^s \left[\frac{s(s+1)(s-2a-1)(s-2a)}{(2b-1)\,2b\,(2b+1)(2c-1)\,2c\,(2c+1)} \right]^{1/2},$$

$$\begin{Bmatrix} a & b & c \\ 1 & c-1 & b \end{Bmatrix} = (-1)^s \left[\frac{2(s+1)(s-2a)(s-2b)(s-2c+1)}{2b(2b+1)(2b+2)(2c-1)\,2c\,(2c+1)} \right]^{1/2},$$

$$\begin{Bmatrix} a & b & c \\ 1 & c-1 & b+1 \end{Bmatrix} = (-1)^s \left[\frac{(s-2b)(s-2b-1)(s-2c+1)(s-2c+2)}{(2b+1)(2b+2)(2b+3)(2c-1)\,2c\,(2c+1)} \right]^{1/2}, \tag{4.92}$$

$$\begin{Bmatrix} a & b & c \\ 1 & c & b \end{Bmatrix} = (-1)^s \frac{2X}{[2b(2b+1)(2b+2)\,2c\,(2c+1)(2c+2)]^{1/2}},$$

$$\begin{Bmatrix} a & b & c \\ \tfrac{3}{2} & c-\tfrac{3}{2} & b-\tfrac{3}{2} \end{Bmatrix} = (-1)^s \left[\frac{(s-1)\,s\,(s+1)(s-2a-2)(s-2a-1)(s-2a)}{(2b-2)(2b-1)\,2b\,(2b+1)(2c-2)(2c-1)\,2c\,(2c+1)} \right]^{1/2},$$

$$\begin{Bmatrix} a & b & c \\ \tfrac{3}{2} & c-\tfrac{3}{2} & b-\tfrac{1}{2} \end{Bmatrix} = (-1)^s \left[\frac{3s(s+1)(s-2a-1)(s-2a)(s-2b)(s-2c+1)}{(2b-1)\,2b\,(2b+1)(2b+2)(2c-2)(2c-1)\,2c\,(2c+1)} \right]^{1/2},$$

$$\begin{Bmatrix} a & b & c \\ \tfrac{3}{2} & c-\tfrac{3}{2} & b+\tfrac{1}{2} \end{Bmatrix} = (-1)^s \left[\frac{3(s+1)(s-2a)(s-2b-1)(s-2b)(s-2c+1)(s-2c+2)}{2b(2b+1)(2b+2)(2b+3)(2c-2)(2c-1)\,2c\,(2c+1)} \right]^{1/2}, \tag{4.93}$$

4.2 Angular Momentum Vector Addition Coefficients

$$\begin{Bmatrix} a & b & c \\ \frac{3}{2} & c-\frac{3}{2} & b+\frac{3}{2} \end{Bmatrix}$$
$$= (-1)^s \left[\frac{(s-2b-2)(s-2b-1)(s-2b)(s-2c+1)(s-2c+2)(s-2c+3)}{(2b+1)(2b+2)(2b+3)(2b+4)(2c-2)(2c-1)2c(2c+1)} \right]^{1/2},$$

$$\begin{Bmatrix} a & b & c \\ \frac{3}{2} & c-\frac{1}{2} & b-\frac{1}{2} \end{Bmatrix}$$
$$= (-1)^s \frac{[2(s-2b)(s-2c)-(s+2)(s-2a-1)][(s+1)(s-2a)]^{1/2}}{[(2b-1)2b(2b+1)(2b+2)(2c-1)2c(2c+1)(2c+2)]^{1/2}},$$

$$\begin{Bmatrix} a & b & c \\ \frac{3}{2} & c-\frac{1}{2} & b+\frac{1}{2} \end{Bmatrix}$$
$$= (-1)^s \frac{[(s-2b-1)(s-2c)-2(s+2)(s-2a)][(s-2b)(s-2c+1)]^{1/2}}{[2b(2b+1)(2b+2)(2b+3)(2c-1)2c(2c+1)(2c+2)]^{1/2}},$$

$$\begin{Bmatrix} a & b & c \\ 2 & c-2 & b-2 \end{Bmatrix}$$
$$= (-1)^s \left[\frac{(s-2)(s-1)s(s+1)(s-2a-3)(s-2a-2)(s-2a-1)(s-2a)}{(2b-3)(2b-2)(2b-1)2b(2b+1)(2c-3)(2c-2)(2c-1)2c(2c+1)} \right]^{1/2},$$

$$\begin{Bmatrix} a & b & c \\ 2 & c-2 & b-1 \end{Bmatrix}$$
$$= (-1)^s 2 \left[\frac{(s-1)s(s+1)(s-2a-2)(s-2a-1)(s-2a)(s-2b)(s-2c+1)}{(2b-2)(2b-1)2b(2b+1)(2b+2)(2c-3)(2c-2)(2c-1)2c(2c+1)} \right]^{1/2},$$

$$\begin{Bmatrix} a & b & c \\ 2 & c-2 & b \end{Bmatrix} = (-1)^s (s-2b-1)^{1/2}$$
$$\times \left[\frac{6s(s+1)(s-2a-1)(s-2a)(s-2b)(s-2c+1)(s-2c+2)}{(2b-1)2b(2b+1)(2b+2)(2b+3)(2c-3)(2c-2)(2c-1)2c(2c+1)} \right]^{1/2},$$

$$\begin{Bmatrix} a & b & c \\ 2 & c-2 & b+1 \end{Bmatrix} = (-1)^s 2$$
$$\times \left[\frac{(s+1)(s-2a)(s-2b-2)(s-2b-1)(s-2b)(s-2c+1)(s-2c+2)(s-2c+3)}{2b(2b+1)(2b+2)(2b+3)(2b+4)(2c-3)(2c-2)(2c-1)2c(2c+1)} \right]^{1/2},$$

(4.94)

$$\begin{Bmatrix} a & b & c \\ 2 & c-2 & b+2 \end{Bmatrix} = (-1)^s (s-2b-3)^{1/2}$$
$$\times \left[\frac{(s-2b-2)(s-2b-1)(s-2b)(s-2c+1)(s-2c+2)(s-2c+3)(s-2c+4)}{(2b+1)(2b+2)(2b+3)(2b+4)(2b+5)(2c-3)(2c-2)(2c-1)2c(2c+1)} \right]^{1/2},$$

$$\begin{Bmatrix} a & b & c \\ 2 & c-1 & b-1 \end{Bmatrix} = (-1)^s$$

$$\times \frac{4[(a+b)(a-b+1)-(c-1)(c-b+1)][s(s+1)(s-2a-1)(s-2a)]^{1/2}}{[(2b-2)(2b-1)2b(2b+1)(2b+2)(2c-2)(2c-1)2c(2c+1)(2c+2)]^{1/2}},$$

$$\begin{Bmatrix} a & b & c \\ 2 & c-1 & b \end{Bmatrix} = (-1)^s$$

$$\times \frac{2[(a+b+1)(a-b)-c^2+1][6(s+1)(s-2a)(s-2b)(s-2c+1)]^{1/2}}{[(2b-1)2b(2b+1)(2b+2)(2b+3)(2c-2)(2c-1)2c(2c+1)(2c+2)]^{1/2}},$$

$$\begin{Bmatrix} a & b & c \\ 2 & c-1 & b+1 \end{Bmatrix} = (-1)^s$$

$$\times \frac{4[(a+b+2)(a-b-1)-(c-1)(b+c+2)][(s-2b-1)(s-2b)(s-2c+1)(s-2c+2)]^{1/2}}{[2b(2b+1)(2b+2)(2b+3)(2b+4)(2c-2)(2c-1)2c(2c+1)(2c+2)]^{1/2}}$$

$$\begin{Bmatrix} a & b & c \\ 2 & c & b \end{Bmatrix}$$

$$= (-1)^s \frac{2[3X(X+1)-4b(b+1)c(c+1)]}{[(2b-1)2b(2b+1)(2b+2)(2b+3)(2c-1)2c(2c+1)(2c+2)(2c+3)]^{1/3}}.$$

In (4.92–94)

$$s = a+b+c, \tag{4.95}$$

$$X = a(a+1) - b(b+1) - c(c+1). \tag{4.96}$$

We also give two formulas for W coefficients which are especially important

$$W(abab; c1) = (-1)^{a+b+c+1} \frac{a(a+1)+b(b+1)-c(c+1)}{2[a(a+1)(2a+1)b(b+1)(2b+1)]^{1/2}} \tag{4.97}$$

$W(abab; c2)$

$$=(-1)^{a+b+c} \frac{2[3C(C-1)-4a(a+1)b(b+1)]}{[(2a-1)2a(2a+1)(2a+2)(2a+3)(2b-1)2b(2b+1)(2b+2)(2b+3)]^{1/2}}, \tag{4.98}$$

$$C = a(a+1) + b(b+1) - c(c+1). \tag{4.99}$$

4.2.5 9j Symbols

Let us consider the transitions between the two schemes of addition of four angular momenta

$$j_1 j_2 [J_{12}]; \; j_3 j_4 [J_{34}] J, \tag{4.100}$$

$$j_1 j_3 [J_{13}]; \; j_2 j_4 [J_{24}] J. \tag{4.101}$$

This transition can be achieved in three stages, altering each time the order of addition of three angular momenta:

$$j_1 j_2 [J_{12}]; \; j_3 j_4 [J_{34}] J \to j_1; j_2, j_3 j_4 [J_{34}] J' J$$
$$\to j_1; j_3, j_2 j_4 [J_{24}] J' J \to j_1 j_3 [J_{13}]; j_2 j_4 [J_{24}] J.$$

As a result,

$$(j_1 j_2 [J_{12}] j_3 j_4 [J_{34}] J | j_1 j_3 [J_{13}] \, j_2 j_4 [J_{24}] J)$$
$$= \sum_{J'} (j_1 j_2 [J_{12}] J_{34} J | j_1; j_2 J_{34} [J'] J) (j_2; j_3 j_4 [J_{34}] J' | j_3; j_2 j_4 [J_{24}] J')$$
$$\times (j_1; j_3 J_{24} [J'] J | j_1 j_3 [J_{13}] J_{24} J). \tag{4.102}$$

Each of the transformation coefficients on the right-hand side of (4.102) is expressed in terms of W coefficients by (4.71) and (4.73). Replacing the W coefficients by $6j$ symbols in the final formula, we obtain

$$(j_1 j_2 [J_{12}]; j_3 j_4 [J_{34}] J | j_1 j_3 [J_{13}]; j_2 j_4 [J_{24}] J)$$
$$= \sqrt{(2J_{12}+1)(2J_{34}+1)(2J_{13}+1)(2J_{24}+1)} \begin{Bmatrix} j_1 & j_2 & J_{12} \\ j_3 & j_4 & J_{34} \\ J_{13} & J_{24} & J \end{Bmatrix}, \tag{4.103}$$

where

$$\begin{Bmatrix} j_1 & j_2 & J_{12} \\ j_3 & j_4 & J_{34} \\ J_{13} & J_{24} & J \end{Bmatrix}$$
$$= \sum_{J'} (-1)^{2J'}(2J'+1) \begin{Bmatrix} j_1 & j_2 & J_{12} \\ J_{34} & J & J' \end{Bmatrix} \begin{Bmatrix} j_3 & j_4 & J_{34} \\ j_2 & J' & J_{24} \end{Bmatrix} \begin{Bmatrix} J_{13} & J_{24} & J \\ J' & j_1 & j_3 \end{Bmatrix}. \tag{4.104}$$

Expression (4.104) defines the so-called $9j$ symbol. Thus the coefficients of the transition between the two schemes of addition of four angular momenta are expressed in terms of $9j$ symbols.[2]

---

[2] In a similar way the variation of the scheme of addition of five angular momenta leads to $12j$ symbols; of six momenta to $15j$ symbols, and so on, all of which can be expressed in the form of sums of the products of $6j$ symbols [8–10].

Using (4.104), one can obtain the main properties of the $9j$ symbols and, in particular, the symmetry of $9j$ symbols under even transpositions of rows or columns, and under reflection along either diagonal. An odd transposition of the rows or columns of a $9j$ symbol multiplies it by $(-1)^s$, where s is the sum of all arguments. Formula (4.104) is considerably simplified if one of the arguments of the $9j$ symbol vanishes. In this case

$$\begin{Bmatrix} a & b & e \\ c & d & e \\ f & f & 0 \end{Bmatrix} = \begin{Bmatrix} 0 & e & e \\ f & d & b \\ f & c & a \end{Bmatrix} = \begin{Bmatrix} e & 0 & e \\ c & f & a \\ d & f & b \end{Bmatrix} = \begin{Bmatrix} f & f & 0 \\ d & c & e \\ b & a & e \end{Bmatrix}$$

$$= \begin{Bmatrix} f & b & d \\ 0 & e & e \\ f & a & c \end{Bmatrix} = \begin{Bmatrix} a & f & c \\ e & 0 & e \\ b & f & d \end{Bmatrix} = \begin{Bmatrix} b & a & e \\ f & f & 0 \\ d & c & e \end{Bmatrix} = \begin{Bmatrix} e & d & c \\ e & b & a \\ 0 & f & f \end{Bmatrix} = \begin{Bmatrix} c & e & d \\ a & e & b \\ f & 0 & f \end{Bmatrix}$$

$$= \frac{(-1)^{b+c+e+f}}{\sqrt{(2e+1)(2f+1)}} \begin{Bmatrix} a & b & e \\ d & c & f \end{Bmatrix} . \qquad (4.105)$$

In the theory of atomic spectra, $9j$ symbols defining the transformation from LS coupling to jj coupling are of particular interest. Formulas for $9j$ symbols of this type are given in Sect. 5.6. $9j$ symbols obey a number of sum rules. We shall give the simplest of these rules, which we shall need later,

$$\sum_{gh} \begin{Bmatrix} a & b & e \\ c & d & f \\ g & h & k \end{Bmatrix} \begin{Bmatrix} a & b & e' \\ c & d & f' \\ g & h & k \end{Bmatrix} (2g+1)(2h+1) = \frac{\delta_{ee'}\delta_{ff'}}{(2e+1)(2f+1)} . \qquad (4.106)$$

4.3 Irreducible Tensor Operators

4.3.1 Spherical Tensors

In calculating the matrix elements of various operators, it is convenient to classify these operators according to their behavior upon rotation of the system of coordinates. From this point of view the usual definition of a tensor in the Cartesian system of coordinates is unsuitable, because from the components of a tensor of rank $\kappa \geq 2$ a number of linear combinations can be formed which behave differently upon rotation of the coordinate system. It is more convenient to define a tensor in such a way that all its components (and any linear combinations of these components) are transformed in one way. This condition is satisfied by the set of $(2\kappa+1)$ spherical harmonics: $Y_{\kappa q}$; $q = \kappa, \kappa-1, \ldots, -\kappa$. We shall therefore define the tensor of rank κ as a set of $(2\kappa+1)$ quantities which transform as the spherical harmonics $Y_{\kappa q}$ upon a rotation of the system of coordinates. The tensors defined in this way are called spherical tensors or irreducible tensors. In accordance with this definition, the irreducible tensor operator T_κ of rank κ is a set of $(2\kappa+1)$ operators $T_{\kappa q}$

$$q = \kappa, \kappa-1, \ldots, -\kappa \tag{4.107}$$

obeying the same commutation rules with the angular momentum of the system $\boldsymbol{J}$ as $Y_{\kappa q}$. In accordance with (4.3), these commutation rules have the form

$$[(J_x \pm iJ_y), T_{\kappa q}] = \sqrt{(\kappa \mp q)(\kappa \pm q + 1)}\, T_{\kappa, q \pm 1} \tag{4.108}$$

$$[J_z, T_{\kappa q}] = q T_{\kappa q}. \tag{4.109}$$

A very simple example of operators of this type is the set of functions

$$f(r) Y_{\kappa q}(\theta, \varphi), \tag{4.110}$$

where $f(r)$ is an arbitrary function of r.

When $\kappa = 1$ the commutation rules (4.108, 109) coincide with the commutation rules for the spherical components of a vector $\boldsymbol{A}$,

$$A_0 = A_z;\; A_{+1} = -\frac{1}{\sqrt{2}}(A_x + iA_y);\; A_{-1} = +\frac{1}{\sqrt{2}}(A_x - iA_y), \tag{4.111}$$

since these components are expressed in the following way in terms of the spherical harmonics

$$\begin{aligned}
A_0 &= |A|\cos\theta = \sqrt{\frac{4\pi}{3}}\,|A|\,Y_{10} = |A|\,C_0^1, \\
A_{+1} &= -|A|\frac{e^{i\varphi}\sin\theta}{\sqrt{2}} = \sqrt{\frac{4\pi}{3}}\,|A|\,Y_{1,+1} = |A|\,C_{+1}^1, \\
A_{-1} &= |A|\frac{e^{-i\varphi}\sin\theta}{\sqrt{2}} = \sqrt{\frac{4\pi}{3}}\,|A|\,Y_{1,-1} = |A|\,C_{-1}^1.
\end{aligned} \tag{4.112}$$

Thus the spherical components of the vector form an irreducible tensor operator of the first rank

$$T_{10} = A_0;\; T_{1,\pm 1} = A_{\pm 1}. \tag{4.113}$$

We shall also consider how the components of a tensor of the second rank a_{ik} ($i,k = x,y,z$) are expressed in terms of $T_{\kappa q}$. This tensor can be represented in the form

$$\alpha_{ik} = \alpha\,\delta_{ik} + \alpha'_{ik} + \alpha''_{ik}, \tag{4.114}$$

where

$$\alpha = \frac{1}{3}\sum_i \alpha_{ii},$$

$$\alpha'_{ik} = \frac{1}{2}(\alpha_{ik} - \alpha_{ki}),$$

$$\alpha''_{ik} = \frac{1}{2}(\alpha_{ik} + \alpha_{ki} - 2\alpha\delta_{ik}).$$

The trace of the tensor α_{ik} is invariant with respect to a rotation of the system of coordinates; therefore α is an irreducible tensor of rank zero:

$$T_{00} = \alpha. \tag{4.115}$$

From the components of the antisymmetric tensor α'_{ik}, one can construct the irreducible tensor of the first rank

$$T_{10} = \alpha'_{xy},$$
$$T_{1,\pm1} = \mp\frac{1}{\sqrt{2}}(\alpha'_{yz} \pm i\alpha'_{zx}), \tag{4.116}$$

and from the components of the symmetric tensor α''_{ik}, the irreducible tensor of the second rank

$$T_{20} = \alpha''_{zz}, \tag{4.117}$$

$$T_{2,\pm1} = \pm\sqrt{\frac{2}{3}}(\alpha'_{zx} \pm i\alpha''_{zy}), \tag{4.118}$$

$$T_{2,\pm2} = \sqrt{\frac{1}{6}}(\alpha''_{xx} - \alpha''_{yy} \pm 2i\alpha''_{xy}). \tag{4.119}$$

Tensors of higher rank can be expanded into irreducible tensors in a similar way. We shall use in future one of the two notations $T_{\kappa q}$ or T_q^κ for the components of irreducible tensors.

4.3.2 Matrix Elements

It follows from (4.62) that

$$\langle LM|Y_{\kappa q}|L'M'\rangle = (-1)^M \int Y_{L-M}Y_{L'M'}Y_{\kappa q}\,dO \propto (-1)^M \begin{pmatrix} L & L' & \kappa \\ -M & M' & q \end{pmatrix}.$$

4.3 Irreducible Tensor Operators

This relation can also be obtained directly from the commutation rules for the functions $Y_{\kappa q}$ with the orbital angular momentum L. The dependence of the matrix elements $T_{\kappa q}$ on the quantum numbers $MM'q$ can be found in precisely the same way from the commutation rules for $T_{\kappa q}$ and J. In the general case we have

$$\overbrace{\langle \gamma JM|T_{\kappa q}|\gamma' J'M'\rangle}^{\text{transition ME}} = (-1)^{J-M}\, \underbrace{(\gamma J\|T_{\kappa}\|\gamma' J')}_{\text{reduced ME}} \begin{pmatrix} J & \kappa & J' \\ -M & q & M' \end{pmatrix} \qquad (4.120)$$

known as the Wigner–Eckart theorem. The coefficients

$$(\gamma J\|T_{\kappa}\|\gamma' J') \qquad (4.121)$$

do not depend on MM' or q, and are called reduced matrix elements. From the orthogonality condition (4.42) of the $3j$ symbols follows the important sum rule

$$\sum_{MM'} |\langle \gamma JM|T_{\kappa q}|\gamma' J'M'\rangle|^2 = \frac{1}{2\kappa+1} |(\gamma J\|T_{\kappa}\|\gamma' J')|^2. \qquad (4.122)$$

The right-hand side of (4.122) is independent of q; therefore

$$\sum_{q}\sum_{MM'} |\langle \gamma JM|T_{\kappa q}|\gamma' J'M'\rangle|^2 = |(\gamma J\|T_{\kappa}\|\gamma' J')|^2. \qquad (4.123)$$

In a number of problems, the sum (4.122) or (4.123) enters into the final formulas and not the matrix elements themselves. It is therefore sufficient to know the reduced matrix elements. The latter are found from (4.120). For example, in the case $\kappa = 1$, as a rule the matrix element with $M = M' = q = 0$ is easiest to calculate. From (4.120) we have

$$\langle \gamma J0|T_{10}|\gamma' J'0\rangle = (-1)^J (\gamma J\|T_1\|\gamma' J') \begin{pmatrix} J & 1 & J' \\ 0 & 0 & 0 \end{pmatrix}, \qquad (4.124)$$

$$(\gamma J\|T_1\|\gamma' J') = (-1)^J \frac{\langle \gamma J0|T_{10}|\gamma' J'0\rangle}{\begin{pmatrix} J & 1 & J' \\ 0 & 0 & 0 \end{pmatrix}}. \qquad (4.125)$$

Here it is necessary that $\begin{pmatrix} J & 1 & J' \\ 0 & 0 & 0 \end{pmatrix} \neq 0$. Let us note that the reduced matrix elements $(\gamma J\|T_1\|\gamma' J')$ are related in the following way to the quantities $(\gamma J:T_1:\gamma' J')$ used in [1]

$$(\gamma J\|T_1\|\gamma'J) = \sqrt{J(J+1)(2J+1)}(\gamma J \vdots T_1 \vdots \gamma'J),$$
$$(\gamma J\|T_1\|\gamma'J-1) = \sqrt{J(2J-1)(2J+1)}\,(\gamma J \vdots T_1 \vdots \gamma'J-1),$$
$$(\gamma J\|T_1\|\gamma'J+1)$$
$$= -\sqrt{(J+1)(2J+1)(2J+3)}\,(\gamma J \vdots T_1 \vdots \gamma'J+1). \qquad (4.126)$$

For hermitian operators $T_{\kappa q}$ the reduced matrix elements obey the relation

$$(\gamma J\|T_\kappa\|\gamma'J') = (-1)^{J-J'}\,(\gamma'J'\|T_\kappa\|\gamma J)^*. \qquad (4.127)$$

4.3.3 Some Examples of Calculation of Reduced Matrix Elements

We shall begin by calculating the reduced matrix elements of the spherical harmonics $Y_{\kappa q}$. According to (4.120) we have

$$\langle lm|Y_{\kappa q}|l'm'\rangle = (-1)^{l-m}(l\|Y_\kappa\|l')\begin{pmatrix} l & \kappa & l' \\ -m & q & m' \end{pmatrix}. \qquad (4.128)$$

On the other hand, (4.62) gives

$$\int Y_{lm}^* Y_{\kappa q} Y_{l'm'} \sin\theta\, d\theta\, d\varphi = (-1)^m \int Y_{l-m} Y_{\kappa q} Y_{l'm'} \sin\theta\, d\theta\, d\varphi$$
$$= (-1)^m \sqrt{\frac{(2l+1)(2\kappa+1)(2l'+1)}{4\pi}} \begin{pmatrix} l & \kappa & l' \\ 0 & 0 & 0 \end{pmatrix}\begin{pmatrix} l & \kappa & l' \\ -m & q & m' \end{pmatrix}. \qquad (4.129)$$

By comparing (4.128) and (4.129), we obtain for the case $l + \kappa + l' = 2g$, where g is an integer,

$$(l\|Y_\kappa\|l') = (-1)^l \sqrt{\frac{(2l+1)(2\kappa+1)(2l'+1)}{4\pi}} \begin{pmatrix} l & \kappa & l' \\ 0 & 0 & 0 \end{pmatrix}, \qquad (4.130)$$

$$(l\|C^\kappa\|l') = (-1)^l \sqrt{(2l+1)(2l'+1)} \begin{pmatrix} l & \kappa & l' \\ 0 & 0 & 0 \end{pmatrix}. \qquad (4.131)$$

When $\kappa = 0$,

$$(l\|Y_0\|l') = \sqrt{\frac{2l+1}{4\pi}}\,\delta_{ll'}, \qquad (4.132)$$

$$(l\|C^0\|l') = \sqrt{2l+1}\,\delta_{ll'}. \qquad (4.133)$$

As

$$Y_{00} = \frac{1}{\sqrt{4\pi}},\qquad(4.134)$$

$$(l\|1\|l') = \sqrt{2l+1}\,\delta_{ll'}.\qquad(4.135)$$

When $\kappa = 1$,

$$\begin{pmatrix} l & 1 & l' \\ 0 & 0 & 0 \end{pmatrix} = (-1)^g \sqrt{\frac{l_{\max}}{(2l+1)(2l'+1)}},\quad l' = l\pm 1,\qquad(4.136)$$

where $l_{\max}$ is the larger of the numbers l and l'. Therefore

$$(l\|Y_1\|l') = (-1)^{l+g}\sqrt{\frac{3}{4\pi}}\sqrt{l_{\max}}\qquad l' = l\pm 1,\qquad(4.137)$$

$$(l\|C^1\|l') = (-1)^{l+g}\sqrt{l_{\max}}\qquad l' = l\pm 1.\qquad(4.138)$$

For $l' \neq l\pm 1$, the reduced matrix elements of Y_1 and C^1 equal zero. The spherical components of the unit vector $\mathbf{n}$ are expressed in the following way in terms of functions Y_{1m}

$$n_0 = \sqrt{\frac{4\pi}{3}}\,Y_{10};\; n_{\pm 1} = \sqrt{\frac{4\pi}{3}}\,Y_{1,\pm 1}\qquad(4.139)$$

Therefore

$$(l\|n\|l') = (-1)^{l+g}\sqrt{l_{\max}},\; l' = l\pm 1.\qquad(4.140)$$

When $\kappa = 2$,

$$\begin{pmatrix} l & 2 & l \\ 0 & 0 & 0 \end{pmatrix} = (-1)^g \sqrt{\frac{l(l+1)}{(2l+3)(2l+1)(2l-1)}},\qquad(4.141)$$

$$\begin{pmatrix} l & 2 & l-2 \\ 0 & 0 & 0 \end{pmatrix} = (-1)^g \sqrt{\frac{3l(l-1)}{2(2l+1)(2l-1)(2l-3)}},\qquad(4.142)$$

$$\begin{pmatrix} l & 2 & l+2 \\ 0 & 0 & 0 \end{pmatrix} = (-1)^g \sqrt{\frac{3(l+1)(l+2)}{2(2l+5)(2l+3)(2l+1)}}.\qquad(4.143)$$

Hence it is not difficult to obtain expressions for the reduced matrix elements of Y_2 and C^2. For example,

$$(l\|Y_2\|l) = -\sqrt{\frac{5}{4\pi}} \sqrt{\frac{l(l+1)(2l+1)}{(2l+3)(2l-1)}}, \qquad (4.144)$$

$$(l\|C^2\|l) = -\sqrt{\frac{l(l+1)(2l+1)}{(2l+3)(2l-1)}}. \qquad (4.145)$$

Let us calculate now the reduced matrix elements of the angular momentum. The eigenvalue of the z component of the angular momentum $J_z \equiv J_0$ is M. Thus

$$\langle JM|J_0|J'M'\rangle = M\delta_{JJ'}\delta_{MM'}, \qquad (4.146)$$

whereas (4.120) gives

$$\langle JM|J_0|J'M'\rangle = (J\|J\|J)\frac{M}{\sqrt{J(J+1)(2J+1)}}\delta_{JJ'}\delta_{MM'}, \qquad (4.147)$$

$$(J\|J\|J') = \sqrt{J(J+1)(2J+1)}\,\delta_{JJ'}. \qquad (4.148)$$

In the particular cases of the orbital angular momentum and spin of an electron, (4.148) assumes the forms

$$(l\|l\|l') = \sqrt{l(l+1)(2l+1)}\,\delta_{ll'}, \qquad (4.149)$$

$$(s\|s\|s') = \sqrt{\frac{3}{2}}\,\delta_{ss'}. \qquad (4.150)$$

4.3.4 Tensor Product of Operators

From the two irreducible tensors T^k and U^r one can construct the irreducible tensor Q^s of rank s with the components

$$Q_\sigma^s = \sum_{q,\lambda}(krq\lambda|krs\sigma)\,T_q^k U_\lambda^r, \qquad (4.151)$$

where

$$s = k+r, k+r-1, \ldots, |k-r| \qquad (4.152)$$

and $(krq\lambda|krs\sigma)$ are the Clebsch–Gordan coefficients. The tensor product of the operators T^k and U^r is defined by (4.151), and this will be denoted below as

$$Q^s = [T^k \times U^r]^s;\quad Q_\sigma^s = [T^k \times U^r]_\sigma^s. \qquad (4.153)$$

With the aid of (4.151), one can construct $(2k+1)$ of the operators $[T^k \times U^r]$

4.3 Irreducible Tensor Operators

if $k \leq r$; or $(2r + 1)$ if $k > r$. If $k = r$, then among the possible values of s is zero. Thus from two tensors of the same rank one can construct the scalar

$$[T^k \times U^k]_0^0 = \sum_q (kkq\ -q|kk\ 00)\ T_q^k U_{-q}^k = \frac{(-1)^k}{\sqrt{(2k+1)}} \sum_q (-1)^q T_q^k U_{-q}^k. \tag{4.154}$$

It is more convenient, however, to define this scalar by the relation

$$(T^k U^k) = \sum_q (-1)^q T_q^k U_{-q}^k = \sum_q (-1)^q T_{-q}^k U_q^k. \tag{4.155}$$

Expression (4.155) is called the scalar product of the tensor operators T^k and U^k.

A simple example of a scalar product of tensor operators is the theorem for addition of spherical harmonics (4.9)

$$\begin{aligned}(C^k(\theta_1, \varphi_1) \cdot C^k(\theta_2, \varphi_2)) &= \sum_q C_q^k(\theta_1, \varphi_1)^* C_q^k(\theta_2, \varphi_2) \\ &= \sum_q (-1)^q C_{-q}^k(\theta_1, \varphi_1)\ C_q^k(\theta_2, \varphi_2). \end{aligned} \tag{4.156}$$

A second example is the ordinary scalar product of the two vectors A and B written in spherical components (4.111)

$$A \cdot B = \sum_m (-1)^m A_m B_{-m}. \tag{4.157}$$

We shall also give an example of the tensor product of irreducible tensor operators. It will be shown in (6.42) that the interaction of the magnetic moment of the nucleus with the spin magnetic moment of an electron has the form

$$W = a_l[3(\mathbf{s} \cdot \mathbf{n})\mathbf{n} - \mathbf{s}] \cdot \mathbf{I} = a_l \mathbf{K} \cdot \mathbf{I}, \tag{4.158}$$

where s is the spin of the electron, I is the spin of the nucleus, and a_l is a constant. The α component of the vector $\mathbf{K}$ can be written in the following way

$$K_\alpha = \sum_\beta D_{\alpha\beta} s_\beta, \tag{4.159}$$

$$D_{\alpha\beta} = (3n_\alpha n_\beta - \delta_{\alpha\beta} n^2). \tag{4.160}$$

Since the tensor $D_{\alpha\beta}$ is symmetric and has a trace equal to zero, one can construct from the components of $D_{\alpha\beta}$ a spherical tensor of the second rank [see (4.117–119)]. The components of this tensor D_m^2 are proportional to the spherical functions $C_m^2(\theta, \varphi)$. The spherical components s_m^1 of the vector s form a tensor of the first rank S^1. According to what has been said above, the tensor product

$$[D^2 \times S^1]^1 \tag{4.161}$$

represents a tensor of the first rank and thus the q component of (4.161),

$$[D^2 \times S^1]^1_q = \sum_{mm'} (21\,mm'|21\,1q) D^2_m S^1_{m'}, \tag{4.162}$$

must coincide with the spherical component K_q of vector $\mathbf{K}$ multiplied by some constant

$$K_q = \text{const} \sum_{mm'} (21mm'|21\,1q)\, C^2_m(\theta_1, \varphi_1)\, S^1_{m'}. \tag{4.163}$$

To determine the constant in (4.163) we shall equate K_z from (4.159) with K_0 from (4.163)

$$K_z = D_{zx}s_x + D_{zy}s_y + D_{zz}s_z, \tag{4.164}$$

$$K_0 = \text{const} \sum_m (21\,m-m|2110)\, C^2_m S^1_{-m}. \tag{4.165}$$

The component $s_z = S^1_0$ enters into only the last term of (4.164); therefore,

$$D_{zz}s_z = \text{const}\,(2100|2110)\, C^2_0 S^1_0.$$

Taking into account that

$$D_{zz} = 3\cos^2\theta - 1, \quad C^2_0 = \sqrt{\frac{1}{4}}\,(3\cos^2\theta - 1),$$

$(2100|2110) = -\sqrt{2/5}$, we obtain

$$K_q = -\sqrt{10}\sum_{mm'}(21mm'|21\,1q)\, C^2_m(\theta,\varphi)\, S^1_{m'} = -\sqrt{10}\,[C^2 \times S^1]^1_q, \tag{4.166}$$

$$W = a_l \mathbf{K}\cdot\mathbf{I} = a_l \sum_q (-1)^q K_q I_{-q}$$

$$= -a_1\sqrt{10}\sum_q (-1)^q [C^2 \times S^1]^1_q I_{-q}. \tag{4.167}$$

The matrix element of the scalar product (4.155) can be calculated with the help of the general formula (4.120)

$$\langle \gamma JM|(T^k U^k)|\gamma' J'M \rangle$$
$$= \sum_{\gamma''J''M''}\sum_q (-1)^q \langle \gamma JM|T^k_q|\gamma''J''M''\rangle \langle \gamma''J''M''|U^k_{-q}|\gamma'J'M\rangle$$
$$= \sum_{\gamma''J''} (-1)^{J+J''-2M}(\gamma J\|T^k\|\gamma''J'')(\gamma''J''\|U^k\|\gamma'J')$$

$$\times \sum_{qM''} \begin{pmatrix} J & k & J'' \\ -M & q & M'' \end{pmatrix} \begin{pmatrix} J'' & k & J' \\ -M'' & -q & M' \end{pmatrix}.$$

Summing with respect to M'' and q with the aid of (4.42) and taking into account that $2J - 2M$ is even, we obtain

$$\langle \gamma JM|(T^k U^k)|\gamma' J' M'\rangle = \sum_{\gamma'' J''} (-1)^{J-J''} (\gamma J\|T^k\|\gamma'' J'')(\gamma'' J''\|U^k\|\gamma' J') \frac{\delta_{JJ'}\delta_{MM'}}{2J+1}. \quad (4.168)$$

If the operators T_q^k and U_q^k act on the coordinates of two different noninteracting systems with angular momenta $\mathbf{J}_1$ and $\mathbf{J}_2$, then T_q^k satisfies (4.108) and (4.109) with respect to the angular momenta $\mathbf{J}_1$ and $\mathbf{J} = \mathbf{J}_1 + \mathbf{J}_2$ and commutes with $\mathbf{J}_2$, but U_q^k, on the other hand, satisfies relations (4.108) and (4.109) with respect to $\mathbf{J}_2$ and $\mathbf{J}$ and commutes with $\mathbf{J}_1$. It can be shown that in this case

$$\langle \gamma J_1 J_2 JM|(T^k U^k)|\gamma' J_1' J_2' JM\rangle = (-1)^{J_1'+J_2+J} \sum_{\gamma''} (\gamma J_1\|T^k\|\gamma'' J_1')(\gamma'' J_2\|U^k\|\gamma' J_2') \begin{Bmatrix} J_1 & J_2 & J \\ J_2' & J_1' & k \end{Bmatrix}. \quad (4.169)$$

For example, for the scalar product of the operators

$$(C_1^k \cdot C_2^k) = \sum_q (-1)^q C_q^k(\theta_1 \varphi_1) C_{-q}^k(\theta_2 \varphi_2),$$

$$\langle l_1 l_2 LM_L|(C_1^k C_2^k)|l_1' l_2' LM_L\rangle = (-1)^{l_1+l_2+L} (l_1\|C^k\|l_1')(l_2\|C^k\|l_2') \begin{Bmatrix} l_1 & l_2 & L \\ l_2' & l_1' & k \end{Bmatrix}. \quad (4.170)$$

For the scalar product of the angular momenta $\mathbf{J}_1 \cdot \mathbf{J}_2$ it follows from (4.169) that

$$\langle J_1 J_2 JM|\mathbf{J}_1 \cdot \mathbf{J}_2|J_1' J_2' JM\rangle = \delta_{J_1 J_1'}\delta_{J_2 J_2'} (J_1\|J_1\|J_1)(J_2\|J_2\|J_2)(-1)^{2J_1+2J_2}$$

$$\times \begin{Bmatrix} J_1 & J_2 & J \\ J_2 & J_1 & 1 \end{Bmatrix} = \frac{1}{2}[J(J+1) - J_1(J_1+1) - J_2(J_2+1)]. \quad (4.171)$$

The matrix elements of the tensor product of the operators T^k and U^r acting on the coordinates of different systems are calculated by the general formula (4.120) in which it is necessary to substitute the following expression for the reduced matrix element

$$(\gamma J_1 J_2 J\|[T^k \times U^r]^t\|\gamma' J_1' J_2' J')$$

$$= \sum_{\gamma''} (\gamma J_1 \| T^k \| \gamma'' J_1') (\gamma'' J_2 \| U^r \| \gamma' J_2') \sqrt{(2J+1)(2J'+1)(2s+1)}$$

$$\times \begin{Bmatrix} J_1 & J_1' & k \\ J_2 & J_2' & r \\ J & J' & s \end{Bmatrix}. \quad (4.172)$$

Thus matrix elements of this type are expressed in terms of 9j symbols. In the example considered above

$$(slj\|[C^2 \times S^1]^1\|slj) = (l\|C^2\|l)(s\|s\|s)(2j+1)\sqrt{3} \begin{Bmatrix} s & s & 1 \\ l & l & 2 \\ j & j & 1 \end{Bmatrix}. \quad (4.173)$$

4.3.5 Matrix Elements with Coupled Angular Momenta

We shall now consider the matrix elements of the operator T^k, which commutes with J_2 in the representation $J_1 J_2 JM$. From the general formula (4.120) we have

$$\langle \gamma J_1 J_2 JM | T_q^k | \gamma' J_1' J_2 J' M' \rangle$$
$$= (-1)^{J-M} (\gamma J_1 J_2 J \| T^k \| \gamma' J_1' J_2 J') \times \begin{pmatrix} J & k & J' \\ -M & q & M' \end{pmatrix}. \quad (4.174)$$

The expression for the reduced matrix element of (4.174) can be obtained from (4.172) by taking $r = 0$ and $U_0^0 = 1$. In this case

$$[T^k \times U^0]_q^k = \sum_{q'} (k0q'0 | k0kq) T_{q'}^k = T_q^k,$$

$$(\gamma'' J_2 \| U^0 \| \gamma J_2') = \sqrt{2J_2 + 1}\, \delta_{J_2 J_2'} \delta_{\gamma''\gamma'},$$

$$\begin{Bmatrix} J_1 & J_1' & k \\ J_2 & J_2' & 0 \\ J & J' & k \end{Bmatrix} = \frac{\delta_{J_2 J_2'}(-1)^{J_1+J'+k+J_2}}{\sqrt{(2k+1)(2J_2+1)}} \begin{Bmatrix} J_1' & J_1 & k \\ J & J' & J_2 \end{Bmatrix}$$

[see (4.105, 135)].

For the final formula, we obtain

$$(\gamma J_1 J_2 J \| T^k \| \gamma' J_1' J_2 J')$$
$$= (-1)^{J_1+J_2+J'+k} (\gamma J_1 \| T^k \| \gamma' J_1') \sqrt{(2J+1)(2J'+1)} \begin{Bmatrix} J_1 & J & J_2 \\ J' & J_1' & k \end{Bmatrix}. \quad (4.175)$$

Similarly for the operator U^k, which commutes with J_1,

$$(\gamma J_1 J_2 J \| U^k \| \gamma' J_1 J_2' J')$$
$$= (-1)^{J_1+J_2'+k+J} (\gamma J_2 \| U^k \| \gamma' J_2') \sqrt{(2J+1)(2J'+1)} \begin{Bmatrix} J_2 & J & J_1 \\ J' & J_2' & k \end{Bmatrix}. \quad (4.176)$$

From (4.175, 176), it follows

$$(\gamma J_2 J_1 J \| T^k \| \gamma' J_2 J_1' J') = (-1)^{J_1 - J_1 + J - J'} (\gamma J_1 J_2 J \| T^k \| \gamma' J_1' J_2 J') . \quad (4.177)$$

Let us consider a number of examples. For the reduced matrix element of J_1 in the representation $J_1 J_2 JM$, we obtain from (4.148, 175)

$$(J_1 J_2 J \| J_1 \| J_1 J_2 J) = (J_1 \| J_1 \| J_1) (-1)^{J_2 + 1 - J_1 - J} (2J + 1) W(J_1 J J_1 J; J_2 1)$$
$$= \sqrt{J(J+1)(2J+1)} \frac{J(J+1) + J_1(J_1+1) - J_2(J_2+1)}{2J(J+1)}$$
$$= (J \| J \| J) \frac{J(J+1) + J_1(J_1+1) - J_2(J_2+1)}{2J(J+1)} , \quad (4.178)$$

and also

$$\langle J_1 J_2 JM | J_{1z} | J_1 J_2 JM \rangle = \frac{J(J+1) + J_1(J_1+1) - J_2(J_2+1)}{2J(J+1)} M . \quad (4.179)$$

The last expression is not difficult to obtain from the obvious quasiclassical picture, according to which the mean value of J_1 in the state $J_1 J_2 J$ is directed along J,

$$\langle J_1 \rangle = \frac{\langle J_1 \cdot J \rangle}{J(J+1)} J = \frac{J(J+1) + J_1(J_1+1) - J_2(J_2+1)}{2J(J+1)} J . \quad (4.180)$$

For the orbital angular momentum l and spin s in the representation $sljm$ we have

$$(slj \| s \| slj) = \sqrt{j(j+1)(2j+1)} \frac{j(j+1) + l(l+1) - s(s+1)}{2j(j+1)} , \quad (4.181)$$

$$(slj \| l \| slj) = \sqrt{j(j+1)(2j+1)} \frac{j(j+1) + s(s+1) - l(l+1)}{2j(j+1)} . \quad (4.182)$$

We also give the formulas for the reduced matrix elements of C^k in the representation $sljm$

$$j = l \pm \frac{1}{2}, \ j' = l' \pm \frac{1}{2} : \left(\frac{1}{2} lj \| C^k \| \frac{1}{2} l'j' \right)$$
$$= (-1)^{\frac{l'+k-l}{2}} \sqrt{\frac{(j+j'-k)!(j+k-j')!(j'+k-j)!}{(j+j'+k+1)!}}$$
$$\times \frac{(j+j'+k+j)!!}{(j+j'-k-1)!!(j+k-j')!!(j'+k-j)!!} ; \quad (4.183)$$

$$j = l \pm \frac{1}{2}, \ j' = l' \mp \frac{1}{2} : \left(\frac{1}{2}lj \|C^k\| \frac{1}{2}l'j'\right)$$

$$= (-1)^{\frac{l'+k-j-1}{2}} \sqrt{\frac{(j+j'-k)!\,(j+k-j')!\,(j'+k-j)!}{(j+j'+k+1)!}}$$

$$\times \frac{(j+j'+k)!!}{(j+j'-k)!!\,(j+k-j'-1)!!\,(j+k-j-1)!!}\,; \qquad (4.184)$$

where $k!! = 2 \times 4 \times 6 \ldots k$ if k is even, and $k!! = 1 \times 3 \times 5 \ldots k$ if k is odd. It is significant that (4.183, 184) do not contain l,l'. For $k=1$, $j=j'$ and for $k=2$, $j=j'$ from (4.183, 184) we obtain

$$\left(\frac{1}{2}lj \|C^1\| \frac{1}{2}l'j\right) = \sqrt{\frac{l_{\max}}{2j(j+1)}}, \qquad l = l' \pm 1, \qquad (4.185)$$

$$\left(\frac{1}{2}lj \|C^2\| \frac{1}{2}lj\right) = -\frac{1}{4} \sqrt{\frac{(2j-1)(2j+1)(2j+3)}{j(j+1)}}. \qquad (4.186)$$

4.3.6 Direct Product of Operators

By multiplying in all possible ways the components of the irreducible tensor operators T^k and U^r, we obtain the set $(2k+1)(2r+1)$ of operators $T^k_q U^r_\lambda$. This set is called the direct product of the operators T^k and U^r. Let the operators T^k_q obey the commutation rules (4.108) and (4.109) with the angular momentum $\boldsymbol{J}_1$ and commute with the angular momentum $\boldsymbol{J}_2$, and the operators U^r_λ, on the other hand, commute with $\boldsymbol{J}_1$ and obey (4.108) and (4.109) with respect to $\boldsymbol{J}_2$. Then the operator R^{kr} with the components $R^{kr}_{q\lambda} = T^k_q U^r_\lambda$ behaves as an irreducible tensor of order k with respect to $\boldsymbol{J}_1$ and as an irreducible tensor of order r with respect to $\boldsymbol{J}_2$.[3] We shall therefore call the operator R^{kr} an irreducible tensor operator of rank kr. The matrix elements of the components of this operator in the representation $J_1 J_2 M_1 M_2$ have the form

$$\langle J_1 J_2 M_1 M_2 | R^{kr}_{q\lambda} | J'_1 J'_2 M'_1 M'_2 \rangle = (-1)^{J_1+J_2-M_1-M_2}(J_1 J_2 \|R^{kr}\| J'_1 J'_2)$$

$$\times \begin{pmatrix} J_1 & k & J'_1 \\ -M_1 & q & M'_1 \end{pmatrix} \begin{pmatrix} J_2 & r & J'_2 \\ -M_2 & \lambda & M'_2 \end{pmatrix}, \qquad (4.187)$$

$$(J_1 J_2 \|R^{kr}\| J'_1 J'_2) = (J_1 \|T^k\| J'_1)(J_2 \|U^r\| J'_2). \qquad (4.188)$$

---

[3] Although each of the operators T^k_q and U^r_λ separately obeys the commutation rules (4.108) and (4.109) with the total angular momentum $\boldsymbol{J} = \boldsymbol{J}_1 + \boldsymbol{J}_2$, their product $T^k_q U^r_\lambda$ does not possess this property. Only definite linear combinations of these products, namely (4.151), satisfy (4.108, 109) with $\boldsymbol{J}$.

The case $J_1 = L$, $J_2 = S$ is of particular interest for various applications. Formulas (4.187) and (4.188) are direct generalizations of (4.120). All the other relations generalize in a similar way. Thus the scalar product of the operators R^{kr} and Q^{kr} is defined as

$$(R^{kr} \cdot Q^{kr}) = \sum_{q,\lambda} (-1)^{q+\lambda} R^{kr}_{q\lambda} Q^{kr}_{-q,-\lambda}. \tag{4.189}$$

If the operator R^{kr} satisfies (4.108, 109) with respect to the angular momenta L_1, S_1 and commutes with L_2, S_2, and the operator Q^{kr} commutes with L_1, S_1 and satisfies (4,108. 109) with respect to L_2, S_2, then

$$\langle \gamma L_1 S_1 L_2 S_2 LSM_L M_S | (R^{kr} \cdot Q^{kr}) | \gamma' L_1' S_1' L_2' S_2' LSM_L M_S \rangle$$
$$= (-1)^{L_1' + S_1' + L_2 + S_2 + L + S} \sum_{\gamma''} (\gamma L_1 S_1 \| R^{kr} \| \gamma'' L_1' S_1') (\gamma'' L_2 S_2 \| Q^{kr} \| \gamma' L_2' S_2')$$
$$\times \begin{Bmatrix} L_1 & L_2 & L \\ L_2' & L_1' & k \end{Bmatrix} \begin{Bmatrix} S_1 & S_2 & S \\ S_2' & S_1' & r \end{Bmatrix}. \tag{4.190}$$

An example of a scalar product of this type is the operator

$$s_1 \cdot s_2 \sum_q (-1)^q C_q^k(\theta_1 \varphi_1) C_{-q}^k(\theta_2 \varphi_2) = (s_1 \cdot s_2)(C_1^k C_2^k), \tag{4.191}$$

where s_1, s_2 are the spins of two electrons and θ_1, φ_1 and θ_2, φ_2 their angular coordinates. According to (4.157)

$$(C_1^k C_2^k)(s_1 \cdot s_2) = \sum_{q,\lambda} (-1)^{q+\lambda} V^{k1}_{q\lambda}(1) V^{k1}_{-q,-\lambda}(2) = (V_1^{k1} V_2^{k1}), \tag{4.192}$$

where

$$V^{k1}_{q\lambda}(1) = C_q^k(\theta_1, \varphi_1)(S_1)^1_\lambda; \quad V^{k1}_{q\lambda}(2) = C_q^k(\theta_2, \varphi_2)(S_2)^1_\lambda. \tag{4.193}$$

The matrix elements $V^{k1}_{q\lambda}$ in the representation $lsm\mu$ are defined by (4.187) and (4.188), which in this case take the form

$$\langle lsm\mu | V^{k1}_{q\lambda} | l'sm'\mu' \rangle$$
$$= (-1)^{l+s-m-\mu}(ls\| V^{k1} \| l's) \begin{pmatrix} l & k & l' \\ -m & q & m' \end{pmatrix} \begin{pmatrix} s & 1 & s \\ -\mu & \lambda & \mu' \end{pmatrix}, \tag{4.194}$$

$$(ls\| V^{k1} \| l's) = \sqrt{\frac{3}{2}} (l\| C^k \| l'). \tag{4.195}$$

Substituting these expressions into (4.190), we obtain

$$\langle l_1 s_1 l_2 s_2 LSM_L M_S | (C_1^k \cdot C_2^k) \, \mathbf{s}_1 \cdot \mathbf{s}_2 | l'_1 s_1 l'_2 s_2 LSM_L M_S \rangle$$
$$= (-1)^{l'_1 + l_2 + 1 + L + S} \frac{3}{2} (l_1 \| C^k \| l'_1)(l_2 \| C^k \| l'_2) \begin{Bmatrix} l_1 & l_2 & L \\ l'_2 & l'_1 & k \end{Bmatrix} \begin{Bmatrix} s_1 & s_2 & S \\ s_2 & s_1 & 1 \end{Bmatrix}. \quad (4.196)$$

We also give the formulas which are a generalization of (4.175–177):

$$(\gamma L_1 S_1 L_2 S_2 LS \| R^{kr} \| \gamma' L'_1 S'_1 L_2 S_2 L' S')$$
$$= (-1)^{L_2 + S_2 + k + r + L_1 + S_1 + L' + S'} (\gamma L_1 S_1 \| R^{kr} \| \gamma' L'_1 S'_1)$$
$$\times \sqrt{(2L+1)(2L'+1)(2S+1)(2S'+1)} \begin{Bmatrix} L_1 & L & L_2 \\ L' & L'_1 & k \end{Bmatrix} \begin{Bmatrix} S_1 & S & S_2 \\ S' & S'_1 & r \end{Bmatrix}, \quad (4.197)$$

$$(\gamma L_1 S_1 L_2 S_2 LS \| Q^{kr} \| \gamma' L_1 S_1 L'_2 S'_2 L' S')$$
$$= (-1)^{L_1 + S_1 + k + r + L'_2 + S'_2 + L + S} (\gamma L_2 S_2 \| Q^{kr} \| \gamma' L'_2 S'_2)$$
$$\times \sqrt{(2L+1)(2L'+1)(2S+1)(2S'+1)} \begin{Bmatrix} L_2 & L & L_1 \\ L' & L'_2 & k \end{Bmatrix} \begin{Bmatrix} S_2 & S & S_1 \\ S' & S'_2 & r \end{Bmatrix}, \quad (4.198)$$

$$(\gamma L_2 S_2 L_1 S_1 LS \| R^{kr} \| \gamma L_2 S_2 L'_1 S'_1 L' S')$$
$$= (-1)^{L'_1 + S'_1 - L_1 - S_1 + L + S - L' - S'} (\gamma L_1 S_1 L_2 S_2 LS \| R^{kr} \| \gamma L'_1 S'_1 L_2 S_2 L' S'). \quad (4.199)$$

The matrix elements of the operator T_q^k which commutes with $\mathbf{S}$ in the representation $LSM_L M_S$ can be obtained by taking $T_q^k = R_{q0}^{k0}$, $U_0^0 = 1$. Thus instead of (4.187, 190, 197), we obtain

$$\langle LSM_L M_S | T_q^k | L' SM'_L M_S \rangle = (-1)^{L-M} (L \| T^k \| L') \begin{pmatrix} L & k & L' \\ -M_L & q & M'_L \end{pmatrix}, \quad (4.200)$$

$$\langle \gamma L_1 S_1 L_2 S_2 LSM_L M_S | (T_1^k \cdot T_2^k) | \gamma' L'_1 S_1 L'_2 S_2 LSM_L M_S \rangle$$
$$= (-1)^{L'_1 + L_2 + L} \sum_{\gamma''} (\gamma L_1 \| T_1^k \| \gamma'' L'_1)(\gamma'' L_2 \| T_2^k \| \gamma' L'_2) \begin{Bmatrix} L_1 & L_2 & L \\ L'_2 & L'_1 & k \end{Bmatrix}, \quad (4.201)$$

$$(\gamma L_1 S_1 L_2 S_2 LS \| T^k \| \gamma' L'_1 S_1 L_2 S_2 L' S)$$
$$= (-1)^{L_2 + k + L_1 + L'} (\gamma L_1 \| T^k \| \gamma' L'_1) \sqrt{(2L+1)(2L'+1)} \begin{Bmatrix} L_1 & L & L_2 \\ L' & L'_1 & k \end{Bmatrix}. \quad (4.202)$$

Chapter 5 Systematics of the States of Multielectron Atoms

Systematics of the levels of multielectron atoms, term structure, and fine splitting are treated in detail using the Racah techniques in the theory of angular momenta and the Racah method of fractional parentage coefficients. There are numerous examples throughout the text which enable one to use these very effective methods in different problems of atomic physics.

5.1 Wave Functions

5.1.1 Central Field Approximation

The wave function Ψ of a system consisting of N noninteracting electrons can be built from the single-electron functions $\psi_a(\xi)$ where ξ is the set of the three spatial coordinates and the spin variable λ. For such a wave function, however, one cannot simply take the product

$$\Psi = \psi_{a_1}(\xi_1)\psi_{a_2}(\xi_2) \ldots \psi_{a_N}(\xi_N) \tag{5.1}$$

because the wave function of a systems of electrons must be antisymmetric with respect to exchange of electrons. The determinant

$$\Psi = \frac{1}{\sqrt{N!}} \begin{vmatrix} \psi_{a_1}(\xi_1) & \psi_{a_1}(\xi_2) & \ldots & \psi_{a_1}(\xi_N) \\ \psi_{a_2}(\xi_1) & \psi_{a_2}(\xi_2) & \ldots & \psi_{a_2}(\xi_N) \\ \cdots & \cdots & & \cdots \\ \psi_{a_N}(\xi_1) & \psi_{a_N}(\xi_2) & \ldots & \psi_{a_N}(\xi_N) \end{vmatrix}, \tag{5.2}$$

which is a linear combination of functions (5.1), satisfies this condition. Exchange of the two electrons i, k corresponds to exchange of the corresponding columns of the determinant, as a result of which the determinant changes sign. In the particular case of $N = 2$,

$$\Psi = \frac{1}{\sqrt{2}} [\psi_{a_1}(\xi_1)\psi_{a_2}(\xi_2) - \psi_{a_1}(\xi_2)\psi_{a_2}(\xi_1)]. \tag{5.3}$$

If any of the states $a_1, a_2, \ldots, a_N$ are identical, then the corresponding rows of the determinant will prove to be identical, and the determinant will vanish. Thus, the function (5.2) complies with the Pauli principle.

The state of an electron in a central field is described by the quantum numbers n, l, m, μ (m is the z component of the orbital angular momentum; μ is the z component of the spin); therefore, the wave function of a system of N electrons in a central field has the form (5.2), if it is assumed (Sect. 4.1) that

$$\psi_a(\xi) = \psi_{nlm\mu}(\xi) = \psi_{nlm}(r)\,\delta_{\mu\lambda}. \tag{5.4}$$

In the wave function (5.2) it sometimes proves to be convenient to distinguish one of the states, for example the state a_N. From the general properties of determinants it follows that

$$\Psi = \frac{1}{\sqrt{N}} \sum_i (-1)^{i-N} \psi_{a_N}(\xi_i)\, \Psi', \tag{5.5}$$

where

$$\Psi' = \frac{1}{\sqrt{(N-1)!}} \begin{vmatrix} \psi_{a_1}(\xi_1) & \cdots & \psi_{a_1}(\xi_{i-1}) & \psi_{a_1}(\xi_{i+1}) & \cdots & \psi_{a_1}(\xi_N) \\ \cdots & \cdots & \cdots & \cdots & \cdots & \cdots \\ \psi_{a_{N-1}}(\xi_1) & \cdots & \psi_{a_{N-1}}(\xi_{i-1}) & \psi_{a_{N-1}}(\xi_{i+1}) & \cdots & \psi_{a_{N-1}}(\xi_N) \end{vmatrix}. \tag{5.6}$$

5.1.2 Two-Electron Wave Functions in LSM_LM_S Representation

We shall now consider how one can construct from the functions $\psi_{nlm\mu}$ and $\psi_{n'l'm'\mu'}$ the wave function of the two-electron system $\Psi_{SLM_SM_L}$ describing a state with prescribed values of the angular momenta, L, S and their z components M_L, M_S.

Using the general rule of addition of angular momenta, namely (4.15,18), we obtain

$$\Psi_{SLM_SM_L}(l_1 l_2') = \sum C^L_{mm'} C^S_{\mu\mu'} \psi_{nlm\mu}(\xi_1)\, \psi_{n'l'm'\mu'}(\xi_2), \tag{5.7}$$

$$\Psi_{SLM_SM_L}(l_2 l_1') = \sum C^L_{mm'} C^S_{\mu\mu'} \psi_{nlm\mu}(\xi_2)\, \psi_{n'l'm'\mu'}(\xi_1). \tag{5.8}$$

The wave functions (5.7, 8) differ in that in the first case the first electron is in the state with angular momentum l, and in the second case, the second electron. This is indicated by the indices 1 and 2 on the angular momenta l and l'. This notation will also be used below. The function $\Psi_{SLM_SM_L}$ can be obtained by constructing an antisymmetric combination of the functions (5.7, 8)

$$\Psi_{SLM_SM_L} = \frac{1}{\sqrt{2}} [\Psi_{SLM_SM_L}(l_1 l_2') - \Psi_{SLM_SM_L}(l_2 l_1')]. \tag{5.9}$$

The factor $1/\sqrt{2}$ is introduced for normalization. Substituting (5.7) and (5.8) into (5.9), one can easily verify that (5.9) is expressed in terms of antisymmetric combinations of the products of single-electron wave functions of the type (5.3).

Thus a two-electron function which is an eigenfunction of the operators L^2, S^2, L_z, S_z ($L = l + l'$; $S = s + s'$) can be constructed by the general rule of addition of angular momenta, with the condition of subsequent antisymmetrization. From the symmetry properties of Clebsch–Gordan coefficients it follows that

$$(ll'mm' | ll'LM_L) = (-1)^{l+l'-L} (l'lm'm | l'lLM_L). \tag{5.10}$$

$$\left(\tfrac{1}{2}\tfrac{1}{2}\mu\mu' \Big| \tfrac{1}{2}\tfrac{1}{2} SM_S\right) = (-1)^{1-S} \left(\tfrac{1}{2}\tfrac{1}{2}\mu'\mu \Big| \tfrac{1}{2}\tfrac{1}{2} SM_S\right). \tag{5.11}$$

Therefore,

$$\Psi_{SLM_SM_L}(l_2 l_1') = (-1)^{l+l'+1-L-S} \Psi_{SLM_SM_L}(l_1' l_2)$$

and (5.9) can be rewritten in the following form

$$\Psi_{SLM_SM_L} = \frac{1}{\sqrt{2}} [\Psi_{SLM_SM_L}(l_1 l_2') + (-1)^{l+l'-L-S} \Psi_{SLM_SM_L}(l_1' l_2)], \tag{5.12}$$

where

$$\Psi_{SLM_SM_L}(l_1' l_2) = \sum C^L_{m'm} C^S_{\mu'\mu} \psi_{n'l'm'\mu'}(\xi_1) \psi_{nlm\mu}(\xi_2). \tag{5.13}$$

The function (5.13) differs from (5.8) by the exchange of states.

We shall now consider the case of equivalent electrons: $n = n'$, $l = l'$. In this case, as is easy to verify, the normalization factor equals $1/2$ and not $1/\sqrt{2}$. Taking this into consideration, and also using the obvious relation

$$\Psi_{SLM_SM_L}(l_1 l_2') = \Psi_{SLM_SM_L}(l_1' l_2) = \Psi_{SLM_SM_L}(l_1 l_2), \tag{5.14}$$

we obtain

$$\begin{aligned}\Psi_{SLM_SM_L} &= \Psi_{SLM_SM_L}(l_1 l_2), \quad L + S \text{ even}; \\ \Psi_{SLM_SM_L} &= 0 \qquad\qquad\qquad, \quad L + S \text{ odd}.\end{aligned} \tag{5.15}$$

Thus the wave function describing the state SLM_SM_L of two equivalent electrons for even values of $L + S$ equals simply the function $\psi_{SLM_LM_S}(l_1 l_2)$ obtained by the general rule of addition of angular momenta, and for odd values of $L + S$ it vanishes. Thus only terms with even values of $L + S$ are allowed for two equivalent electrons. ^1S, ^3P, and ^1D will be such terms for the configuration p^2, and $^1S\,^3P\,^1D\,^3F\,^1G$ for the configuration d^2. In the general case of the configuration l^2, the terms $^1S\,^3P\,^1D\,^3F\ldots\,^1L = 2l$ are allowed.

It is convenient in a number of cases to represent the wave function $\Psi_{SLM_SM_L}$ in the form of the product of the independent coordinate and spin functions

$$\Psi_{SLM_SM_L} = \Phi_{LM_L}Q_{SM_S}. \tag{5.16}$$

Each of the functions Φ_{LM_L} and Q_{SM_S} separately does not have to be antisymmetric. It is sufficient that the total function $\Psi_{SLM_SM_L}$ be antisymmetric. Therefore, the two cases

$$\Psi_{SLM_SM_L} = \Phi^+_{LM_L}Q^-_{SM_S}, \tag{5.17}$$

$$\Psi_{SLM_SM_L} = \Phi^-_{LM_L}Q^+_{SM_S} \tag{5.18}$$

are possible. The symmetric and antisymmetric functions are denoted by the indices $+$ and $-$, respectively in (5.17,18). Using again the general rule of addition of angular momenta and taking (5.10,11) into account, we obtain

$$\Phi_{LM_L}(l_1 l'_2) = \sum C^L_{mm'}\varphi_{lm}(r_1)\,\varphi_{l'm'}(r_2), \tag{5.19}$$

$$\Phi_{LM_L}(l_2 l'_1) = \sum C^L_{mm'}\varphi_{lm}(r_2)\,\varphi_{l'm'}(r_1), \tag{5.20}$$

$$\Phi^+_{LM_L} = \frac{1}{\sqrt{2}}[\Phi_{LM_L}(l_1 l'_2) + \Phi_{LM_L}(l_2 l'_1)]$$

$$= \frac{1}{\sqrt{2}}[\Phi_{LM_L}(l_1 l'_2) + (-1)^{l+l'-L}\Phi_{LM_L}(l'_1 l_2)], \tag{5.21}$$

$$\Phi^-_{LM_L} = \frac{1}{\sqrt{2}}[\Phi_{LM_L}(l_1 l'_2) - \Phi_{LM_L}(l_2 l'_1)]$$

$$= \frac{1}{\sqrt{2}}[\Phi_{LM_L}(l_1 l'_2) - (-1)^{l+l'-L}\Phi_{LM_L}(l'_1 l_2)]. \tag{5.22}$$

Functions $Q^+_{SM_S}$ and $Q^-_{SM_S}$ can be constructed in a similar way. In this case it is necessary to take into account the fact that the spins of the electrons cannot be different

$$Q^+_{SM_S} = \frac{1}{\sqrt{2}}[Q_{SM_S}(s_1 s_2) + (-1)^{1-s}Q_{SM_S}(s_1 s_2)], \tag{5.23}$$

$$Q^-_{SM_S} = \frac{1}{\sqrt{2}}[Q_{SM_S}(s_1 s_2) - (-1)^{1-s}Q_{SM_S}(s_1 s_2)]. \tag{5.24}$$

From (5.23, 24) it follows that when $S = 0$, $Q^+_{SM_S} = 0$ and $Q^-_{SM_S} \neq 0$, and when $S = 1$, $Q^+_{SM_S} \neq 0$ and $Q^-_{SM_S} = 0$.

Thus an antisymmetric spin function corresponds to singlet states ($S = 0$) and a symmetric one to triplet states ($S = 1$). Collecting all these formulas together, we obtain

$S = 0$, $\Psi_{SLM_S M_L}$
$$= \frac{1}{\sqrt{2}} [\Phi_{LM_L}(l_1 l_2') + (-1)^{l+l''-L} \Phi_{LM_L}(l_1' l_2)] Q^-_{SM_S} ; \qquad (5.25)$$

$S = 1$, $\Psi_{SLM_S M_L}$
$$= \frac{1}{\sqrt{2}} [\Phi_{LM_L}(l_1 l_2') - (-1)^{l+l''-L} \Phi_{LM_L}(l_1' l_2)] Q^+_{SM_S} . \qquad (5.26)$$

In the case of equivalent electrons $l = l'$, these expressions take the form

$$S = 0, \quad \Psi_{SLM_S M_L} = \Phi_{LM_L}(l_1 l_2) Q^-_{SM_S}, \quad L \text{ even} ; \qquad (5.27)$$

$$S = 1, \quad \Psi_{SLM_S M_L} = \Phi_{LM_L}(l_1 l_2) Q^+_{SM_S}, \quad L \text{ odd} . \qquad (5.28)$$

In accordance with (5.15), $L + S$ is even in both cases.

5.1.3 Two-Electron Wave Functions in $mm'SM_S$ Representation

In some applications it is convenient to use the functions $\Psi_{mm'SM_S}$. These functions are eigenfunctions of the operators $l^2, l_z, l'^2; l_z'$ and S^2, S_z. In constructing these functions it is sufficient to sum up only the spin angular momenta of the electrons. It is not necessary to sum up the orbital angular momenta. The coordinate functions $\Psi_{mm'}$ can be constructed directly from the functions $\psi_{nlm}(\mathbf{r})$ and $\psi_{n'l'm'}(\mathbf{r})$. By summing the spins of the electrons, we obtain the symmetric and antisymmetric spin functions $Q^+_{SM_S}$ and $Q^-_{SM_S}$. Taking, therefore, the requirement of antisymmetry of the total wave function into account, we obtain

$S = 0$,
$$\psi_{mm'SM_S} = \frac{1}{\sqrt{2}} [\psi_{nlm}(\mathbf{r}_1) \psi_{n'l'm'}(\mathbf{r}_2) + \psi_{nlm}(\mathbf{r}_2) \psi_{n'l'm'}(\mathbf{r}_1)] Q^-_{SM_S} ; \qquad (5.29)$$

$S = 1$,
$$\psi_{mm'SM_S} = \frac{1}{\sqrt{2}} [\psi_{nlm}(\mathbf{r}_1) \psi_{n'l'm'}(\mathbf{r}_2) - \psi_{nlm}(\mathbf{r}_2) \psi_{n'l'm'}(\mathbf{r}_1)] Q^+_{SM_S} . \qquad (5.30)$$

5.1.4 Multielectron Wave Functions in a Parentage Scheme Approximation

As a rule, several identical terms arise in the case of multielectron configurations. For example, for the configuration $np\,n'p\,n''p$ we have the following terms

$np\,n'p\,[^1S]\,n''p\,^2P$,
$np\,n'p\,[^3S]\,n''p\,^2P\,^4P$,

$np\ n'p\ [^1P]\ n''p\ ^2SPD$,

$np\ n'p\ [^3P]\ n''p\ ^2SPD\ ^4SPD$,

$np\ n'p\ [^1D]\ n''p\ ^2PDF$,

$np\ n'p\ [^3D]\ n''p\ ^2PDF\ ^4PDF$,

among which there are six ^2P terms, four ^2D terms, two ^2F terms, and so on. We shall characterize each of these terms by the assignment of the initial term, i.e., the term of configuration $np\ n'p$. In the general case, by the initial term of an atom we mean that term of the ion which gives, upon adding an electron, a particular term of the atom. The assignment of an initial term is usually spoken of as the assignment of the origin or parentage of the term.

The concept of a parentage of a term has meaning only if the interaction between the added electron and the electrons of the initial ion is considerably less than the interaction of the latter electrons with each other. In this case the energy of the atom is formed from the energy of the unperturbed ion and the energy of the valence electron moving in the field of the ion. In exactly the same way, the orbital and spin angular momenta of the atom L and S are formed from the angular momenta L_1, S_1 of the initial ion and the angular momenta l, s of the valence electron, conservation of the absolute magnitudes of L_1 and S_1 occurring together with conservation of L and S. Precisely this circumstance enables one to associate the term of the atom with a definite initial term of the ion. In the general case, terms observed in reality may not have definite initial terms.

We shall denote the wave functions of states relating to the term LS obtained by adding an electron with angular momentum l to the initial term L_1S_1 by means of $\Psi_{SLM_SM_L}(S_1L_1,l)$. The wave functions $\Psi_I = \Psi_{SLM_SM_L}(S_1L_1 l)$ and Ψ_{II} $\Psi_{SLM_SM_L}(S_2L_2,l)$ obviously correspond to essentially different states. In the case when the energy of interaction of the added electron with the electrons of the initial ion is of the same order of magnitude as the interaction of the latter electrons with each other, the off-diagonal matrix elements $U_{I\ II}$ of the interaction are not small in comparison with $U_{I\ I}$ and $U_{II\ II}$. This means that in this case only the total angular momenta S and L are conserved, and conservation of S_1, L_1 does not occur. To determine the energy of electrostatic splitting of two identical terms, it is necessary to find the roots of the secular equation

$$\begin{vmatrix} U_{I\ I} - \varepsilon & U_{I\ II} \\ U_{II\ I} & U_{II\ II} - \varepsilon \end{vmatrix} = 0. \tag{5.31}$$

To these roots ε_1 and ε_2, defining the energies of the terms, correspond the wave functions Ψ_1 and Ψ_2 which are linear combinations of the functions Ψ_I and Ψ_{II}. Thus it is necessary to attribute to the terms observed in reality not the states S_1L_1,lSL or S_2L_2,lSL but a mixture of these states. The true terms do not have a definite initial term in the general case.

The question of the applicability of the parentage of the terms can easily be solved in each actual case if the relative arrangement of the terms is known. Systems of terms corresponding to different initial terms are similar, and are displaced relative to each other by approximately the energy difference between the initial terms. We have already encountered such a situation in analyzing the terms of atoms with p and d optical electrons. The atom of oxygen is a typical example. Among the terms of this atom there are three systems of terms converging to different ionization limits corresponding to the three ground state terms of the oxygen ion, ^2P, ^2D, and ^4S. Identical terms of each of these systems are displaced relative to each other by approximately the same amount as the corresponding initial terms of the oxygen ion. For example, the difference between the terms $2s^2 2p^3 [^2D] np\,^1P$ and $2s^2 2p^3 [^2P] np\,^1P$ of the oxygen atom approximately coincides with the difference between the initial terms $2s^2 2p^3 [^2D]$, $2s^2 2p^3 [^2P]$ of the oxygen ion.

It is sometimes convenient to attribute the term of an atom to a definite initial term also in the case when the interaction of the valence electron with the electrons of the initial ion is comparable with but nevertheless smaller than the interaction of the latter between themselves. In this case there is no strict similarity of terms of different parentage. Violation of this similarity is usually spoken of as the interaction of terms. Essentially this means that it is impossible to neglect off-diagonal matrix elements in the secular equation (5.31).

Let us pass to the construction of wave functions in the parentage approximation. We shall denote by means of $\Psi_{SLM_S M_L}(S'L', l_i)$ the wave function of the state $[S'L']l_i SLM_S M_L$ in which the electrons $1, 2, \ldots, i-1, i+1, \ldots, N$ belong to the initial ion, and electron i is in the state with the angular momentum l. The function $\Psi_{SLM_S M_L}(S'L', l_i)$ can be constructed by the general rule of addition of angular momenta

$$\Psi_{SLM_S M_L}(S'L', l_i) = \sum C^{L}_{M'_L m} C^{S}_{M'_S \mu} \Psi_{S'L'M'_S M'_L} \psi_{nlm\mu}(\xi_i) \,. \tag{5.32}$$

The wave function of the initial ion $\Psi_{S'L'M'_S M'_L}$ is antisymmetric with respect to exchange of electrons $1, 2, \ldots, i-1, i+1, \ldots, N$. Therefore, the wave function (5.32) is also antisymmetric with respect to the electrons $1, 2, \ldots, i-1, i+1, \ldots, N$ but not antisymmetric with respect to all N electrons.

The wave function $\Psi_{SLM_S M_L}(S'L', l)$, antisymmetric with respect to all the electrons of an atom, can be represented in the form of a linear combination of functions (5.32)

$$\Psi_{SLM_S M_L}(S'L', l) = \frac{1}{\sqrt{N}} \sum_i^N (-1)^{N-i} \Psi_{SLM_S M_L}(S'L', l_i). \tag{5.33}$$

Function (5.33) has the same structure as function (5.9) and is the natural generalization of (5.9) for the case of a large number of electrons. When $N = 2$, (5.33) coincides with (5.9).

5.1.5 Fractional Parentage Coefficients

In the case of equivalent electrons, the parentage scheme does not make sense even as a first approximation, since the interaction with the remaining electrons is not small for one of the equivalent electrons. The wave function $\Psi_{SLM_SM_L}(l^n)$ describing the state SLM_SM_L of a group l^n of equivalent electrons is the linear combination of the functions $\Psi_{SLM_SM_L}(l^{n-1}[S'L']l)$ corresponding to the different initial terms $S'L'$ of configuration l^{n-1}. Here, however, it is necessary to take into account the fact that among the states $l^{n-1}[S'L']lSLM_SM_L$ obtained by the general rule of addition of angular momenta there will also be some which are forbidden by the Pauli principle. Only well-defined linear combinations of the functions $\Psi_{SLM_SM_L}(l^{n-1}[S'L']l)$,

$$\Psi_{SLM_SM_L}(l^n) = \sum_{S'L'} G^{SL}_{S'L'} \Psi_{SLM_SM_L}(l^{n-1}[S'L']l) , \tag{5.34}$$

will comply with the Pauli principle. The coefficients $G^{SL}_{S'L'}$ are called fractional parentage coefficients. In the following, following Racah, we shall also denote these coefficients by means of $(l^{n-1}[S'L']lSL\} l^n SL)$.[1] The general method of calculation of fractional parentage coefficients has been developed by *Racah* [7, 12]. The idea of the method is as follows. It has been shown above that in the case of two equivalent electrons the wave functions $\Psi_{S'L'M_S'M_L'}(l_1 l_2)$, constructed by the general rule of addition of angular momenta, are normalized and antisymmetric functions of the configuration l^2 for even values of $S' + L'$. Let us add to the configuration l^2 a third l electron and construct the function

$$\Psi_{SLM_SM_L}(l_1 l_2 [S'L'] l_3), \quad S' + L' \text{ is even},$$

again using the general rule of addition of angular momenta. This function is obviously antisymmetric with respect to exchange of electrons 1 and 2, but it is not antisymmetric with respect to exchange of these electrons and electron 3. Altering the scheme of addition of angular momenta, we obtain

$$\Psi_{SLM_SM_L}(l_1 l_2 [S'L'] l_3) = \sum_{S''L''} (ll[S'L'] lSL | l, ll[S''L''] SL)\, \Psi_{SLM_SM_L}(l_1; l_2 l_3 [S''L'']) .$$

The functions $\Psi_{SLM_SM_L}(l_1; l_2 l_3 [S''L''])$ are also constructed by the general rule of addition of angular momenta from the functions $\Psi_{l_1 m_1 s_1 \mu_1}$ and $\Psi_{S''L''M_{S''}M_{L''}}(l_2 l_3)$. Among these functions there are those for which $S'' + L''$ is an even number and those for which $S'' + L''$ is odd. Only the former correspond to states

---

[1] As there can be several terms with identical values of SL among the terms of the configuration l^n, it is necessary to introduce additional quantum numbers. In the general case, fractional parentage coefficients have to be written in the form $G^{\gamma SL}_{\gamma' S'L'} = (l^{n-1}[\gamma' S'L']lSL\} l^n \gamma SL)$. However, below, when this will not lead to misunderstandings, the additional quantum numbers γ, γ' will be omitted.

5.1 Wave Functions

which are antisymmetric with respect to exchange of electrons 2 and 3. We shall therefore construct linear combinations

$$\sum_{S'L'} (l^2 [S'L'] lSL\} \, l^3\gamma SL) \, \Psi_{SLM_SM_L} (l_1l_2 [S'L'] l_3) \, ,$$

which do not contain functions $\Psi_{SLM_SM_L} (l_1, l_2l_3 [S''L''])$ with an odd value of $S'' + L''$. This is fulfilled provided

$$\sum_{S'L'} (ll[S'L']lSL \mid l, ll[S''L'']SL) \cdot (l^2[S'L']lSL)\}l^3 \gamma SL) = 0 \, .$$

The system of equations we have obtained enables us to find the unknown coefficients $(l^2 [S'L'] lSL\} \, l^3\gamma SL)$.

Since a function which is antisymmetric with respect to exchange of electrons 1,2 and 2,3 is antisymmetric with respect to all three electrons, we finally obtain

$$\Psi_{\gamma SLM_SM_L} (l^3) = \sum_{S'L'} (l^2[S'L'] lSL\} \, l^3\gamma SL) \, \Psi_{\gamma SLM_SM_L} (l^2 [S'L'] l) \, .$$

In adding a fourth electron to the configuration l^3, all the reasoning can be repeated in a similar way and a system of equations can be obtained for determining the fractional parentage coefficients $(l^3[\gamma'S'L'] lSL\} \, l^4\gamma SL)$ and so on.

This method enables us to calculate relatively simply the fractional parentage coefficients for the simplest configurations l^n, notably for p^n and d^n. Considerably more general group theoretical methods of calculating these coefficients are discussed in [12, 13].

Fractional parentage coefficients for the configurations p^n and d^n and also for terms of maximum multiplicity of configurations f^n ($n \leqslant 7$) are given at the end of this section in Tables 5.1–16. All these coefficients are real.[2]

The following relation occurs between the coefficients $G^{SL}_{S'L'}$ for the configurations l^{n+1} and l^{4l+2-n}:

$$(-1)^{-S-L} \sqrt{(N - n)(2S + 1)(2L + 1)} \, G^{SL}_{S'L'} (l^{N-n})$$
$$= (-1)^{S'+L'-l-1/2} \sqrt{(n + 1)(2S' + 1)(2L' + 1)} \, G^{S'L'}_{SL} (l^{n+1}) \, . \qquad (5.35)$$

Here

$$N = 2(2l + 1) \, .$$

Thus, it is sufficient to calculate coefficients $G^{SL}_{S'L'}$ for configurations l^n with $n \leqslant 2l + 1$, i.e., for shells less than half-filled. In the following we shall need, in addition, the following property of the coefficients $G^{SL}_{S'L'}$:

---

[2] Tables 5.1–7 are taken from [7], Tables 5.8–11 from [14], Tables 5.12–16 from [15].

$$(l, l^{n-1}[S'L']SL\}l^n SL) = (-1)^{L+S+L'+S'-l-1/2}(l^{n-1}[S'L']lSL\}l^n SL). \quad (5.36)$$

In the case $n = 2$, (5.34) transforms into (5.15) if it is assumed that $(llSL\}l^2 SL) = 1$ for even $L+S$ and zero for odd $L+S$. In exactly the same way $(l^{4l+1}[1/2\ l] l00\}l^{4l+2}\ 00) = 1$. The wave functions $\Psi_{SLM_S M_L}(l^{n-1}[S'L']l)$ in the right-hand part of (5.34) are eigenfunctions of the operators $L'^2, S'^2, l^2, L^2, S^2, L_z, S_z$ and are constructed by the general rule of addition of angular momenta without taking into account equivalence of electrons. For applications, it is necessary to be able to separate one of the electrons. This is achieved by the following formula

$$\Psi_{SLM_S M_L}(l^n) = \sum_{S'L'} G^{SL}_{S'L'} (-1)^{n-i}\ \Psi_{SLM_S M_L}(l^{n-1}[S'L']\ l_i), \quad (5.37)$$

where $i = 1, 2, \ldots, n$. This follows directly from the definition of the G coefficients and from the method of calculation given above. We also give the generalization of (5.37) for the case of two groups of equivalent electrons

$$\Psi_{SLM_S M_L}(l^n S_1 L_1, l'^p S_2 L_2)$$
$$= \sqrt{\frac{n}{n+p}}(-1)^{n-i}\sum_{S_1'L_1'} G^{S_1 L_1}_{S_1'L_1'}\ \Psi_{SLM_S M_L}(l^{n-1}[S_1'L_1']\ l_i S_1 L_1, l'^p [S_2 L_2])$$
$$+ \sqrt{\frac{p}{n+p}}(-1)^{n+p-i}\sum_{S_2'L_2'} G^{S_2 L_2}_{S_2'L_2'}\ \Psi_{SLM_S M_L}(l^n S_1 L_1, l'^{p-1}[S_2'L_2']\ l_i'S_2 L_2).$$

(5.38)

Generalization for several groups of equivalent electrons is conducted in a similar way.

5.1.6 Classification of Identical Terms of l^n Configuration According to Seniority (Seniority Number)

Identical terms (Table 2.1) usually appear among terms of the configuration l^n when $n > 2$. Additional quantum numbers are necessary, therefore, for full description of the states $SLM_S M_L$ of a system. In this case the angular momenta $S'L'$ of the initial ion cannot be these additional quantum numbers because it is impossible to attribute terms of the configuration l^n to definite terms of the configuration l^{n-1}. It proves to be possible, however, to classify the terms S, L of the configuration l^n through their relationship to terms of the same type (i.e., with the same values of S, L) in the configuration l^{n-2}. This classification was proposed by *Racah*. We shall briefly enumerate below these principal results which are most important for the systematics of spectra [7, 13]. According to Racah, all identical terms S, L of the configuration l^n divide into two classes. The states $SLM_S M_L$, belonging to terms of the first class, can be obtained from states of the same type in the configuration l^{n-2} by the addition of two l electrons forming the closed pair $l^2; L = 0, S = 0$. Terms of the second class cannot be obtained thusly from definite SL terms of the configuration l^{n-2}, and in this sense they appear for the first time in the given configuration. Some of the SL terms of the configuration l^{n-2} can be obtained in turn from definite terms of the same type in the configuration l^{n-4} by the addition of the closed pair l^2 and so on.

Continuing this reasoning, we arrive at the configuration l^v in which the term SL appears for the first time since it cannot be obtained from any definite term of the configuration l^{v-2} by the addition of the pair l^2 [00]. The assignment of the number v uniquely determines the whole chain of terms generated by the term SL of the configuration l^v. It thus becomes possible to classify the terms of the configuration l^n by ascribing to them different values of the number v, which indicates in what configuration a given term appears for the first time. In accordance with the above, $(n-v)/2$ closed pairs l^2 [00] correspond to the states vSL of configuration l^n.

If the function $\Psi_{vSLM_SM_L}(l^n)$ with $v \neq n$ is represented in the form of an expansion in terms of wave functions $\Psi_{vSM_SM_L}(l^{n-2}[v_1S_1L_1], l^2[S_2L_2])$, then of all the possible functions $\Psi_{vSLM_SLM_L}(l^{n-2}[v_1SL]l^2[00])$ there enters into this expansion only one which corresponds to the value $v_1 = v$. The term vSL of configuration l^n with $v \neq n$ is generated by the term vSL of configuration l^{n-2} precisely in this sense.

Racah proposed the name "seniority number" for the number v. In accordance with this terminology, the numbers v classify terms according to their seniority. The value v is shown before and below the value of the term $^{2S+1}_vL$.

Let us consider, as an example, the d^n configurations. When $n = 1$, only one term, ^2D, is possible. It is necessary to assign the value $v = 1$ to this term. Thus we obtain the term ^2_1D. A chain of terms in the configurations d^3, d^5 is generated by this term (it is sufficient to consider l^n configurations with $n \leq 2l+1$). When $n = 2$, the terms $^1S\,^1D\,^1G\,^3P\,^3F$ appear.

The term ^1S can be obtained by the addition of the pair l^2 [00] to the configuration l^0. The value $v = 0$ is therefore assigned to the term ^1S. The other terms appear for the first time in the configuration d^2, and the value $v = 2$ therefore has to be assigned to them; we obtain the terms $^1_2D\,^1_2G\,^3_2P\,^3_2F$. When $n = 3$, two ^2D terms are possible. One of these terms is the term ^2_1D as it is generated by the term ^2_1D of configuration d^1. The second term ^2D appears for the first time and thus corresponds to the value $v = 3$. This term is denoted ^2_3D.[3] The other terms of the configuration d^3 also appear for the first time; therefore $v = 3$ for them also. The terms of the configurations d^4 and d^5 can be classified in a similar way. In accordance with this classification the notation

$$(l^{n-1}[v'S'L']\,lSL\}\,l^nvSL)$$

is adopted in Tables 5.5 – 11 for the fractional parentage coefficients G. The set

---

[3] The terms ^2_1D and ^2_3D correspond to the terms ^2_aD and ^2_bD in old notations.

of three numbers vSL uniquely defines a term of the configuration d^n. In the case of the configuration f^n, the situation is more complicated because several terms can arise which correspond to one and the same set of numbers vSL. For the separation of these terms it is necessary to introduce additional quantum numbers. A detailed investigation of this question is contained in [12], see also [11].

Matrix elements of the symmetric single-electron operators T^{rk}, of rank r with respect to spin S and of rank k with respect to orbital angular momentum L, will be met later in various applications. For reduced matrix elements of T^{rk} diagonal in v, we have the relations

$k+r$ is odd

$$(l^nvSL\|T^{rk}\|l^nvS'L') = (l^{n-2}vSL\|T^{rk}\|l^{n-2}vS'L') = \ldots$$
$$\ldots = (l^vSL\|T^{rk}\|l^vS'L'); \qquad (5.39)$$

$k + r$ is even

$$(l^nvSL\|T^{rk}\| l^nvS'L') = \frac{2l + 1 - n}{2l + 1 - v}(l^vSL\|T^{rk}\|l^vS'L'). \qquad (5.40)$$

In addition, for odd values of $r + k$ the matrix of T^{rk} is diagonal with respect to v.

Table 5.1

$(p^2SL\{p[^2P]\,pSL)$

p^2	p
	^2P
^1S	1
^3P	1
^1D	1

Table 5.2

$(p^3SL\{p^2[S'L']\,pSL)$

p^3	p^2		
	^1S	^3P	^1D
^4S	0	1	0
^2P	$\frac{\sqrt{2}}{3}$	$-\frac{1}{\sqrt{2}}$	$-\sqrt{\frac{5}{18}}$
^2D	0	$\frac{1}{\sqrt{2}}$	$-\frac{1}{\sqrt{2}}$

Table 5.3

$(p^4SL\{p^3[S'L']\,pSL)$

p^4	p^3		
	^4S	^2P	^2D
^1S	0	1	0
^3P	$-\frac{1}{\sqrt{3}}$	$-\frac{1}{2}$	$\sqrt{\frac{5}{12}}$
^1D	0	$-\frac{1}{2}$	$\sqrt{\frac{3}{4}}$

Table 5.4

$(p^5\,^2P\{p^4[S'L']\,p^2P)$

p^5	p^4		
	^1S	^3P	^1D
^2P	$\frac{1}{\sqrt{15}}$	$\sqrt{\frac{3}{5}}$	$\sqrt{\frac{1}{3}}$

Table 5.5

d^3	N^a	$(d^3vSL \{|d^2 [u'S'L']dSL)$				
		d^2				
		^1_0S	^3_2P	^1_2D	^3_2F	^1_2G
^2_3P	$30^{-1/2}$	0	$7^{1/2}$	$15^{1/2}$	$-8^{1/2}$	0
^4_3P	$15^{-1/2}$	0	$-8^{1/2}$	0	$-7^{1/2}$	0
^2_1D	$60^{-1/2}$	4	-3	$-5^{1/2}$	$-21^{1/2}$	-3
^2_3D	$140^{-1/2}$	0	-7	$45^{1/2}$	$21^{1/2}$	-5
^2_3F	$70^{-1/2}$	0	$28^{1/2}$	$-10^{1/2}$	$7^{1/2}$	-5
^4_3F	$5^{-1/2}$	0	-1	0	2	0
^2_3G	$42^{-1/2}$	0	0	$-10^{1/2}$	$21^{1/2}$	$11^{1/2}$
^2_3H	$2^{-1/2}$	0	0	0	-1	1

[a] Here and below, N is the normalization factor. The numbers given in the table must be multiplied by N.

Table 5.6

d^4	N	$(d^4vSL \{|d^3 [v'S'L'] dSL)$							
		d^3							
		^2_3P	^4_3P	^2_1D	^2_3D	^2_3F	^4_3F	^2_3G	^2_3H
^1_0S	1	0	0	1	0	0	0	0	0
^1_4S	1	0	0	0	0	1	0	0	0
^3_2P	$360^{-1/2}$	$-14^{1/2}$	-8	$135^{1/2}$	$-35^{1/2}$	$-56^{1/2}$	$-56^{1/2}$	0	0
^3_4P	$90^{-1/2}$	5	$-14^{1/2}$	0	$10^{1/2}$	-5	4	0	0
^1_2D	$280^{-1/2}$	$-42^{1/2}$	0	$105^{1/2}$	$45^{1/2}$	$28^{1/2}$	0	$-60^{1/2}$	0
^1_4D	$140^{-1/2}$	42	0	0	$20^{1/2}$	$63^{1/2}$	0	$15^{1/2}$	0
^3_2D	$210^{-1/2}$	$-14^{1/2}$	7	0	$60^{1/2}$	$-21^{1/2}$	$-21^{1/2}$	$45^{1/2}$	0
^5_2D	$10^{-1/2}$	0	$3^{1/2}$	0	0	0	7	0	0
^1_4F	$560^{-1/2}$	$120^{1/2}$	0	0	$200^{1/2}$	$-105^{1/2}$	0	$-3^{1/2}$	$-132^{1/2}$
^3_2F	$840^{-1/2}$	4	$-56^{1/2}$	$315^{1/2}$	$15^{1/2}$	$-14^{1/2}$	$224^{1/2}$	$90^{1/2}$	$110^{1/2}$
^3_4F	$1680^{-1/2}$	$-200^{1/2}$	$-448^{1/2}$	0	$120^{1/2}$	$-175^{1/2}$	$-112^{1/2}$	$-405^{1/2}$	$220^{1/2}$
^1_2G	$504^{-1/2}$	0	0	$189^{1/2}$	-5	$70^{1/2}$	0	$66^{1/2}$	$-154^{1/2}$
^1_4G	$1008^{-1/2}$	0	0	0	$88^{1/2}$	$385^{1/2}$	0	$-507^{1/2}$	$-28^{1/2}$
^3_2G	$1680^{-1/2}$	0	0	0	$200^{1/2}$	$315^{1/2}$	$-560^{1/2}$	$297^{1/2}$	$308^{1/2}$
^3_2H	$60^{-1/2}$	0	0	0	0	$5^{1/2}$	$20^{1/2}$	-3	$26^{1/2}$
^1_4I	$10^{-1/2}$	0	0	0	0	0	0	$3^{1/2}$	$7^{1/2}$

Table 5.7

$$(d^5 vSL\{|d^4 [v'S'L']dSL)$$

d^5	N	^1_0S	^1_4S	^3_2P	^3_4P	^1_2D	^1_4D	^3_2D	^5D	^1_4F	^3_2F	^3_4F	^1_2G	^1_4G	^3_4G	^3_4H	^1_4I
^2S	$5^{1/2}$	0	0	0	0	0	$-2^{1/2}$	$3^{1/2}$	0	0	0	0	0	0	0	0	0
^6S	1	0	0	0	0	0	0	0	1	0	0	0	0	0	0	0	0
^2P	$150^{-1/2}$	0	0	$14^{1/2}$	5	$30^{1/2}$	$15^{1/2}$	$10^{1/2}$	0	$-15^{1/2}$	-4	-5	0	0	0	0	0
^4P	$300^{-1/2}$	0	0	-8	$14^{1/2}$	0	0	$35^{1/2}$	$-75^{1/2}$	0	$-56^{1/2}$	$56^{1/2}$	0	0	0	0	0
^2_1D	$50^{-1/2}$	$6^{1/2}$	0	-3	0	$-5^{1/2}$	0	0	0	0	$-21^{1/2}$	0	-3	0	0	0	0
^2_2D	$350^{-1/2}$	0	$-14^{1/2}$	-7	$-14^{1/2}$	$45^{1/2}$	$-10^{1/2}$	$60^{1/2}$	0	$35^{1/2}$	$21^{1/2}$	$-21^{1/2}$	-5	$-11^{1/2}$	$45^{1/2}$	0	0
^2_3D	$700^{-1/2}$	0	$-56^{1/2}$	0	$126^{1/2}$	0	$90^{1/2}$	$60^{1/2}$	0	$35^{1/2}$	0	$189^{1/2}$	0	$99^{1/2}$	$45^{1/2}$	0	0
^4D	$700^{-1/2}$	0	0	0	$126^{1/2}$	0	0	$-135^{1/2}$	$-175^{1/2}$	0	0	$-84^{1/2}$	0	0	$180^{1/2}$	0	0
^2_1F	$2800^{-1/2}$	0	0	$448^{1/2}$	$-200^{1/2}$	$-160^{1/2}$	$180^{1/2}$	$120^{1/2}$	0	$105^{1/2}$	$112^{1/2}$	$-175^{1/2}$	-20	$275^{1/2}$	$-405^{1/2}$	$220^{1/2}$	0
^2_2F	$2800^{-1/2}$	0	0	0	$360^{1/2}$	0	-10	$600^{1/2}$	0	$-525^{1/2}$	0	$-315^{1/2}$	0	$495^{1/2}$	-3	$-396^{1/2}$	0
^4F	$700^{-1/2}$	0	0	$-56^{1/2}$	-4	0	0	$-15^{1/2}$	$-175^{1/2}$	0	$224^{1/2}$	$14^{1/2}$	0	0	$-90^{1/2}$	$-110^{1/2}$	0
^2_1G	$8400^{-1/2}$	0	0	0	0	$-800^{1/2}$	-10	$600^{1/2}$	0	$-7^{1/2}$	$1680^{1/2}$	$945^{1/2}$	$880^{1/2}$	$845^{1/2}$	$891^{1/2}$	$924^{1/2}$	$728^{1/2}$
^2_2G	$18480^{-1/2}$	0	0	0	0	0	$1452^{1/2}$	$968^{1/2}$	0	$2541^{1/2}$	0	$4235^{1/2}$	0	$-1215^{1/2}$	$-5577^{1/2}$	$-308^{1/2}$	$-2184^{1/2}$
^4G	$420^{-1/2}$	0	0	0	0	0	0	5	$-105^{1/2}$	0	0	$-70^{1/2}$	0	0	$-66^{1/2}$	$154^{1/2}$	0
^2H	$1100^{-1/2}$	0	0	0	0	0	0	0	0	$33^{1/2}$	$-220^{1/2}$	$55^{1/2}$	$220^{1/2}$	$-5^{1/2}$	$-99^{1/2}$	$286^{1/2}$	$172^{1/2}$
^2I	$550^{-1/2}$	0	0	0	0	0	0	0	0	0	0	0	0	$-45^{1/2}$	$99^{1/2}$	$231^{1/2}$	$-175^{1/2}$

Table 5.8

$$(d^6 vSL\{|d^5 (v_1 S_1 L_1)dSL)$$

d^5 \ d^6	^1_0S	^1_2S	^3_2P	^3_4P	^1_2D	^1_4D	^3_2D	^5_0D	^1_2F	^3_2F	^3_4F	^1_2G	^1_4G	^3_2G	^3_4H	^1_2I
^2S	0	0	0	0	0	$-280^{1/2}$	$42^{1/2}$	0	0	0	0	0	0	0	0	0
^6S	0	0	0	0	0	0	0	$6^{1/2}$	0	0	0	0	0	0	0	0
^4_1S	0	0	$-14^{1/2}$	-5	$-42^{1/2}$	$-210^{1/2}$	$14^{1/2}$	0	$-120^{1/2}$	0	$200^{1/2}$	0	0	0	0	0
^3_2P	0	0	-8	$14^{1/2}$	0	0	-7	$-3^{1/2}$	0	$-224^{1/2}$	$448^{1/2}$	0	0	0	0	0
^4P	1	0	$-45^{1/2}$	0	$-35^{1/2}$	0	0	0	0	$-420^{1/2}$	0	$-63^{1/2}$	0	0	0	0
^1_2D	0	-1	$-35^{1/2}$	$-10^{1/2}$	$45^{1/2}$	-10	$-60^{1/2}$	0	$-200^{1/2}$	$60^{1/2}$	$-120^{1/2}$	-5	$-968^{1/2}$	$-200^{1/2}$	0	0
^3_2D	0	$-2^{1/2}$	0	$45^{1/2}$	0	$450^{1/2}$	$-30^{1/2}$	0	-10	0	$540^{1/2}$	0	$4356^{1/2}$	-10	0	0
^3_4D	0	0	0	$-90^{1/2}$	0	0	$-135^{1/2}$	$5^{1/2}$	0	0	$480^{1/2}$	0	0	$800^{1/2}$	0	0
^5D	0	0	$-56^{1/2}$	5	$28^{1/2}$	$-315^{1/2}$	$21^{1/2}$	0	$105^{1/2}$	$-56^{1/2}$	$175^{1/2}$	$70^{1/2}$	$-4235^{1/2}$	$-315^{1/2}$	$-55^{1/2}$	0
^3_2F	0	0	0	$-45^{1/2}$	0	$175^{1/2}$	$105^{1/2}$	0	$-525^{1/2}$	0	$315^{1/2}$	0	$-7623^{1/2}$	$-7^{1/2}$	$99^{1/2}$	0
^3_4F	0	0	$-56^{1/2}$	-4	$-60^{1/2}$	$-75^{1/2}$	$21^{1/2}$	$-7^{1/2}$	$3^{1/2}$	$896^{1/2}$	$112^{1/2}$	0	0	$560^{1/2}$	$-220^{1/2}$	0
^3_4G	0	0	0	0	0	0	$-45^{1/2}$	0	$495^{1/2}$	$360^{1/2}$	$405^{1/2}$	$66^{1/2}$	$5577^{1/2}$	$-297^{1/2}$	$99^{1/2}$	$-33^{1/2}$
^3_2G	0	0	0	0	$-60^{1/2}$	$495^{1/2}$	$-33^{1/2}$	0	0	0	$825^{1/2}$	0	$-3645^{1/2}$	$845^{1/2}$	$-15^{1/2}$	$-45^{1/2}$
^1_4G	0	0	0	0	0	0	$75^{1/2}$	3	0	0	$1200^{1/2}$	0	0	$-880^{1/2}$	$-660^{1/2}$	0
^3_2H	0	0	0	0	0	0	0	0	$132^{1/2}$	$440^{1/2}$	$-220^{1/2}$	$-154^{1/2}$	$308^{1/2}$	$-308^{1/2}$	$-286^{1/2}$	$-77^{1/2}$
^3_2I	0	0	0	0	0	0	0	0	0	0	0	0	$-6552^{1/2}$	$-728^{1/2}$	$546^{1/2}$	$-175^{1/2}$
N	1	$3^{-1/2}$	$270^{-1/2}$	$270^{-1/2}$	$210^{-1/2}$	$2100^{-1/2}$	$630^{-1/2}$	$30^{-1/2}$	$1680^{-1/2}$	$2520^{-1/2}$	$5040^{-1/2}$	$387^{-1/2}$	$33264^{-1/2}$	$5040^{-1/2}$	$1980^{-1/2}$	$330^{-1/2}$

104   5. Systematics of the Levels of Multielectron Atoms

Table 5.9

d^7 d^6	^2_3P	^4_3P	^2_1D	^2_3D	^2_3F	^4_3F	^2_3G	^2_3H
^1_0S	0	0	$8^{1/2}$	0	0	0	0	0
^1_4S	0	0	0	$56^{1/2}$	0	0	0	0
^2_2P	$7^{1/2}$	$-16^{1/2}$	$27^{1/2}$	$-49^{1/2}$	$112^{1/2}$	$-14^{1/2}$	0	0
^4_3P	$-50^{1/2}$	$-14^{1/2}$	0	$56^{1/2}$	$200^{1/2}$	$16^{1/2}$	0	0
^2_1D	$15^{1/2}$	0	$15^{1/2}$	$45^{1/7}$	$-40^{1/2}$	0	$-200^{1/2}$	0
^2_3D	$-30^{1/2}$	0	0	$40^{1/2}$	$-180^{1/2}$	0	$100^{1/2}$	0
^2_3D	$-20^{1/2}$	$-35^{1/2}$	0	$-240^{1/2}$	$-120^{1/2}$	$15^{1/2}$	$-600^{1/2}$	0
^2_3D	0	$75^{1/2}$	0	0	0	$175^{1/2}$	0	0
^2_1F	$30^{1/2}$	0	0	$-140^{1/2}$	$-105^{1/2}$	0	$7^{1/2}$	$-33^{1/2}$
^2_3F	$-8^{1/2}$	$-14^{1/2}$	$63^{1/2}$	$21^{1/2}$	$28^{1/2}$	$56^{1/2}$	$420^{1/2}$	$-55^{1/2}$
^4_3F	$50^{1/2}$	$-56^{1/2}$	0	$84^{1/2}$	$175^{1/2}$	$-14^{1/2}$	$-945^{1/2}$	$-55^{1/2}$
^2_1G	0	0	$27^{1/2}$	$-25^{1/2}$	$-100^{1/2}$	0	$220^{1/2}$	$55^{1/2}$
^2_1G	0	0	0	$44^{1/2}$	$-275^{1/2}$	0	$-845^{1/2}$	$5^{1/2}$
^2_3G	0	0	0	$-180^{1/2}$	$405^{1/2}$	$90^{1/2}$	$-891^{1/2}$	$99^{1/2}$
^2_3H	0	0	0	0	$-220^{1/2}$	$110^{1/2}$	$-924^{1/2}$	$-286^{1/2}$
^1_4I	0	0	0	0	0	0	$728^{1/2}$	$-182^{1/2}$
N	$210^{-1/2}$	$210^{-1/2}$	$140^{-1/2}$	$980^{-1/2}$	$1960^{-1/2}$	$490^{-1/2}$	$5880^{-1/2}$	$770^{-1/2}$

Table 5.10

d^8 d^7	^1_0S	^3_2P	^1_2D	^3_2F	^1_2G
^2_3P	0	$-14^{1/2}$	$-126^{1/2}$	4	0
^4_3P	0	$-8^{1/2}$	0	$-56^{1/2}$	0
^2_1D	1	$-15^{1/2}$	$-35^{1/2}$	$-35^{1/2}$	$-21^{1/2}$
^2_3D	0	$-35^{1/2}$	$135^{1/2}$	$15^{1/2}$	-5
^2_3F	0	$-56^{1/2}$	$84^{1/2}$	$-14^{1/2}$	$70^{1/2}$
^4_3F	0	$-56^{1/2}$	0	$224^{1/2}$	0
^2_3G	0	0	$-180^{1/2}$	$90^{1/2}$	$66^{1/2}$
^2_3H	0	0	0	$110^{1/2}$	$-154^{1/2}$
N	1	$240^{-1/2}$	$560^{-1/2}$	$560^{-1/2}$	$336^{-1/2}$

Table 5.11

d^8 d^9	^1S	^3P	^1D	^3F	^1G
^2D	$\dfrac{1}{\sqrt{45}}$	$\dfrac{1}{\sqrt{5}}$	$\dfrac{1}{3}$	$\sqrt{\dfrac{7}{15}}$	$\sqrt{\dfrac{1}{5}}$

$(d^9\,{}^2D\,\{d^8\,[S'L']\,d^2D)$

$(d^7vSL\,\{d^6\,(v_1S_1L_1)\,dSL)$

$(d^8vSL\,\{d^7\,(v_1S_1L_1)\,dSL)$

5.1 Wave Functions

Table 5.12

	$(f^3\,{}^4L\{|f^2\,[{}^3L_1]\}f^4L)$		
${}^4L$ ${}^3L_1$	${}^3P$	${}^3F$	${}^3H$
${}^4S$	0	1	0
${}^4D$	$\sqrt{\frac{3}{7}}$	$\frac{\sqrt{2}}{3}$	$\frac{1}{3}\sqrt{\frac{2\cdot 11}{7}}$
${}^4F$	$\frac{1}{\sqrt{2\cdot 7}}$	$-\sqrt{\frac{2}{3}}$	$\sqrt{\frac{11}{2\cdot 3\cdot 7}}$
${}^4G$	$\sqrt{\frac{11}{2\cdot 3\cdot 7}}$	$\frac{\sqrt{2}}{3}$	$\frac{1}{3}\sqrt{\frac{5\cdot 13}{2\cdot 7}}$
${}^4I$	0	$\frac{\sqrt{2}}{3}$	$-\frac{\sqrt{7}}{3}$

Table 5.13

	$(f^4\,{}^5L\{|f^3\,[{}^4L_1]\}f^5L)$				
${}^5L$ ${}^4L_1$	${}^4S$	${}^4D$	${}^4F$	${}^4G$	${}^4I$
${}^5S$	0	0	1	0	0
${}^5D$	0	$\sqrt{\frac{2}{7}}$	$-\frac{1}{2\sqrt{2}}$	$-\frac{1}{2}\sqrt{\frac{3\cdot 11}{2\cdot 7}}$	0
${}^5F$	$\frac{1}{\sqrt{7}}$	$-\frac{1}{2}\sqrt{\frac{5}{2\cdot 7}}$	$\frac{1}{2}\sqrt{\frac{3}{2}}$	$-\frac{3}{2\sqrt{2\cdot 7}}$	$-\frac{1}{2}\sqrt{\frac{13}{2\cdot 7}}$
${}^5G$	0	$\frac{1}{2}\sqrt{\frac{5\cdot 11}{2\cdot 3\cdot 7}}$	$\frac{1}{2\sqrt{2}}$	$-\frac{5}{2}\sqrt{\frac{5}{2\cdot 7\cdot 11}}$	$\frac{1}{2}\sqrt{\frac{7\cdot 13}{2\cdot 3\cdot 11}}$
${}^5I$	0	0	$\frac{1}{2\sqrt{2}}$	$\frac{1}{2}\sqrt{\frac{3\cdot 7}{2\cdot 11}}$	$\sqrt{\frac{7}{11}}$

Table 5.14

	$(f^5\,{}^6L\{|f^4\,[{}^5L_1]\}f^6L)$				
${}^6L$ ${}^5L_1$	${}^5S$	${}^5D$	${}^5F$	${}^5G$	${}^5I$
${}^6P$	0	$\sqrt{\frac{3}{7}}$	$\frac{1}{\sqrt{2\cdot 5}}$	$\sqrt{\frac{3\cdot 11}{2\cdot 5\cdot 7}}$	0
${}^6F$	$\sqrt{\frac{3}{5\cdot 7}}$	$\sqrt{\frac{2}{3\cdot 7}}$	$-\sqrt{\frac{2}{5}}$	$-\sqrt{\frac{2\cdot 3}{5\cdot 7}}$	$-\sqrt{\frac{2\cdot 13}{3\cdot 5\cdot 7}}$
${}^6H$	0	$\sqrt{\frac{2}{3\cdot 7}}$	$\frac{1}{\sqrt{2\cdot 5}}$	$-\sqrt{\frac{3\cdot 13}{2\cdot 7\cdot 11}}$	$\sqrt{\frac{7\cdot 13}{3\cdot 5\cdot 11}}$

Table 5.15

	$(f^6\,{}^7L\{|f^5\,[{}^6L_1]\}f^7L)$		
${}^7L$ ${}^6L_1$	${}^6P$	${}^6F$	${}^6H$
${}^1F$	$\frac{1}{\sqrt{7}}$	$\frac{1}{\sqrt{3}}$	$\sqrt{\frac{11}{3\cdot 7}}$

Table 5.16

	$(f^7\,{}^8L\{|f^6\,[{}^7L_1]\}f^8L)$
${}^8L$ ${}^7L_1$	${}^7F$
${}^8S$	1

5.2 Matrix Elements of Symmetric Operators

5.2.1 Statement of the Problem

In various applications one meets matrix elements of operators of two types

$$F = \sum_i f_i, \tag{5.41}$$

$$Q = \frac{1}{2} \sum_{i,k}' q_{ik} = \sum_{i>k} q_{ik}. \tag{5.42}$$

The operators F and Q are symmetric with respect to all the electrons of an atom. The first of these operators is a sum of single-electron operators because each of the operators f_i acts only on the variables of the ith electron. Operators of this type are, for example, the electric dipole moment of an atom

$$\boldsymbol{D} = - e \sum_i \boldsymbol{r}_i \tag{5.43}$$

and also the Coulomb interaction of the atomic electrons with the nucleus

$$U = - e^2 \sum_i Z/r_i. \tag{5.44}$$

The operator Q is a sum of two-electron operators q_{ik}. Summation in (5.42) is carried out over all possible pairs, i,k ($i \neq k$). The number of such pairs is $N(N-1)/2$. An example of an operator of this type is the electrostatic interaction between electrons

$$U = e^2 \sum_{i>k} \frac{1}{|\boldsymbol{r}_i - \boldsymbol{r}_k|}. \tag{5.45}$$

Before proceeding to the consideration of specific problems, it is useful to establish a number of general relations for the matrix elements of the operators F and Q between antisymmetric states of a system, i.e., states described by antisymmetric wave functions.

Owing to the indistinguishability of electrons, the integrals

$$\int \Psi_\gamma^* f_i \Psi_{\gamma'} d\tau, \quad \int \Psi_\gamma^* q_{ik} \Psi_{\gamma'} d\tau,$$

where Ψ_γ are antisymmetric wave functions, do not depend on the indices i and i, k, respectively. Therefore,

$$\int \Psi_\gamma^* F \Psi_{\gamma'} d\tau = N \int \Psi_\gamma^* f_i \Psi_{\gamma'} d\tau = N \int \Psi_\gamma^* f_N \Psi_{\gamma'} d\tau, \tag{5.46}$$

$$\int \Psi_\gamma^* Q \Psi_{\gamma'} \, d\tau = \frac{N(N-1)}{2} \int \Psi_\gamma^* q_{ik} \Psi_{\gamma'} \, d\tau = \frac{N(N-1)}{2} \int \Psi_\gamma^* q_{N-1N} \Psi_{\gamma'} \, d\tau. \tag{5.47}$$

The operator f_N acts only on the variables ξ_N. Consequently, to do the integration in (5.46), it is necessary to separate the variables of the electron N from the variables of all the other electrons. In exactly the same way it is necessary to separate the variables ξ_{N-1}, ξ_N in the integrals (5.47).

We shall explain what has been said by calculating, as an example, the diagonal matrix element of the operator q_{12} in the case of a two-electron configuration. We shall limit ourselves to the central field approximation. Putting the wave functions in the form (5.3)

$$\Psi_{aa'} = \frac{1}{\sqrt{2}} [\psi_a(\xi_1) \psi_{a'}(\xi_2) - \psi_a(\xi_2) \psi_{a'}(\xi_1)], \tag{5.48}$$

we find

$$\langle aa' | q_{12} | aa' \rangle = \frac{1}{2} \int [\psi_a^*(\xi_1) \psi_{a'}^*(\xi_2) q_{12} \psi_a(\xi_1) \psi_{a'}(\xi_2) \\
+ \psi_a^*(\xi_2) \psi_{a'}^*(\xi_1) q_{12} \psi_a(\xi_2) \psi_{a'}(\xi_1) - \psi_a^*(\xi_1) \psi_{a'}^*(\xi_2) q_{12} \psi_a(\xi_2) \psi_{a'}(\xi_1) \\
- \psi_a^*(\xi_2) \psi_{a'}^*(\xi_1) q_{12} \psi_a(\xi_1) \psi_{a'}(\xi_2)] \, d\xi_1 \, d\xi_2$$

or

$$\langle aa' | q_{12} | aa' \rangle = \langle a_1 a_2' | q_{12} | a_1 a_2' \rangle - \langle a_1 a_2' | q_{12} | a_2 a_1' \rangle. \tag{5.49}$$

In this expression the lower indices on the quantum numbers a, a' indicate which of the electrons is in the given state. A similar notation will be used everywhere in this and subsequent sections of this chapter. The matrix elements in the right-hand part of (5.49) are calculated with the aid of the nonantisymmetrized functions

$$\Psi_{a_1 a_2'} = \psi_a(\xi_1) \psi_{a'}(\xi_2), \quad \Psi_{a_2 a_1'} = \psi_a(\xi_2) \psi_{a'}(\xi_1). \tag{5.50}$$

The matrix element entering into (5.49) with a minus sign is called an exchange matrix element. This is because in the right-hand part of the corresponding matrix element, a transposition (exchange) of electrons between the states a and a' is carried out. The physical meaning of an exchange matrix element will be explained in Sect. 5.3. Let us introduce the exchange operator P_{12}, which we shall define by the relation

$$P_{12} \Psi_{a_1 a_2'} = \Psi_{a_2 a_1'}. \tag{5.51}$$

By means of this operator, (5.49) can be written in the more compact form

$$\langle aa'|q_{12}|aa'\rangle = \langle a_1 a'_2|q_{12}(1-P_{12})|a_1 a'_2\rangle. \tag{5.52}$$

The problem of reducing the matrix elements F and Q to the matrix elements of the operators f_N and $q_{N-1,N}$, calculated with the aid of nonantisymmetrized wave functions of the type (5.50), is a typical problem which one has to encounter in considering multielectron configurations. The general methods of calculating matrix elements stated in Sect. 4.3 can be used only after this problem has been solved.

5.2.2 F Matrix Elements. Parentage Scheme Approximation

We shall begin by considering the matrix elements for the transitions

$$[\gamma_1 S_1 L_1] lSLM_S M_L \to [\gamma_1 S_1 L_1] lS'L'M'_S M'_L,$$

in which neither the initial term nor the quantum numbers of the optical electron vary. Diagonal matrix elements are a particular case of matrix elements of this type. We shall represent the wave functions $\Psi_{SLM_S M_L}(S_1 L_1, l)$ in the form of (5.33), i.e., in the form of an expansion in functions $\tilde{\Psi}_{SLM_S M_L}(S_1 L_1, l_i)$. We recall that these functions are constructed by the general rule of addition of angular momenta on the assumption that the electrons $1, 2, \ldots, i-1, i+1, \ldots, N$ relate to the initial ion and the electron i is in the state with angular momentum l_i. Thus the functions under consideration are antisymmetric with respect to exchange of the electrons $1, 2, \ldots, i-1, i+1, \ldots, N$, but are not antisymmetric with respect to exchange of these electrons with the electron i. Taking this into account, we obtain

$$\langle \gamma_1 S_1 L_1, lSLM_S M_L|F|\gamma_1 S_1 L_1, lS'L'M'_S M'_L\rangle$$
$$= N \frac{1}{N} \sum_{i,k} (-1)^{i+k} \langle \gamma_1 S_1 L_1, l_i SLM_S M_L|f_N|\gamma_1 S_1 L_1, l_k S'L'M'_S M'_L\rangle. \tag{5.53}$$

Only the terms $i = k$ are nonzero in the sum, all the terms $i \neq N$ being equal. This enables us to write the right-hand part of (5.53) in the form

$$\langle \gamma_1 S_1 L_1, l_N SLM_S M_L|f_N|\gamma_1 S_1 L_1, l_N S'L'M'_S M'_L\rangle$$
$$+ \langle \gamma_1 S_1 L_1, l_i SLM_S M_L|(N-1)f_N|\gamma_1 S_1 L_1, l_i S'L'M'_S M'_L\rangle. \tag{5.54}$$

In the second term of (5.54), one can substitute $(N-1)f_N = \sum_{p \neq i} f_p$ and then replace the index i by N. After this

$$\langle \gamma_1 S_1 L_1, lSLM_S M_L | F | \gamma_1 S_1 L_1, lS'L'M'_S M'_L \rangle$$
$$= \langle \gamma_1 S_1 L_1, l_N SLM_S M_L | f_N + \sum_{p \neq N} f_p | \gamma_1 S_1 L_1, l_N S'L'M'_S M'_L \rangle$$
$$= \langle \gamma_1 S_1 L_1, l_N SLM_S M_L | F | \gamma_1 S_1 L_1, l_N S'L'M'_S M'_L \rangle . \quad (5.55)$$

The F matrix elements off-diagonal in the quantum numbers of the optical electron are nonzero only in the case when the state of the initial ion does not change, i.e., for transitions

$$\gamma_1 S_1 L_1, nlSLM_S M_L \to \gamma_1 S_1 L_1, n'l'S'L'M'_S M'_L .$$

By using (5.33) again for the wave functions, it is easy to obtain an expression analogous to (5.53). Only one term $i = k = N$ will now remain in the sum over i, k. Thus

$$\langle \gamma_1 S_1 L_1, nlSLM_S M_L | F | \gamma_1 S_1 L_1, n'l'S'L'M'_S M'_L \rangle$$
$$= \langle \gamma_1 S_1 L_1, nl_N SLM_S M_L | f_N | \gamma_1 S_1 L_1, n'l'_N S'L'M'_S M'_L \rangle . \quad (5.56)$$

It is easy to see that (5.55, 56) coincide with those expressions for the F matrix elements that would be obtained if the state l were ascribed to the electron N from the very beginning. In other words, one can use the functions $\Psi_{SLM_S M_L}(S_1 L_1, l_N)$ instead of the antisymmetric functions $\Psi_{SLM_S M_L}(S_1 L_1, l)$ when calculating F matrix elements.

By calculating in exactly the same way F matrix elements in the central field approximation, it is not difficult to obtain

$$\langle a^1 \ldots a^N | F | a^1 \ldots a^N \rangle = \sum_k \langle a_N^k | f_N | a_N^k \rangle = \sum_k \langle a_k^k | f_k | a_k^k \rangle , \quad (5.57)$$

$$\langle a^1 \ldots a^k \ldots a^N | F | a^1 \ldots b^k \ldots a^N \rangle = \langle a_N^k | f_N | b_N^k \rangle = \langle a_k^k | f_k | b_k^k \rangle , \quad (5.58)$$

In this case again, the result has the same form as when the system is described by nonantisymmetrized functions

$$\Psi = \psi_{a1}(\xi_1) \psi_{a2}(\xi_2) \ldots \psi_{aN}(\xi_N) . \quad (5.59)$$

5.2.3 F Matrix Elements. Equivalent Electrons

We shall proceed from (5.34) for the wave function of the states of the configuration l^n. For the transition between the states of this configuration it follows from (5.34) that

$$\langle l^n\gamma SLM_SM_L|F|l^n\gamma'S'L'M'_SM'_L\rangle = n\sum_{\gamma_1S_1L_1} G^{\gamma SL}_{\gamma_1S_1L_1} G^{\gamma'S'L'}_{\gamma_1S_1L_1}$$
$$\times \langle l^{n-1}[\gamma_1S_1L_1], l_nSLM_SM_L|f_n|l^{n-1}[\gamma_1S_1L_1], l_nS'L'M'_SM'_L\rangle. \qquad (5.60)$$

In the case of the transition $l^n\gamma SLM_SM_L \to l^{n-1}[\gamma_1S_1L_1]l'S'L'M'_SM'_L$, it is necessary to take the wave function of the initial state in the form (5.34) and the wave function of the final state in the form (5.33). In this case

$$\langle l^n\gamma SLM_SM_L|F|\, l^{n-1}[\gamma_1S_1L_1]\, l'S'L'M'_SM'_L\rangle = \sqrt{n}\sum_{\gamma_2S_2L_2} G^{\gamma SL}_{\gamma_2S_2L_2}$$
$$\times \sum_i (-1)^{n-i} \langle l^{n-1}[\gamma_2S_2L_2]\, l_nSLM_SM_L|f_n|\, l^{n-1}[\gamma_1S_1L_1]\, l'_iS'L'M'_SM'_L\rangle$$
$$= \sqrt{n}\, G^{\gamma SL}_{\gamma_1S_1L_1}\, \langle l^{n-1}[\gamma_1S_1L_1]\, l_nSLM_SM_L|f_n|\, l^{n-1}[\gamma_1S_1L_1]\, l'_nS'L'M'_SM'_L\rangle. \qquad (5.61)$$

In the particular case of the configuration l^2, (5.60) and (5.61) take the form

$$\langle l^2SLM_SM_L|F|\, l^2S'L'M'_SM'_L\rangle = 2\langle l_1l_2SLM_SM_L|f_2|\, l_1l_2S'L'M'_SM'_L\rangle, \qquad (5.62)$$

$$\langle l^2SLM_SM_L|F|\, l,l'S'L'M'_SM'_L\rangle = \sqrt{2}\langle l_1l_2SLM_SM_L|f_2|\, l_1l'_2S'L'M'_SM'_L\rangle. \qquad (5.63)$$

We shall also consider the transition $l^n[\gamma_1S_1L_1]\, l'^p[\gamma_2S_2L_2]\, SLM_SM_L \to l^{n-1}[\gamma'_1S'_1L'_1]\, l'^{p+1}[\gamma'_2S'_2L'_2]\, S'L'M'_SM'_L$, in which two groups of equivalent electrons take part. In this case the functions of both the initial and final states must be given in the form of (5.38).

Using these functions, it is not difficult to obtain

$$\langle l^n[\gamma_1S_1L_1]\, l'^p[\gamma_2S_2L_2]\, SLM_SM_L|F|\, l^{n-1}[\gamma'_1S'_1L'_1], l'^{p+1}[\gamma'_2S'_2L'_2],$$
$$S'L'M'_SM'_L\rangle = \sqrt{n(p+1)}\,(-1)^p\, G^{\gamma_1S_1L_1}_{\gamma'_1S'_1L'_1}\, G^{\gamma_2S_2L_2}_{\gamma'_2S'_2L'_2}$$
$$\times \langle l^{n-1}[\gamma'_1S'_1L'_1]\, l_N[\gamma_1S_1L_1], l'^p[\gamma_2S_2L_2]\, SLM_SM_L|f_N|\, l^{n-1}[\gamma'_1S'_1L'_1],$$
$$l'^p[\gamma_2S_2L_2]\, l'_N[\gamma'_2S'_2L'_2]\, S'L'M'_SM'_L\rangle = \sqrt{n(p+1)}\, G^{\gamma_1S_1L_1}_{\gamma'_1S'_1L'_1}\, G^{\gamma_2S_2L_2}_{\gamma'_2S'_2L'_2}$$
$$\times (-1)^{L_2+S_2+L'_2+S'_2-l'-1/2}\, \langle l^{n-1}[\gamma'_1S'_1L'_1]\, l_N[\gamma_1S_1L_1]\, l'^p,$$
$$[\gamma_2S_2L_2]\, SLM_SM_L|f_N|\, l^{n-1}[\gamma'_1S'_1L'_1], l'_Nl'^p[\gamma_2S_2L_2]\, \gamma'_2S'_2L'_2S'L'M'_SM'_L\rangle. \qquad (5.64)$$

All other transitions with the participation of groups of equivalent electrons can be reduced to the three considered above.

5.2.4. Q Matrix Elements. Parentage Scheme Approximation

We shall begin by considering the diagonal matrix element of Q for the states $\gamma_1 S_1 L_1 \, l S L M_S M_L$. Using (5.33) again, we obtain

$$\langle \gamma_1 S_1 L_1, l S L M_S M_L | Q | \gamma_1 S_1 L_1, l S L M_S M_L \rangle = \tfrac{1}{2} N(N-1) \tfrac{1}{N}$$

$$\times \sum_{i,k} (-1)^{i+k} \langle \gamma_1 S_1 L_1, l_i S L M_S M_L | q_{N-1,N} | \gamma_1 S_1 L_1, l_k S L M_S M_L \rangle . \qquad (5.65)$$

In the sum of (5.65), the only nonvanishing terms are of two types

1) $i = N,\ N-1\ ;\ k = N,\ N-1$,

2) $i = k \neq N,\ N-1$,

the terms $i = N$, $k = N$ and $i = N-1$, $k = N-1$, and also $i = N$, $k = N-1$ and $i = N-1$, $k = N$ being equal. The contribution of terms of the first type to the matrix element (5.65) equals

$$(N-1)\,\{\langle \gamma_1 S_1 L_1 l_N S L M_S M_L | q_{N-1,N} | \gamma_1 S_1 L_1 l_N S L M_S M_L \rangle$$
$$- \langle \gamma_1 S_1 L_1 l_N S L M_S M_L | q_{N-1,N} | \gamma_1 S_1 L_1 l_{N-1} S L M_S M_L \rangle\} . \qquad (5.66)$$

The terms of the second type give

$$\tfrac{1}{2}(N-1) \sum_i \langle \gamma_1 S_1 L_1 l_i S L M_S M_L | q_{N-1,N} | \gamma_1 S_1 L_1 l_i S L M_S M_L \rangle$$
$$= \tfrac{1}{2}(N-1)(N-2) \langle \gamma_1 S_1 L_1 l_i S L M_S M_L | q_{N-1,N} | \gamma_1 S_1 L_1 l_i S L M_S M_L \rangle . \qquad (5.67)$$

In this expression one can replace $\tfrac{1}{2}(N-1)(N-2) q_{N-1,N}$ by $\sum_{r>k} q_{rk}(r, k \neq i)$ and then replace the index i by N. Similarly in (5.66), one can substitute $\sum_{P}^{N-1} q_{PN}$ for $(N-1)q_{N-1,N}$. As a result we obtain

$$\langle \gamma_1 S_1 L_1, l S L M_S M_L | Q | \gamma_1 S_1 L_1 l S L M_S N_L \rangle$$
$$= \langle \gamma_1 S_1 L_1 l_N S L M_S M_L \left| \sum_{r>k}^{N-1} q_{rk} + \sum_p q_{pN}(1-P_{pN}) \right| \gamma_1 S_1 L_1, l_N S L M_S M_L \rangle . \qquad (5.68)$$

Formula (5.68) has a simple physical meaning. The two terms in (5.68) correspond to the interactions of the electrons of the initial ion and the interaction of the electron N with the electrons of the initial ion. It follows form (5.68) that when calculating diagonal matrix elements of Q, one can use the nonantisymmetrized functions $\Psi_{SLM_SM_L}(\gamma_1 S_1 L_1, l_N)$, ascribing the state l to the electron N. Here it is necessary to add exchange terms to the interaction of electron N with the remaining electrons.

If the states of the initial ion can also be given in the parentage scheme approximation, then in the matrix element

$$\langle \gamma_2 S_2 L_2 l' S_1 L_1 l_N SLM_SM_L | \sum_{r>k}^{N-1} q_{rk} + \sum_r q_{rN}(1 - P_{rN}) | \gamma_2 S_2 L_2,$$

$$l' S_1 L_1 l_N SLM_SM_L \rangle \tag{5.69}$$

it is easy to separate one further electron, having assigned to it the state l'. By repeating the derivation of (5.68), we obtain

$$\langle \gamma_2 S_2 L_2 l' S_1 L_1 lSLM_SM_L | Q | \gamma_2 S_2 L_2, l' S_1 L_1 lSLM_SM_L \rangle$$
$$= \langle \gamma_2 S_2 L_2, l'_{N-1} S_1 L_1 l_N SLM_SM_L | W | \gamma_2 S_2 L_2, l'_{N-1} S_1 L_1 l_N SLM_SM_L \rangle , \tag{5.70}$$

$$W = \sum_{r>k}^{N-2} q_{rk} + \sum_{r \neq N} q_{rN}(1 - P_{rN}) + \sum_{r \neq N-1} q_{r,N-1}(1 - P_{r,N-1})$$
$$+ q_{N,N-1}(1 - P_{N,N-1}) . \tag{5.71}$$

The first term in (5.71) describes the interaction between the electrons of a double ion; the others describe the interaction of the electrons N, $N-1$ with each other and with the electrons of the initial ion.

For two electrons, (5.71) takes the form

$$\langle ll'SLM_SM_L | q_{12} | ll'SLM_SM_L \rangle$$
$$= \langle l_1 l'_2 SLM_SM_L | q_{12}(1 - P_{12}) | l_1 l'_2 SLM_SM_L \rangle . \tag{5.72}$$

A treatment of the same type can also be used for off-diagonal matrix elements of Q. We shall quote the final results. Off-diagonal matrix elements Q are nonzero only for transitions to which there corresponds a change of one or two electron states. These matrix elements have the form

$$\langle \gamma_1 S_1 L_1 lSLM_SM_L | Q | \gamma_1 S_1 L_1 l'S'L'M'_SM'_L \rangle$$
$$= \langle \gamma_1 S_1 L_1 l_N SLM_SM_L | \sum_{i \neq N} q_{iN}(1 - P_{iN}) | \gamma_1 S_1 L_1 l'_N S'L'M'_SM'_L \rangle , \tag{5.73}$$

$$\langle \gamma_2 S_2 L_2 l' S_1 L_1 lSLM_SM_L | Q | \gamma_2 S_2 L_2 l'' S'_1 L'_1 l''' S'L'M_SM'_L \rangle$$
$$= \langle \gamma_2 S_2 L_2 l'_{N-1} S_1 L_1 l_N SLM_SM_L | q_{N-1,N}(1 - P_{N-1,N}) | \gamma_2 S_2 L_2$$
$$l'''_{N-1} S'_1 L'_1 l'''_N S'L'M'_SM'_L \rangle . \tag{5.74}$$

5.2 Matrix Elements of Symmetric Operators

In the first case the change of the state of electron N is caused by the interaction of this electron with all the other electrons. In the second case, only the interaction of the electrons $N-1$, N plays a part. In assigning definite states to the electrons $N-1$, N it is necessary, as in (5.70), to add the corresponding exchange terms.

The matrix of Q in the central field approximation is easy to obtain either directly or from (5.70–74):

$$\langle a^1 \ldots a^N | Q | a^1 \ldots a^N \rangle = \sum_{i>k} \langle a^i_{N-1} a^k_N | q_{N-1,N} (1 - P_{N-1,N}) | a^i_{N-1} a^k_N \rangle$$
$$= \sum_{i>k} \langle a^i_i a^k_k | q_{ik} (1 - P_{ik}) | a^i_i a^k_k \rangle, \quad (5.75)$$

$$\langle a^1 \ldots a^k \ldots a^N | Q | a^1 \ldots b^k \ldots a^N \rangle$$
$$= \sum_i \langle a^i_{N-1} a^k_N | q_{N-1 N} (1 - P_{N-1 N}) | a^i_{N-1} b^k_N \rangle$$
$$= \sum_i \langle a^i_i a^k_k | q_{ik} (1 - P_{ik}) | a^i_i b^k_k \rangle, \quad (5.76)$$

$$\langle a' \ldots a^i \ldots a^k \ldots a^N | Q | a' \ldots b^i \ldots b^k \ldots a^N \rangle$$
$$= \langle a^i_{N-1} a^k_N | q_{N-1,N} (1 - P_{N-1,N}) | b^i_{N-1} b^k_N \rangle = \langle a^i_i a^k_k | q_{ik} (1 - P_{ik}) | b^i_i b^k_k \rangle. \quad (5.77)$$

5.2.5 Q Matrix Elements. Equivalent Electrons

In this section we shall restrict ourselves to considering diagonal matrix elements of Q for l^n and $l^n l'$ configurations. Results can be obtained by means of analogous methods in all other cases.

A double application of (5.34) gives

$$\Psi_{\gamma SLM_SM_L}(l^n) = \sum_{\substack{\gamma_1 S_1 L_1 \\ \gamma_2 S_2 L_2}} G^{\gamma SL}_{\gamma_1 S_1 L_1} G^{\gamma_1 S_1 L_1}_{\gamma_2 S_2 L_2} \Psi_{\gamma SLM_SM_L} (l^{n-2}[\gamma_2 S_2 L_2], l_{n-1}[S_1 L_1] l_n), \quad (5.78)$$

whence it follows that

$$\langle l^n \gamma SLM_S M_L | Q | l^n \gamma SLM_S M_L \rangle$$
$$= \tfrac{1}{2} n(n-1) \sum_{\substack{\gamma_1 S_1 L_1 \\ \gamma_2 S_2 L_2 \\ \gamma_1' S_1' L_1'}} G^{\gamma SL}_{\gamma_1 S_1 L_1} G^{\gamma_1 S_1 L_1}_{\gamma_2 S_2 L_2} G^{\gamma SL}_{\gamma_1' S_1' L_1'} G^{\gamma_1' S_1' L_1'}_{\gamma_2 S_2 L_2}$$
$$\times \langle l^{n-2} [\gamma_2 S_2 L_2] l_{n-1} [S_1 L_1] l_n SLM_S M_L | q_{n-1, n} | l^{n-2} [\gamma_2 S_2 L_2],$$
$$l_{n-1} [S_1' L_1'] l_n SLM_S M_L \rangle. \quad (5.79)$$

In the particular case $n = 2$, (5.79) takes the form

$$\langle l^2 SLM_S M_L | q_{12} | l^2 SLM_S M_L \rangle = \langle l_1 l_2 SLM_S M_L | q_{12} | l_1 l_2 SLM_S M_L \rangle . \quad (5.80)$$

Expression (5.80) coincides with the matrix element of the same type for two nonequivalent electrons (5.72) if in this matrix element it is assumed $l = l'$ and the exchange term is omitted.

We shall now pass to the configuration $l^n l'$. In this case the expression for the matrix element has the same form as (5.68) because in the derivation of (5.68) no assumptions were made about the structure of the electron shells of the initial ion; namely

$$\langle l^n [\gamma_1 S_1 L_1] l' SLM_S M_L | Q | l^n [\gamma_1 S_1 L_1] l' SLM_S M_L \rangle$$
$$= \langle l^n [\gamma_1 S_1 L_1] l'_N SLM_S M_L \Big| \sum_{i>k}^{N-1} q_{ik} + \sum_{i\neq N} q_{iN} (1 - P_{iN}) \Big|$$
$$l^n [\gamma_1 S_1 L_1] l'_N SLM_S M_L \rangle . \quad (5.81)$$

We shall show in conclusion that for the diagonal matrix elements

$$\langle l^n \gamma SLM_S M_L | Q | l^n \gamma SLM_S M_L \rangle$$

of an operator Q which commutes with the angular momenta S, L, there is a simple recurrence formula. The operator $Q = \sum_{i>k}^{n} q_{ik}$ contains $n(n-1)/2$ terms, and the operator $Q' = \sum_{i>k}^{n-1} q_{ik}$ contains $(n-1)(n-2)/2$ terms; therefore,

$$\langle l^n \gamma SLM_S M_L | Q | l^n \gamma SLM_S M_L \rangle$$
$$= \frac{n}{n-2} \langle l^n \gamma SLM_S M_L | Q' | l^n \gamma SLM_S M_L \rangle . \quad (5.82)$$

We shall write the wave function $\Psi_{\gamma SLM_S M_L}(l^n)$ in the form

$$\Psi_{\gamma SLM_S M_L}(l^n) = \sum_{\gamma_1 S_1 L_1} G_{\gamma_1 S_1 L_1}^{\gamma SL} \Psi_{\gamma SLM_S M_L}(l^{n-1} \gamma_1 S_1 L_1, l_n)$$
$$= \sum_{\gamma_1 S_1 L_1} G_{\gamma_1 S_1 L_1}^{\gamma SL} \sum C_{M_L 1 m}^{L} C_{M_S 1 \mu}^{S} \Psi_{\gamma_1 S_1 L_1 M_{S1} M_{L1}}(l^{n-1}) \psi_{m\mu}(l_n) . \quad (5.83)$$

The operator Q' does not act on variables of the electron n. With the aid of (5.83) this enables us to isolate from the matrix element in the right-hand part of (5.82) the integral

$$\int \psi_{m\mu}^*(\xi_n) \psi_{m'\mu'}(\xi_n) d\xi_n = \delta_{mm'} \delta_{\mu\mu'} ,$$

after which this matrix element becomes

$$\sum_{\gamma_1 \gamma_1' S_1 L_1} G^{\gamma SL}_{\gamma_1 S_1 L_1} G^{\gamma SL}_{\gamma_1' L_1} \sum_{M_{L_1} M_{S_1}} |C^L_{M_{L_1}, M_L - M_{L_1}}|^2 |C^S_{M_{S_1}, M_S - M_{S_1}}|^2$$
$$\times \langle l^{n-1} \gamma_1 S_1 L_1 M_{S_1} M_{L_1} | Q' | l^{n-1} \gamma_1' S_1 L_1 M_{S_1} M_{L_1} \rangle .$$

Taking into account that the matrix element of the operator Q' does not depend on the quantum numbers M_S, M_L, we finally obtain

$$\langle l^n \gamma SL M_S M_L | Q | l^n \gamma SL M_S M_L \rangle = \frac{n}{n-2} \sum_{\gamma_1 \gamma_1' S_1 L_1} G^{\gamma SL}_{\gamma_1 S_1 L_1} G^{\gamma SL}_{\gamma_1' S_1 L_1}$$
$$\times \langle l^{n-1} \gamma_1 S_1 L_1 M_{S_1} M_{L_1} | Q' | l^{n-1} \gamma_1' S_1 L_1 M_{S_1} M_{L_1} \rangle . \tag{5.84}$$

5.2.6 Summary of Results

The results obtained above can be briefly formulated in the following way:

1) When calculating the matrix elements of operators of type F, one can proceed from nonantisymmetrized wave functions, ascribing to each electron, or few electrons, definite states (5.55–58).

2) When calculating the matrix elements of operators of type Q, one can also proceed from nonantisymmetrized wave functions. In this case, however, in assigning to the electron i a definite state, it is necessary to replace each of the operators $q_{ik}, k = 1, 2, \ldots i-1, i+1, \ldots N$ by $q_{ik}(1 - P_{ik})$ which is equivalent to the addition of the exchange interaction (5.68, 70, 77).

An exception to these rules is the case of equivalent electrons. Thus in the case of the configuration $l^n l'$, one can ascribe to a particular electron the state l', but at the same time it is impossible to ascribe the l states to particular electrons. Therefore configurations containing equivalent electrons require special treatment (5.60, 61, 64, 79, 84).

5.3 Electrostatic Interaction in LS Coupling. Two-Electron Configuration

5.3.1 Coulomb and Exchange Integrals

In the approximation of the self-consistent centrally symmetric field the energy of an atom is determined by the set of quantum numbers $nl, n'l' \ldots$. Let us take into account the omitted noncentral part of the electrostatic interaction U in the framework of perturbation theory. Despite the fact that part of the electrostatic interaction between the electrons is already included in the self-consistent field, everywhere below we shall assume that U is the total expression for this interaction, i.e.,

$$U = \frac{1}{2} e^2 \sum_{i \neq k} \frac{1}{r_{ik}} = e^2 \sum_{i > k} \frac{1}{r_{ik}} . \tag{5.85}$$

This is because only the splitting will interest us, i.e., the relative position of the terms. The centrally symmetric part of U is not essential for the splitting and appears only in a total shift of all the terms. The interaction U, like any scalar quantity, is invariant with respect to a rotation of the system of coordinates. Hence, it follows that U commutes with $\mathbf{L}$ and the matrix U is diagonal with respect to the quantum numbers L and M_L. Moreover, the matrix U is diagonal with respect to S and M_S since U does not depend on the spins of the electrons.

We shall proceed to calculate the matrix element U. We shall first of all use the expansion of $1/r_{12}$ in a series of Legendre polynomials, which permits us to separate the radial and angular variables

$$\frac{1}{r_{12}} = (r_1^2 + r_2^2 - 2r_1 r_2 \cos\omega)^{-1/2} = \sum_{k=0}^{\infty} \frac{r_<^k}{r_>^{k+1}} P_k(\cos\omega), \qquad (5.86)$$

where the smaller and greater of the magnitudes of the vectors $\mathbf{r}_1$ and $\mathbf{r}_2$ are denoted by $r_<$ and $r_>$ and ω is the angle between the vectors $\mathbf{r}_1$ and $\mathbf{r}_2$, i.e., between the directions θ_1, φ_1 and θ_2, φ_2. Such notation means that

$$\int \frac{r_<^k}{r_>^{k+1}} dr_1 dr_2 = \int_0^\infty dr_1 \left(\int_0^{r_1} \frac{r_2^k}{r_1^{k+1}} dr_2 + \int_{r_1}^\infty \frac{r_1^k}{r_2^{k+1}} dr_2 \right).$$

By using the theorem of addition for spherical functions, we obtain

$$\frac{1}{r_{12}} = \sum_{k=0}^{\infty} \frac{4\pi}{2k+1} \frac{r_<^k}{r_>^{k+1}} \sum_q Y_{kq}(\theta_1 \varphi_1) Y^*_{kq}(\theta_2 \varphi_2). \qquad (5.87)$$

The wave functions $\Psi_{SLM_S M_L}$ for the configuration $nl, n'l'$ have the form

$$\Psi_{SLM_S M_L} = \frac{1}{\sqrt{2}} [\Phi_{LM_L}(l_1 l_2') \pm (-1)^{l+l'-L} \Phi_{LM_L}(l_1' l_2)] Q^{\pm}_{SM_S}. \qquad (5.88)$$

The upper sign corrsponds to singlet terms $S = 0$ and the lower sign to triplet terms $S = 1$. Hence,

$$\langle SLM_S M_L | U | SLM_S M_L \rangle = \langle l_1 l_2' LM_L | U | l_1 l_2' LM_L \rangle$$
$$\pm (-1)^{l+l''-L} \langle l_1 l_2' LM_L | U | l_1' l_2 LM_L \rangle. \qquad (5.89)$$

By using (5.87), it is not difficult to obtain

$$\langle SLM_S M_L | U | SLM_S M_L \rangle = \sum_k (f_k F^k \pm g_k G^k). \qquad (5.90)$$

Here, as in (5.88), the upper sign corresponds to $S = 0$ and the lower sign to $S = 1$. The coefficients f_k and g_k are determined by the following formulas:

5.3 Electrostatic Interaction in LS Coupling. Two-Electron Configuration

$$f_k = (-1)^{l+l'+L} (l||C^k||l) (l'||C^k||l') \begin{Bmatrix} l & l' & L \\ l' & l & k \end{Bmatrix}, \qquad (5.91)$$

$$g_k = (l||C^k||l')^2 \begin{Bmatrix} l & l' & L \\ l & l' & k \end{Bmatrix}. \qquad (5.92)$$

The radial integrals F^k and G^k have the form

$$F^k = e^2 \int \frac{r_<^k}{r_>^{k+1}} R_{nl}^2(r_1) R_{n'l'}^2(r_2) r_1^2 \, dr_1 r_2^2 dr_2, \qquad (5.93)$$

$$G^k = e^2 \int \frac{r_<^k}{r_>^{k+1}} R_{nl}(r_1) R_{n'l'}(r_1) R_{nl}(r_2) R_{n'l'}(r_2) r_1^2 dr_1 r_2^2 \, dr_2. \qquad (5.94)$$

The integrals G^k define the so-called exchange part of the electrostatic interaction. The integrals F^k are usually called the Coulomb integrals and G^k the exchange integrals. The terms of direct and exchange Coulomb interaction are also used.

The formulas (5.90–92) give the terms for any two-electron configuration l,l'. For two equivalent electrons (configuration l^2), using (5.80), we obtain

$$\langle l^2 SLM_S M_L | U | l^2 SLM_S M_L \rangle = \sum_k f_k F^k, \qquad (5.95)$$

$$f_k = (-1)^L (l||C^k||l)^2 \begin{Bmatrix} l l L \\ l l k \end{Bmatrix}. \qquad (5.96)$$

The reduced matrix elements $(l||C^k||l')$ are nonzero only if $k+l+l' = 2g$, where g is integer, and if the triangle condition $\Delta(k,l,l')$ is fulfilled. Therefore the number of terms in the sums with respect to k in (5.90, 95) is usually not great. Let us consider, as an example, the configurations pp' and p^2. For the six possible terms in the first case we obtain

$$(^1S), (^3S) = F^0 + \frac{10}{25} F^2 \pm \left(G^0 + \frac{10}{25} G^2\right),$$

$$(^1P), (^3P) = F^0 - \frac{5}{25} F^2 \pm \left(G^0 - \frac{5}{25} G^2\right), \qquad (5.97)$$

$$(^1D), (^3D) = F^0 + \frac{F^2}{25} \pm \left(G^0 + \frac{1}{25} G^2\right).$$

In the second case there are three possible terms

$$(^1S) = F^0 + \frac{10}{25} F^2, \quad (^3P) = F^0 - \frac{5}{25} F^2, \quad (^1D) = F^0 + \frac{1}{25} F^2. \qquad (5.98)$$

118   5. Systematics of the Levels of Multielectron Atoms

In full agreement with Hund's rule, the lowest term is the term with the greatest multiplicity.

Eliminating F^0 and F^2, it is easy to obtain from (5.98)

$$R \equiv \frac{(^1S) - (^1D)}{(^1D) - (^3P)} = \frac{3}{2}. \tag{5.99}$$

It is important that this ratio does not depend on the numerical values of F^0 and F^2 and can be directly compared with experiment. The number of parameters F^k and G^k is always less than the number of terms. This enables one to eliminate the parameters F^k and G^k and to obtain for the distances between terms a series of relations not depending on numerical values of F^k and G^k, similar to that in (5.99); see Table 5.17. The discrepancy between theoretical and experimental values of R will be discussed in Sect. 5.4.4.

Table 5.17 Experimental values of R

spectrum	C I	N II	O III	F IV	Ne V	Na VI	Mg VII	Al VIII	Si IX	P X
R	1.14	1.14	1.14	1.14	1.14	1.14	1.14	1.13	1.13	1.13

5.3.2 Configuration Mixing

In the above, when analyzing electrostatic splitting, we did not consider the connection between terms of different configurations. Let us denote by I and II the configurations $n_\mathrm{I} l_\mathrm{I} n'_\mathrm{I} l'_\mathrm{I}$ and $n_\mathrm{II} l_\mathrm{II} n'_\mathrm{II} l'_\mathrm{II}$ for which the matrix element

$$\langle n_\mathrm{I} l_\mathrm{I} n'_\mathrm{I} l'_\mathrm{I} SLM_SM_L | U | n_\mathrm{II} l_\mathrm{II} n'_\mathrm{II} l'_\mathrm{II} SLM_SM_L \rangle = U_\mathrm{I\,II} \tag{5.100}$$

is nonvanishing. This matrix element defines the corrections $\Delta E_{LS}^{(\mathrm{I})}$ and $\Delta E_{LS}^{(\mathrm{II})}$ to the terms

$$\Delta E_{LS}^{(\mathrm{I})} = \frac{|U_\mathrm{I\,II}|^2}{E_{LS}^{(\mathrm{I})} - E_{LS}^{(\mathrm{II})}}, \quad \Delta E_{LS}^{(\mathrm{II})} = \frac{|U_\mathrm{I\,II}|^2}{E_{LS}^{(\mathrm{II})} - E_{LS}^{(\mathrm{I})}}. \tag{5.101}$$

According to (5.101), the corrections to the terms have different signs; therefore, taking into account the off-diagonal matrix elements $U_\mathrm{I\,II}$ leads to an increase in the distance between the terms. This effect is usually spoken of as repulsion and interaction of terms or interaction of configurations. The term configuration mixing is also used. In some cases the corrections (5.101) are of the same order of magnitude as the diagonal matrix elements $U_\mathrm{I\,I}$ and $U_\mathrm{II\,II}$ or even greater than them. This means that a single-configurational approximation becomes too rough. To determine the terms, it is necessary to solve the secular equation

5.3 Electrostatic Interaction in LS Coupling. Two-Electron Configuration

$$\begin{vmatrix} U_{\text{I I}}-\varepsilon & U_{\text{I II}} \\ U_{\text{II I}} & U_{\text{II II}}-\varepsilon \end{vmatrix} = 0 \ . \quad \varepsilon = 1/2 \, [(U_{\text{II}}-U_{\text{II II}}) \\ \pm \sqrt{(U_{\text{II}}-U_{\text{II II}})^2+4|U_{\text{I II}}|^2}]$$

The wave functions corresponding to the roots of this equation ε_1, ε_2 are linear combinations of the functions Ψ_{I}, Ψ_{II}. Thus there is no sense in relating the real terms in this case to any specific configuration.

The different effects of configuration mixing will be discussed in Sect. 5.4. Here we shall deal only with the calculation of off-diagonal matrix elements of the type U_{III}.

Just like single-configurational matrix elements $U_{\text{I I}}$ and $U_{\text{II II}}$, two-configurational matrix elements $U_{\text{I II}}$ are diagonal with respect to the quantum numbers SLM_SM_L. Moreover, from the invariance of U with respect to inversion of coordinates it follows that the matrix elements $U_{\text{I II}}$ are nonzero only for configurations I, II of the same parity.

Calculation of the matrix elements $U_{\text{I II}}$ is carried out by the same methods that were used previously. Thus

$$\langle nl, n'l' SLM_SM_L | U | n''l'', n'''l''' SLM_SM_L \rangle$$
$$= \sum_k [R_k(nln'l'; n''l'' n'''l''') \alpha_k \pm R_k(nln'l'; n'''l''' n''l'') \beta_k] , \quad (5.102)$$

$$R_k(nln'l'; n''l'' n'''l''')$$
$$= \int \frac{r_<^k}{r_>^{k+1}} R_{nl}(r_1) R_{n''l''}(r_1) R_{n'l'}(r_2) R_{n'''l'''}(r_2) r_1^2 dr_1 r_2^2 dr_2 , \quad (5.103)$$

$$R_k(nl\, n'l'; n'''l''' n''l'')$$
$$= \int \frac{r_<^k}{r_>^{k+1}} R_{nl}(r_1) R_{n'''l'''}(r_1) R_{n'l'}(r_2) R_{n''l''}(r_2) r_1^2 dr_1 r_2^2 dr_2 , \quad (5.104)$$

$$\alpha_k = \langle l_1 l'_2 LM_L | P_k(\cos \omega) | l''_1 l'''_2 LM_L \rangle$$
$$= (-1)^{l'+l'''+L} (l\|C^k\|l'') (l'\|C^k\|l''') \begin{Bmatrix} l & l' & L \\ l''' & l'' & k \end{Bmatrix} , \quad (5.105)$$

$$\beta_k = \langle l_1 l'_2 LM_L | P_k(\cos \omega) | l''_2 l'''_1 LM_L \rangle$$
$$= (-1)^{l+l''-L} \langle l_1 l'_2 LM_L | P_k(\cos \omega) | l'''_1 l''_2 LM_L \rangle$$
$$= (-1)^{l+l'''} (l\|C^k\|l''') (l'\|C^k\|l'') \begin{Bmatrix} l & l' & L \\ l'' & l''' & k \end{Bmatrix} . \quad (5.106)$$

The + sign in (5.102) corresponds to singlet terms and the − sign to triplet terms. Similarly, we obtain for the interaction of the configurations l^2 and l'^2

$$\langle l^2 SLM_S M_L | \sum_k \frac{r_<^k}{r_>^{k+1}} P_k(\cos\omega) | l'^2 SLM_S M_L \rangle$$

$$= \sum_k \alpha_k \int R_{nl}(r_1) R_{nl}(r_2) \frac{r_<^k}{r_>^{k+1}} R_{n'l'}(r_1) R_{n'l'}(r_2) r_1^2 dr_1 r_2^2 dr_2 , \qquad (5.107)$$

$$\alpha_k = \langle l_1 l_2 LM_L | P_k(\cos\omega) | l'_1 l'_2 LM_L \rangle$$

$$= (-1)^{l+l'+L} (l\|C^k\|l')^2 \begin{Bmatrix} l & l & L \\ l' & l' & k \end{Bmatrix} . \qquad (5.108)$$

The radial integral in (5.107) is none other than the exchange integral $G^k(nl; n'l')$. Therefore,

$$\langle l^2 SLM_S M_L | \frac{e^2}{r_{12}} | l'^2 SLM_S M_L \rangle = \sum_k \alpha_k G^k(nl; n'l') . \qquad (5.109)$$

Let us consider in conclusion the interaction of the configurations l^2, $l'l''$ (the case l^2, ll' is most frequent)

$$\langle l^2 SLM_S M_L | \sum_k \frac{r_<^k}{r_>^{k+1}} P_k(\cos\omega) | l'l'' SLM_S M_L \rangle$$

$$= \frac{1}{\sqrt{2}} \sum_k R_k(nl\,nl; n'l'n''l'') (\alpha_k \pm \beta_k) , \qquad (5.110)$$

$$\alpha_k = \langle l_1 l_2 LM_L | P_k(\cos\omega) | l'_1 l''_2 LM \rangle$$

$$= (-1)^{l+l'+L} (l\|C^k\|l') (l\|C^k\|l'') \begin{Bmatrix} l & l & L \\ l'' & l' & k \end{Bmatrix} , \qquad (5.111)$$

$$\beta_k = (-1)^{-L} \langle l_1 l_2 LM_L | P_k(\cos\omega) | l''_1 l'_2 LM_L \rangle$$

$$= (-1)^{l+l''} (l\|C^k\|l'') (l\|C^k\|l') \begin{Bmatrix} l & l & L \\ l' & l'' & k \end{Bmatrix} . \qquad (5.112)$$

5.4 Electrostatic Interaction in *LS* Coupling. Multielectron Configuration

5.4.1 Configurations l^n and $l^n l'$

The electrostatic interaction between electrons $U = \sum_{i>j} e^2/r_{ij}$ is a symmetric two-electron operator of the type (5.42). Therefore, using general methods of calculating the matrix elements of such operators (Sects. 4.3 and 5.2) it is possible to express the matrix elements U in terms of radial integrals F^k and obain formulas similar to (5.95). The final result is

5.4 Electrostatic Interaction in LS Coupling: Multielectron Configuration

$$\langle l^n \gamma SL|U|l^n \gamma SL\rangle = \sum_k f_k F^k ,\qquad (5.113)$$

$$f_k = \frac{1}{2}|(l||C^k||l)|^2 \left[\frac{1}{2L+1}\sum_{\gamma'L'}|(\gamma SL||U^k||\gamma' SL')|^2 - \frac{n}{2l+1}\right]. \qquad (5.114)$$

Here $U^k = \sum_i u_i^k$ is a symmetric one-electron operator, defined in such a way that

$$(l||u^k||l') = \delta_{ll'} . \qquad (5.115)$$

Summing in (5.114) is carried out with respect to terms of the configuration l^n. Matrix elements U^k are given by the formula

$$(l^n \gamma SL||U^k||l^n \gamma' SL')$$
$$= n \sum_{\gamma_1 S_1 L_1} G^{\gamma SL}_{\gamma_1 S_1 L_1} G^{\gamma' SL'}_{\gamma_1 S_1 L_1} (-1)^{L_1+k-l+L} \sqrt{(2L+1)(2L'+1)} \begin{Bmatrix} l & L & L_1 \\ L' & l & k \end{Bmatrix}, \qquad (5.116)$$

where $G^{\gamma SL}_{\gamma_1 S_1 L_1}$ are coefficients of fractional parentage

$$(l^n \gamma SL||U^0||l^n \gamma SL') = n\sqrt{\frac{2L+1}{2l+1}}\delta_{LL'} . \qquad (5.117)$$

The last term in (5.114) is the same for all terms SL. This term gives only a shift common to all terms and can be omitted in calculating the relative position of terms.

The parentage scheme approximation is, as a rule, applicable to the configuration $l^n l'$. In this approximation the energy of electrostatic interaction in the state $l^n[\gamma_1 S_1 L_1]l'\gamma SL$ is the sum of two parts — the energy of the group l^n in the state $\gamma_1 S_1 L_1$ and the energy of interaction of the electron l' with the group l^n (Sect. 5.2). The latter is defined by the matrix element

$$\langle l^n [\gamma_1 S_1 L_1] l'_N SL \Big| \sum_{p=1}^n \frac{e^2}{r_{pN}}(1 - P_{pN}) \Big| l^n [\gamma_1 S_1 L_1] l'_N SL\rangle$$
$$= \sum_k [\alpha_k F^k(nln'l'; nln'l') - \beta_k G^k(nln'l'; n'l'nl)]; \qquad (5.118)$$

$$\alpha_k = (l||C^k||l)(l'||C^k||l')(-1)^{L_1+l'+L}$$
$$\times (l^n \gamma_1 S_1 L_1||U^k||l^n \gamma_1 S_1 L_1)\begin{Bmatrix} L_1 & l' & L \\ l' & L_1 & k \end{Bmatrix}, \qquad (5.119)$$

$$\beta_k = (l\|C^k\|l')^2 \sum_r (-1)^r (2r+1) \begin{Bmatrix} l & l & r \\ l' & l' & k \end{Bmatrix} \left[\frac{1}{2}(-1)^{L_1+l'+L} \right.$$

$$\times (l^n \gamma_1 S_1 L_1 \| U^r \| l^n \gamma_1 S_1 L_1) \begin{Bmatrix} L_1 & l' & L \\ l' & L_1 & r \end{Bmatrix} + 2(-1)^{L_1+S_1+l'+1/2+L+S} \sqrt{\frac{3}{2}}$$

$$\times (l^n \gamma_1 S_1 L_1 \| V^{1r} \| l^n \gamma_1 S_1 L_1) \begin{Bmatrix} L_1 & l' & L \\ l' & L_1 & r \end{Bmatrix} \begin{Bmatrix} S_1 & 1/2 & S \\ 1/2 & S_1 & 1 \end{Bmatrix} \right]. \quad (5.120)$$

Here V^{1r} is the symmetric electron operator of rank 1 with respect to spin variables and rank r with respect to coordinate variables [see (4.193)]

$$V^{1r} = \sum_i v_i^{1r}; \quad \left(\frac{1}{2} l \| v^{1r} \| \frac{1}{2} l' \right) = (l\|u^r\|l') \left(\frac{1}{2} \|s\| \frac{1}{2} \right) = \delta_{ll'} \sqrt{\frac{3}{2}} ; \quad (5.121)$$

$$(l^n \gamma_1 S_1 L_1 \| V^{1r} \| l^n \gamma_1' S_1 L_1)$$
$$= n \sum_{\gamma_2 S_2 L_2} G^{\gamma_1 S_1 L_1}_{\gamma_2 S_2 L_2} G^{\gamma_1' S_1 L_1}_{\gamma_2 S_2 L_2} (-1)^{L_2+S_2+r+1+l+1/2+S_1+L_1}$$
$$\times \sqrt{\frac{3}{2}} (2L_1+1)(2S_1+1) \begin{Bmatrix} l & L_1 & L_2 \\ L_1 & l & r \end{Bmatrix} \begin{Bmatrix} 1/2 & S_1 & S_2 \\ S_1 & 1/2 & 1 \end{Bmatrix}. \quad (5.122)$$

The values of the matrix elements U^k and V^{1r} are given in Tables 5.18 and 5.19.

Table 5.18 Reduced matrix elements U^2, V^{11}, and V^{12} for configurations p^n

$(p^2SL\|U^2\|p^2S'L')$				$(p^3SL\|U^2\|p^3S'L')$		
	^1S	^3P	^1D		^4S	^2P	^2D
^1S	0	0	$\frac{2}{\sqrt{3}}$	^4S	0	0	0
^3P	0	-1	0	^2P	0	0	$-\sqrt{3}$
^1D	$\frac{2}{\sqrt{3}}$	0	$\sqrt{\frac{7}{3}}$	^2D	0	$\sqrt{3}$	0
$(p^2SL\|\sqrt{6}\,V^{11}\|p^2S'L')$				$(p^3SL\|\sqrt{6}\,V^{11}\|p^3S'L')$		
	^1S	^3P	^1D		^4S	^2P	^2D
^1S	0	$\sqrt{6}$	0	^4S	0	$2\sqrt{3}$	0
^3P	$\sqrt{6}$	3	$-\frac{1}{2}\sqrt{30}$	^2P	$2\sqrt{3}$	0	$\sqrt{15}$
^1D	0	$-\frac{1}{2}\sqrt{30}$	0	^2D	0	$-\sqrt{15}$	0
$(p^2SL\|2V^{12}\|p^2S'L')$				$(p^3SL\|2V^{12}\|p^3S'L')$		
	^1S	^3P	^1D		^4S	^2P	^2D
^1S	0	0	0	^4S	0	0	$-2\sqrt{2}$
^3P	0	$-\sqrt{6}$	-3	^2P	0	$\sqrt{6}$	0
^1D	0	-3	0	^2D	$2\sqrt{2}$	0	$-\sqrt{14}$

5.4 Electrostatic Interaction in LS Coupling: Multielectron Configuration

Table 5.19 Reduced matrix elements $(vSL\|V^{11}\|uSL)$ for configuration d^n

$d^2\,{}^3P$	$d^2\,{}^3F$	$d^3\,{}^2P$	$d^3\,{}^4P$	$d^3\,{}^2D$	$d^3\,{}^2D$	$d^3\,{}^2F$	$d^3\,{}^4F$	$d^3\,{}^2G$	$d^3\,{}^2H$
$\sqrt{\dfrac{3}{10}}$	$\sqrt{\dfrac{21}{5}}$	$\sqrt{\dfrac{2}{15}}$	$\sqrt{\dfrac{1}{3}}$	$\dfrac{1}{2}\sqrt{\dfrac{3}{2}}$	$-\dfrac{1}{2}\sqrt{\dfrac{1}{6}}$	$-\dfrac{1}{2}\sqrt{\dfrac{7}{15}}$	$2\sqrt{\dfrac{7}{6}}$	$\dfrac{9}{10}$	$\dfrac{3}{5}\sqrt{\dfrac{11}{6}}$

$d^4\,{}^3_2P$	$d^4\,{}^1P$	$d^4\,{}^3D$	$d^4\,{}^1D$	$d^4\,{}^3F$	$d^4\,{}^1F$	$d^4\,{}^3G$	$d^4\,{}^3H$
$\sqrt{\dfrac{1}{30}}$	$2\sqrt{\dfrac{1}{30}}$	$-\dfrac{5}{2}\sqrt{\dfrac{1}{6}}$	$\dfrac{3}{2}\sqrt{\dfrac{5}{6}}$	$\sqrt{\dfrac{7}{15}}$	$-\dfrac{1}{2}\sqrt{\dfrac{7}{15}}$	$\dfrac{9}{10}$	$\dfrac{3}{5}\sqrt{\dfrac{11}{6}}$

5.4.2 More Than Half Filled Shells

The reduced matrix elements of the symmetric hermitian operator $T^{kr} = \sum_i t_i^{kr}$ with $k+r \geq 1$ obey the relation $(n < 2l+1)$

$$(l^n \gamma SL \| T^{kr} \| l^n \gamma' S'L')$$
$$= -(-1)^{k+r}\,(l^{4l+2-n}\gamma SL \| T^{kr} \| l^{4l+2-n}\gamma' S'L') . \tag{5.123}$$

Consequently, in passing from the configuration l^n to the configuration l^{4l+2-n}, the reduced matrix elements of $U^1\,V^{12}\,\ldots$ do not change but those of U^2, $V^{11}, \ldots$ do change sign.

For scalar operators

$$(l^{4l+2-n}\gamma SL \| T^{00} \| l^{4l+2-n}\gamma' SL) = \frac{4l+2-n}{n}\,(l^n \gamma SL \| T^{00} \| l^n \gamma' SL) . \tag{5.124}$$

Therefore, the structures of the terms of the configurations l^n and l^{4l+2-n} are identical. The coefficients f_k are related by

$$k \neq 0, \quad f_k(l^n) = f_k(l^{4l+2-n}) + (l\|C^k\|l)^2 ; \tag{5.125}$$

$$k = 0, \quad \frac{f_0(l^n)}{n(n-1)} = \frac{f_0(l^{4l+2-n})}{(4l+2-n)(4l+1-n)} . \tag{5.126}$$

The coefficients α_k in (5.118) for $k \neq 0$ obey the relation

$$\alpha_k(l^n, l') = -\alpha_k(l^{4l+2-n}, l') . \tag{5.127}$$

General relations between $\beta_k(l^n, l')$ and $\beta_k(l^{4l+2-n}, l')$ analogous to (5.127) do not exist.

5.4.3 Filled Shells

For a filled shell

$$\langle l^{4l+2} | U | l^{4l+2} \rangle = \frac{(4l+2)(4l+1)}{2} F^0 - \sum_{k \neq 0} (l\|C^k\|l)^2\, F^k , \tag{5.128}$$

$$\alpha_k(l^{4l+2},l') = (4l+2)\delta_{k0} \ ; \quad \beta_k(l^{4l+2},l') = +\frac{(l||C^k||l')^2}{(2l'+1)} . \tag{5.129}$$

For the interaction of a group l'^n with a filled shell l^{4l+2}

$$\alpha_k(l^{4l+2},l'^n) = n(4l+2)\delta_{k0} \ ; \quad \beta_k(l^{4l+2},l'^n) = +\frac{n(l||C^k||l')^2}{(2l'+1)} . \tag{5.130}$$

When $n = 4l' + 2$, (5.130) gives the coefficients α_k and β_k for interaction of two filled shells.

Let us mention that interaction between the electrons of each of the filled shells, interaction between the electrons of the different filled shells, and interaction between the electrons of unfilled shells with the electrons of filled shells appear only in a shift common to all terms.

5.4.4 Applicability of the Single-Configuration Approximation

It has already been pointed out above that the number of parameters F^k and G^k is usually less than the number of terms. This enables us in some cases to eliminate the parameters F^k and G^k and to obtain for the distances between terms a series of relations not depending on a particular form of the centrally symmetric field and absolute values of F^k and G^k. A typical example is the configuration p^2, the terms of which obey the condition (5.99). Comparison of (5.99) with experimental data provides a test of those assumptions (LS coupling approximation, single-configuration approximation, etc.) which were taken as the basis of the calculation. In the case of more complex configurations, it proves to be convenient not to find relations of the type (5.99) but to adjust the parameters F^k and G^k to experimental data so that divergences are minimized. In this case it is also possible to give a quantitative evaluation of the approximation used.

The main issue, which will be discussed in this section, is the question of the applicability of the single-configuration approximation. This is very important for atomic spectroscopy because a strong interaction between different configurations is by no means a rare exception.

The most fully studied are the configurations p^n, the simplest of the multielectron configurations. The terms of these configurations obey the following relations:

$$p^2 \quad R \equiv \frac{(^1S) - (^1D)}{(^1D) - (^3P)} = \frac{3}{2}, \tag{5.131}$$

$$p^3 \quad R \equiv \frac{(^2P) - (^2D)}{(^2D) - (^4S)} = \frac{2}{3}, \tag{5.132}$$

$$p^4 \quad R \equiv \frac{(^1S) - (^1D)}{(^1D) - (^3P)} = \frac{3}{2}. \tag{5.133}$$

There is a large systematic discrepancy between these formulas and experimental data. According to experimental data for the isoelectronic sequences $2s^2 2p^2$ C(I), the ratio R is $1.12-1.14$ instead of 1.5 as given by (5.131); see Table 5.17. Likewise the ratio for the spectra of the isoelectronic sequence $2s^2 2p^4$ O(I) equals $1.14-1.17$ instead of 1.5 given by (5.133). Similarly, in the isoelectronic sequence N(I) experiment gives R of the order of 0.5 instead of $2/3$. This deviation of experimental data from the theoretical value of R is very regular. In all cases the experimental value of R is less than theoretical one (Table 5.17). This type of deviation can take place as a result of configuration interaction. Since an interaction is possible only between configurations of the same parity, one can expect mutual perturbation (repulsion) of terms of the configurations $2s^2 2p^2$ and $2p^4$. There is a direct indication in a number of cases of the existence of such an interaction. Thus, in the spectrum of O(III), the deviations from theory for configurations $2s^2 2p^2$ and $2p^4$ have different signs.

Vast experimental data show that configuration interaction for atoms with d optical electrons plays a greater role than for atoms with p optical electrons. This is connected with the irregular filling of d shells noted above. Among atoms with d optical electrons the most interesting are the atoms of the iron group, for which deviations from LS coupling are still not very large and, therefore, the conditions for analysis of experimental data are more favorable.

In some cases configuration interaction causes the so-called perturbation of series. This effect arises in the perturbation of terms of one series by the presence of a foreign term. A typical example is the perturbation of the series of terms $3d^{10}$ np $^2P_{3/2\,1/2}$ Cu. The levels $3d^9 4s4p$ $^2P_{1/2}$, $^2P_{3/2}$ are located between the unperturbed positions of the levels $3d^{10} 8p\, ^2P_{3/2\,1/2}$, $3d^{10} 7p\, ^2P_{3/2\,1/2}$, and $3d^{10} 6p\, ^2P_{3/2\,1/2}$. As a result these levels are perturbed especially strongly. In accordance with (5.101) the terms located above and below the perturbing one undergo shifts of different signs. A characteristic feature of the perturbation of a series in this instance is the inversion of the doublet splitting of the terms $3d^{10} 6p\, ^2P_{3/2\,1/2}$ and $3d^{10} 7p\, ^2P_{3/2\,1/2}$. The distance between the unperturbed positions of the levels $3d^{10} 6p\, ^2P_{3/2}$ and $3d^9 4s4p\, ^2P_{3/2}$ is less than that between the levels $3d^{10} 6p\, ^2P_{1/2}$ and $3d^9 4p^2 P_{1/2}$. As a consequence of this, the shift of the level $3d^{10} 6p\, ^2P_{3/2}$ considerably exceeds the total magnitude of the shift of the level $3d^{10} 6p\, ^2P_{1/2}$ and the initial doublet splitting. The inversion of the doublet $3d^{10} 8p\, ^2P_{3/2\,1/2}$ is explained in a similar manner. It is thus evident that the configuration interaction can not only violate series regularities, but also alter the character of multiplet splitting.

5.5 Multiplet Splitting in LS Coupling

5.5.1 Preliminary Remarks

Relativistic effects in the theory of multielectron atoms can be taken into account by including in the Hamiltonian the so-called Breit terms (Sect. 5.5.6). This is the best available approximation at present. The fact is that even for two electrons there does not exist an accurate relativistic equation analogous to the Dirac equation for one electron. A relativistic equation for a two-electron system can be constructed only with an accuracy to terms of the order $(v/c)^2$ inclusive. The Breit equation is such an equation. In addition to effects which occur also in single-electron atoms (dependence of the mass of the electrons on the velocity, spin-orbit interaction proportional to $l_i \cdot s_i$), the Breit equation contains a number of other terms, in particular: the interaction of the spin of one electron with the orbital motion of the other; the interaction between the magnetic moments of the electrons; the retardation in the electromagnetic interaction of the electron charges. All these effects are of the order $(v/c)^2$. Nevertheless, the fine-structure splitting is usually calculated taking into account only the single-electron spin-orbit interaction

$$W = \sum_i a(r_i) l_i \cdot s_i . \qquad (5.134)$$

This is because for elements located in the middle and end of the periodic system, the interaction (5.134) plays the principal role (Sect. 5.5.7). For this reason the simple approximation (5.134) is sufficient in a large number of cases for the purposes of spectrum systematics because it correctly conveys the qualitative features of the splitting. The light atoms are an important exception. For example, (5.134) is completely inadequate for describing the fine structure in the helium spectrum; this problem will be considered in detail below.

5.5.2 Landé Interval Rule

In calculating fine-structure splitting, one can neglect as a first approximation the off-diagonal matrix elements of W connecting the different LS terms and consider the splitting of each term separately. In this case, the magnitude of the splitting is defined by the matrix element

$$\langle \gamma SLJM | W | \gamma SLJM \rangle . \qquad (5.135)$$

Each of the single-electron operators in the sum (5.134) is the scalar product of irreducible tensor operators of rank one, $a(r_i) l_i$ commuting with S, and s_i commuting with L. Therefore the dependence of the matrix element of W on quantum numbers S, L and J is given by the 6-j symbol

$$\langle \gamma SLJM|W|\gamma SLJM\rangle \propto \begin{Bmatrix} S & L & J \\ L & S & 1 \end{Bmatrix} \propto [J(J+1) - L(L+1) - S(S+1)] \tag{5.136}$$

and the expression defining the fine-splitting energy ΔE has the form

$$\Delta E_J = \frac{1}{2} A(\gamma SL)[J(J+1) - L(L+1) - S(S+1)]. \tag{5.137}$$

The constant A is called the fine-structure splitting constant, and depends on the electron configuration and on S and L.

According to (5.137), each term splits into $(2S+1)$ components if $S \leqslant L$ or into $(2L+1)$ components if $S > L$. The separation between adjacent components of a multiplet is

$$\Delta E_J - \Delta E_{J-1} = \Delta E_{J,J-1} = A(\gamma SL) J. \tag{5.138}$$

This relationship is called the Landé interval rule. As already noted in Sect. 2.2, the multiplet splitting constant A can be of either sign, as a result of which normal and inverted multiplets are encountered. It also follows from (5.137) that the energy of splitting does not depend on M; this has a simple physical meaning – the energy of an isolated atom cannot depend on the orientation of its angular momentum J in space. The degree of degeneracy of a level SLJ with respect to M is equal to $2J+1$. It is easy to show that

$$\sum_{|L-S|\leqslant J\leqslant L+S} (2J+1) \Delta E_J = 0. \tag{5.139}$$

This means that the "center of gravity" of a multiplet

$$\bar{E}_{SLJ} = \frac{\Sigma_J(2J+1) E_{SLJ}}{\Sigma_J(2J+1)} \tag{5.140}$$

coincides with the unsplit term. Thus, by the distance between terms one has to understand the distance between the centers of gravity of multiplets.

The distance between the extreme components of a multiplet, $J_{\max} = L+S$ and $J_{\min} = |L-S|$, is

$$\frac{1}{2} A [J_{\max}(J_{\max}+1) - J_{\min}(J_{\min}+1)] = \begin{cases} AS(2L+1), & L \geqslant S, \\ AL(2S+1), & S \geqslant L. \end{cases} \tag{5.141}$$

Thus the total splitting is approximately proportional to LS. The magnitude $[J(J+1) - L(L+1) - S(S+1)]/2$ is the eigenvalue of the operator $\mathbf{L} \cdot \mathbf{S} = (\mathbf{J}^2 - \mathbf{L}^2 - \mathbf{S}^2)/2$ in the state $SLJM$. This indicates that for the term γSL the spin-orbit interaction operator can be written in the form

$$W = A\mathbf{L}\cdot\mathbf{S}. \tag{5.142}$$

For elements in the middle and at the end of the periodic system, even in the case when the LS coupling approximation is applicable, the need to take into account the off-diagonal matrix elements of W frequently arises. Corrections in the second order of perturbation theory to the levels γSLJ equal

$$\Delta E''_{\gamma SLJ} = \sum_{\gamma' S'L'} \frac{|\langle\gamma SLJM|W|\gamma' S'L'JM\rangle|^2}{E_{\gamma SL} - E_{\gamma' S'L'}}. \tag{5.143}$$

These corrections are one of the possible causes of deviations from the Landé interval rule.

5.5.3 One Electron Outside Closed Shells

The operator (5.134) is a symmetric operator of the type (5.41). Therefore, the diagonal matrix element of W in the representation $\gamma S_1 L_1 l SLJM$ is composed of two parts

$$\langle\gamma S_1 L_1; l_N SLJM|W_N|\gamma S_1 L_1; l_N SLJM\rangle \tag{5.144}$$

and

$$\langle\gamma S_1 L_1; l_N SLJM|\sum_{p\neq N} W_p|\gamma S_1 L_1; l_N SLJM\rangle, \tag{5.145}$$

where in this case $L_1 = 0$, $S_1 = 0$ and the energy of the spin-orbit interaction of the initial ion (5.145) is zero. So (5.144) takes the form

$$\langle\gamma 00; l_N sljm|a_N \mathbf{l}_N\cdot\mathbf{s}_N|\gamma 00; l_N sljm\rangle = \langle sljm|a\mathbf{l}\cdot\mathbf{s}|sljm\rangle. \tag{5.146}$$

Thus the problem reduces to calculating the spin-orbit splitting of the levels of the electron in the centrally symmetric field created by the nucleus and the filled shells.

In accordance with (1.21)

$$a(r) = (\hbar^2/2m^2c^2r)\,dU/dr$$

and

$$\Delta E_{nlj} = \zeta_{nl}\frac{1}{2}[j(j+1) - l(l+1) - s(s+1)],$$
$$\zeta_{nl} = \int a(r) R_{nl}^2 r^2\,dr. \tag{5.147}$$

5.5 Multiplet Splitting in LS Coupling

Just as in the case of hydrogen, a level with a given value of l splits into two components $j = l + 1/2$ and $j = l - 1/2$. The shift of these components from the initial level is

$$\Delta E_{nlj=l+1/2} = \zeta_{nl} l/2,$$
$$\Delta E_{nlj=l-1/2} = -\zeta_{nl}(l+1)/2, \tag{5.148}$$

and the distance between the components $j = l \pm 1/2$ is

$$\Delta E_{J,J-1} = \zeta_{nl} j = \zeta_{nl}(l + 1/2). \tag{5.149}$$

To calculate the splitting constant ζ_{nl}, it is necessary to find, by means of some approximate method, the explicit form of the centrally symmetric field $U(r)$ and the radial functions R_{nl}. As a rule, this is an extremely complex task. For this reason, a simple semiempirical formula based on pictorial quasi-classical representations is used to make estimates. The effective field $U(r)$ for an optical electron at large distances becomes the Coulomb field $-Z_a e^2/r$, where $Z_a e$ is the charge of the atomic core, and at small distances can be approximated by the Coulomb field $-Z_i e^2/r$. This enables one to put

$$\left\langle \frac{1}{r} \frac{\partial U}{\partial r} \right\rangle \approx Z_i e^2 \left\langle \frac{1}{r^3} \right\rangle. \tag{5.150}$$

An estimate of the relative time spent by the electron in the fields $-Z_a e^2/r$ and $-Z_i e^2/r$ shows that as a first approximation it is possible to retain for the factor $\langle r^{-3} \rangle$ in (5.150) the same kind of expression as in the case of the hydrogen atom, with the replacement of Z^3/n^3 by $Z_a^2 Z_i / n_*^3$. Thus

$$\zeta_{nl} = \alpha^2 \frac{Z_a^2 Z_i^2}{n_*^3 l(l+1)(l+1/2)} \text{Ry}. \tag{5.151}$$

The number Z_a, defining the effective charge of the atomic core, is 1 for neutral atoms, 2 for singly charged ions, and so on. The effective principal quantum number n_* is determined from experimentally known values of the energy terms (Sect. 3.2). Choosing the value of Z_i is slightly more difficult. Substitution of the experimental values of ζ_{nl} in (5.151) shows that usually for p electrons $Z_i \approx Z - 4$ and for d electrons $Z_i \approx Z - 11$. Table 5.20 gives an idea of the accuracy to which one can make calculations with such a choice of Z_i. This table gives values of Z_i determined from experimental values of the fine structure splitting of levels np [16].

For heavy nuclei it proves to be necessary to introduce in (5.151) the relativistic correction $H_r(lZ_i)$

Table 5.20 Values of the effective charge Z_i

Spectrum	nl	n*	Z_i	Z	Spectrum	nl	n*	Z_i	Z
Li I	2p	1.97	0.94	3	Na I	6p	5.14	7.62	11
Be II	2p	1.95	2.06	4	Mg II	6p	5.30	9.85	13
B I	2p	1.28	3.40	5	Al I	6p	4.71	10.05	13
B III	3p	2.96	3.17	5	Al III	6p	5.40	11.12	15
C II	3p	2.60	4.11	6	K I	7p	5.29	15.10	19
C IV	4p	3.96	4.21	6	Ca II	6p	4.55	17.00	21
N III	3p	2.69	5.06	7	Cu I	4p	1.86	23.4	29
N V	3p	2.96	5.14	7	Rb I	7p	4.33	31.3	37
O IV	3p	2.80	6.30	8	Sr II	6p	3.64	34.5	39
O VI	4p	3.97	6.19	8	Ag I	8p	4.97	42.2	47
F V	3p	2.78	7.12	9	Ba II	8p	4.80	53.6	57
F VII	3p	2.97	7.20	9					

$$\zeta_{nl} = \alpha^2 \frac{Z_a^2 Z_i^2 H_r(lZ_i)}{n_*^3 l(l+1)(l+1/2)} \text{ Ry}. \tag{5.152}$$

This correction begins to be significant in the magnitude of ζ_{nl} only for $Z \geqslant 50$. For small values of Z the correction factor H_r is very close to unity. The relationship of H_r to Z_i for p electrons is shown in Fig. 6.1.

Formula (5.152) can be used for approximate estimates of the factor ζ_{nl}, but of considerably greater interest it its use in determining the effective charge Z_i, because this quantity also enters into the formula for hyperfine splitting (Sect. 6.2).

Formula (5.152) in general correctly gives the basic regularities of the doublet splitting of terms of alkali elements. Individual cases when the approximation becomes inapplicable and the splitting is determined by some additional effects are an exception: for example, when configuration mixing plays a greater role.

5.5.4 Configuration l^n

Before passing on to calculate the constant $A(l^n \gamma SL)$ which defines the splitting of terms of the configuration l^n, let us return to (5.146, 147). From (4.169) it follows that

$$\langle sljm | a(r) \mathbf{l} \cdot \mathbf{s} | sljm \rangle = (-1)^{s+l+j} \zeta_{nl} (s\|s\|s)(l\|l\|l) \begin{Bmatrix} s & l & j \\ l & s & 1 \end{Bmatrix}. \tag{5.153}$$

As

$$\frac{(s\|s\|s)(l\|l\|l)}{\sqrt{l(l+1)(2l+1)}} = (s\|s\|s)(l\|u^1\|l) = (sl\|v^{11}\|sl) \tag{5.154}$$

[see (4.149, 150)], (5.153) can also be written in the form

5.5 Multiplet Splitting in LS Coupling

$$\langle sljm|a\mathbf{l}\cdot\mathbf{s}|sljm\rangle$$

$$= (-1)^{s+l+j}\zeta_{nl}\sqrt{l(l+1)(2l+1)}\,(sl\|v^{11}\|sl)\begin{Bmatrix} s & l & j \\ l & s & 1 \end{Bmatrix}. \tag{5.155}$$

Thus the matrix element (5.153) is expressed in terms of the reduced matrix element of the operator v^{11}. Similarly, the matrix elements

$$\langle l^n\gamma SLJM|\sum_i a(r_i)\,\mathbf{l}_i\cdot\mathbf{s}_i|l^n\gamma SLJM\rangle \tag{5.156}$$

can be expressed in terms of the reduced matrix elements V^{11}

$$\langle l^n\gamma SLM|\sum_i a(r_i)\mathbf{l}_i\cdot\mathbf{s}_i|l^n\gamma SLJM\rangle = (-1)^{S+L+J}\zeta_{nl}n$$

$$\times \sum_{\gamma_1 S_1 L_1}|G^{\gamma SL}_{\gamma_1 S_1 L_1}|^2 (l^{n-1}[\gamma_1 S_1 L_1]l_N L\|l_N\|l^{n-1}[\gamma_1 S_1 L_1]l_N L)$$

$$\times (l^{n-1}[\gamma_1 S_1 L_1]s_N S\|s_N\|l^{n-1}[\gamma_1 S_1 L_1]s_N S)\begin{Bmatrix} S & L & J \\ L & S & 1 \end{Bmatrix}$$

$$= (-1)^{S+L+J}n \sum_{\gamma_1 S_1 L_1}|G^{\gamma SL}_{\gamma_1 S_1 L_1}|^2 \zeta_{nl}\sqrt{l(l+1)(2l+1)}$$

$$\times (l^{n-1}[\gamma_1 S_1 L_1]l_N SL\|v^{11}\|l^{n-1}[\gamma_1 S_1 L_1]l_N SL)\begin{Bmatrix} S & L & J \\ L & S & 1 \end{Bmatrix}. \tag{5.157}$$

Comparing this expression with (5.121, 122, 137), we obtain

$$\langle l^n\gamma SLJM|\sum_i a(r_i)\,\mathbf{l}_i\cdot\mathbf{s}_i|l^n\gamma SLJM\rangle$$

$$= (-1)^{S+L+J}\zeta_{nl}\sqrt{l(l+1)(2l+1)}\,(l^n\gamma SL\|V^{11}\|l^n\gamma SL)\begin{Bmatrix} S & L & J \\ L & S & 1 \end{Bmatrix} \tag{5.158}$$

$$A(l^n\gamma SL) = \zeta_{nl}\sqrt{\frac{l(l+1)(2l+1)}{S(S+1)(2S+1)L(L+1)(2L+1)}}\,(l^n\gamma SL\|V^{11}\|l^n\gamma SL). \tag{5.159}$$

By means of (5.159) and Tables 5.18 and 5.19 given in Sect. 5.4, it is easy to calculate the spin-orbit splitting for any of the configurations p^n and d^n. According to (5.159), for shells less than half filled, $A(l^n\gamma SL) > 0$. For the configurations l^n and l^{4l+2-n}, the reduced matrix elements of V^{11} have opposite sign [see (5.123)]; therefore, for shells more than half filled, A is negative, corresponding to inverted multiplets. When $n = 2l+1$, $A(l^{2l+1}\gamma SL) = 0$, and the matrix elements (5.158) are zero. This does not mean, of course, that multiplet splitting is absent, because, in general, corrections of second order (5.143) are nonvanishing. For matrix elements of W connecting different terms of a configuration l^n, it is easy to obtain

$$\langle l^n SLJM | \sum_i a(r_i) \, \boldsymbol{l}_i \cdot \boldsymbol{s}_i | l^n \gamma' S' L' JM \rangle$$

$$(-1)^{S'+L+J} \zeta_{nl} \sqrt{l(l+1)(2l+1)} \, (l^n \gamma SL \| V^{11} \| l^n \gamma' S' L') \begin{Bmatrix} S & L & J \\ L' & S' & 1 \end{Bmatrix}. \quad (5.160)$$

5.5.5 Parentage Scheme Approximation

The fine-structure splitting constant A of the term

$$\gamma_1 S_1 L_1, \, nlSL$$

can be expressed in terms of the single-electron constant ζ_{nl} and of the fine-structure splitting constant of the initial term $A(\gamma_1 S_1 L_1)$.

We shall average the operator (5.134) over the state with given values of the total angular momenta L_1, S_1 and l,s. This gives

$$A(\gamma_1 S_1 L_1) \, \boldsymbol{L}_1 \cdot \boldsymbol{S}_1 + \zeta_{nl} \, \boldsymbol{l} \cdot \boldsymbol{s}. \quad (5.161)$$

Then, averaging (5.161) over the state with given values of the total angular momenta L and S with the aid of (4.180), we obtain

$$W = A(\gamma_1 S_1 L_1) \frac{(\boldsymbol{L}_1 \cdot \boldsymbol{L})(\boldsymbol{S}_1 \cdot \boldsymbol{S})}{L(L+1) \, S(S+1)} \boldsymbol{L} \cdot \boldsymbol{S} + \zeta_{nl} \frac{(\boldsymbol{l} \cdot \boldsymbol{L})(\boldsymbol{s} \cdot \boldsymbol{S})}{L(L+1) \, S(S+1)} \boldsymbol{L} \cdot \boldsymbol{S}$$

whence

$$A(\gamma SL) = A(\gamma_1 S_1 L_1) \frac{L(L+1) + L_1(L_1+1) - l(l+1)}{2L(L+1)}$$

$$\times \frac{S(S+1) + S_1(S_1+1) - 3/4}{2S(S+1)} + \zeta_{nl} \frac{L(L+1) - L_1(L_1+1) + l(l+1)}{2L(L+1)}$$

$$\times \frac{S(S+1) - S_1(S_1+1) + 3/4}{2S(S+1)}. \quad (5.162)$$

It is easy to generalize (5.162) to configurations containing two groups of equivalent electrons. For the term $l^n \gamma_1 S_1 L_1 l'^p \gamma_2 S_2 L_2, LS$ of such a configuration, we have

$$A = A(l^n \gamma_1 S_1 L_1) \frac{L(L+1) + L_1(L_1+1) - L_2(L_2+1)}{2(2L+1)}$$

$$\times \frac{S(S+1) + S_1(S_1+1) - S_2(S_2+1)}{2S(S+1)}$$

$$+ A(l'^p \gamma_2 S_2 L_2) \frac{L(L+1) - L_1(L_1+1) + L_2(L_2+1)}{2L(L+1)}$$

$$\times \frac{S(S+1) - S_1(S_1+1) + S_2(S_2+1)}{2S(S+1)}. \quad (5.163)$$

5.5.6 Fine-Structure Splitting of Levels of He

In the same approximation in which the fine-structure splitting of levels of hydrogen is calculated, it is possible to obtain the following expression for the Hamiltonian of a two-electron atom [2]

$$H = H_0 + H_1 + H_2 + H_3 + H_4 + H_5, \tag{5.164}$$

where

$$H_0 = \frac{1}{2m}(p_1^2 + p_2^2) - \frac{Ze^2}{r_1} - \frac{Ze^2}{r_2} + \frac{e^2}{r_{12}}, \tag{5.165}$$

$$H_1 = -\frac{1}{8m^3c^2}(p_1^4 + p_2^4), \tag{5.166}$$

$$H_2 = -\frac{e^2}{2m^2c^2}\frac{1}{r_{12}}\left[\boldsymbol{p}_1 \cdot \boldsymbol{p}_2 + \frac{(\boldsymbol{r}_{12}\cdot\boldsymbol{p}_1)(\boldsymbol{r}_{12}\cdot\boldsymbol{p}_2)}{r_{12}^2}\right], \tag{5.167}$$

$$H_3 = \frac{Z\pi e \hbar^2}{2m^2c^2}[(\delta(\boldsymbol{r}_1) + \delta(\boldsymbol{r}_2)] - \frac{\pi e \hbar^2}{m^2c^2}\delta(\boldsymbol{r}_{12}), \tag{5.168}$$

$$H_4 = \frac{e^2\hbar}{2m^2c^2}\left(\frac{Z}{r_1^3}[\boldsymbol{r}_1, \boldsymbol{p}_1] - \frac{1}{r_{12}^3}[\boldsymbol{r}_{12}, \boldsymbol{p}_1] + \frac{2}{r_{12}^3}[\boldsymbol{r}_{12}, \boldsymbol{p}_2]\right)\boldsymbol{s}_1$$
$$+ \frac{e^2\hbar}{2m^2c^2}\left(\frac{Z}{r_2^3}[\boldsymbol{r}_2, \boldsymbol{p}_2] - \frac{1}{r_{12}^3}[\boldsymbol{r}_{21}, \boldsymbol{p}_2] + \frac{2}{r_{12}^3}[\boldsymbol{r}_{21}, \boldsymbol{p}_1]\right)\boldsymbol{s}_2, \tag{5.169}$$

$$H_5 = \frac{e^2\hbar^2}{m^2c^2}\left(-\frac{8\pi}{3}\boldsymbol{s}_1\cdot\boldsymbol{s}_2\delta(\boldsymbol{r}_{12}) + \frac{1}{r_{12}^3}\left[\boldsymbol{s}_1\cdot\boldsymbol{s}_2 - \frac{3(\boldsymbol{s}_1\cdot\boldsymbol{r}_{12})(\boldsymbol{s}_2\cdot\boldsymbol{r}_{12})}{r_{12}^2}\right]\right). \tag{5.170}$$

The Hamiltonian (5.164) corresponds to the Breit approximation, the first term (5.165) being the nonrelativistic Hamiltonian. The remaining terms, (5.166–170), are associated with relativistic effects. The dependence of the mass of the electron on velocity and the retardation in the electromagnetic interaction are taken into account by (5.166) and (5.167). H_1, H_2 and also H_3 do not contain spin operators, i.e., they are purely orbital and are therefore not involved in the splitting. In the following we shall assume that the corrections due to H_1, H_2, H_3 have already been taken into account in the energy of a term.

The splitting of terms is defined by the last two quantities – the spin-orbit interaction (5.169) and the interaction between electron spins (5.170), or rather

(5.169) and the second component in (5.170), since the first component in (5.170) is also not involved in splitting.

It is convenient to isolate terms of the type (5.134) from the more general expression for the spin-orbit interaction (5.169). After this, the operator responsible for the splitting of terms can be written in the form

$$V = H'_{s0} + H''_{s0} + H_{ss},\tag{5.171}$$

$$H'_{s0} = \alpha^2 a_0^3 Z \left(\frac{1}{r_1^3} l_1 \cdot s_1 + \frac{1}{r_2^3} l_2 \cdot s_2\right) \text{Ry},\tag{5.172}$$

$$H''_{s0} = \alpha^2 \frac{a_0^3}{\hbar} \frac{1}{r_{12}^3} [(-[r_{12}, p_1] + 2[r_{12}, p_2]) \cdot s_1 \\ + (-[r_{21}, p_2] + 2[r_{21}, p_1]) \cdot s_2] \text{Ry},\tag{5.173}$$

$$H_{ss} = 2\alpha^2 a_0^3 \frac{1}{r_{12}^3} \left[s_1 \cdot s_2 - \frac{3(s_1 \cdot r_{12})(s_2 \cdot r_{12})}{r_{12}^2}\right] \text{Ry}.\tag{5.174}$$

From hereon we shall refer to the three components in (5.171) as the spin-own orbit, spin-other orbit, and spin-spin interactions.

We need to find the corrections, due to the perturbation (5.171), to the triplet terms of configurations $1snl$. Obviously, singlet terms do not have fine structure. Using the general results of Sect. 5.2 we can assign the state $1s$ to electron 1 and the state l to electron 2. Then it is necessary to replace the two-electron operators H''_{s0} and H_{ss} by $H''_{s0}(1-P_{12})$ and $H_{ss}(1-P_{12})$. In this case, however, the exchange terms, proportional to integrals of the type

$$\int f(r) R_{nl}(r) R_{1s}(r) r^2 dr,$$

are small and can be omitted. In fact, in the region where the function R_{1s} is substantially nonvanishing, R_{nl} is small and vice versa. Neglecting the exchange terms considerably simplifies calculations.

We shall begin by calculating the mean value of V. Since $l'_1 = 0$, $l_2 = L$, and $s_2 = 1/2\, S$, we obtain

$$\langle H'_{s0}\rangle = \alpha^2 Z \left\langle \frac{a_0^3}{r_2^3} l_2 \cdot s_2 \right\rangle \text{Ry} = \frac{1}{2}\alpha^2 Z \left\langle \frac{a_0^3}{r_2^3} L \cdot S \right\rangle \text{Ry}$$

$$= \frac{1}{4}\alpha^2 Z \left\langle \frac{a_0^3}{r_2^3} \right\rangle [J(J+1) - L(L+1) - S(S+1)] \text{Ry}.\tag{5.175}$$

When calculating the corrections due to the interactions H''_{s0} and H_{ss}, we can use the fact that the $1s$ electron is on the average considerably closer to the nucleus than the nl electron. Therefore, $r_2 \gg r_1$ and in the expression for H''_{s0} and H_{ss} one can put

$$r_{12} = r_1 - r_2 \simeq -r_2 ;$$

in which case we obtain

$$H''_{s0} = \alpha^2 \text{ Ry} \frac{a_0^3}{\hbar} \frac{1}{r_2^3} \Big[([r_2, p_1] - 2[r_2, p_2]) \cdot s_1 + (-[r_2, p_2] + 2[r_2, p_1]) \cdot s_2 \Big],$$

(5.176)

$$H_{ss} = 2\alpha^2 \text{ Ry } a_0^3 \frac{1}{r_2^3} \left[s_1 \cdot s_2 - \frac{3(s_1 \cdot r_2)(s_2 \cdot r_2)}{r_2^2} \right].$$

(5.177)

It is easy to see that the simplification of H''_{s0} and H_{ss} which we have effected leads to errors of the same order as neglecting the exchange terms.

From the general expression for the matrix element of a product of operators

$$\langle \gamma | fg | \gamma \rangle = \sum_{\gamma'} \langle \gamma | f | \gamma' \rangle \langle \gamma' | g | \gamma \rangle$$

it follows that

$$\langle [r_2, p_1] \rangle = 0 ,$$

since the matrix elements of r_2 are nonzero for $l_1 = l'_1$ and the matrix elements p_1 are nonzero only for the transitions $l_1 = l'_1 \pm 1$, so

$$\langle H''_{s0} \rangle = -\alpha^2 \text{ Ry} \frac{a_0^3}{\hbar} \left\langle \frac{[r_2, p_2]}{r_2^3} \cdot (2s_1 + s_2) \right\rangle = -\frac{3}{2} \alpha^2 \text{ Ry } a_0^3 \left\langle \frac{1}{r_2^3} L \cdot S \right\rangle$$

$$= -\frac{3}{4} \alpha^2 Z \left\langle \frac{a_0^3}{r_2^3} \right\rangle [J(J+1) - L(L+1) - S(S+1)] \text{ Ry} .$$

(5.178)

It remains only to consider the perturbation (5.177). The expression included in brackets in (5.177)

$$[s_1 \cdot s_2 - 3(s_1 \cdot n)(s_2 \cdot n)] = \sum_{i,k} s_{1i} s_{2k} (\delta_{ik} - 3 n_i n_k)$$

(5.179)

can be represented in the form of the scalar product of irreducible tensors of rank two. The tensor

$$(3 n_i n_k - \delta_{ik}) = D_{ik}$$

(5.180)

is a symmetrical tensor with a trace equal to zero. From the components of this tensor one can construct the spherical tensor of rank two D^2, where

$$D_m^2 = 2 C_m^2 (\theta, \varphi) \qquad (5.181)$$

[see the derivation of (4.167)].

The tensor $s_{1i}s_{2k}$ can be represented in the form (4.114)

$$s_{1i}s_{2k} = \frac{1}{3} s_1 \cdot s_2 \delta_{ik} + \frac{1}{2} (s_{1i}s_{2k} - s_{1k}s_{2i})$$
$$+ \frac{1}{2} \left(s_{1i}s_{2k} + s_{2k}s_{1i} - \frac{2}{3} s_1 \cdot s_2 \delta_{ik} \right), \qquad (5.182)$$

where only the last term having the same symmetry as D_{ik} contributes to (5.179). The products of the first two terms of (5.182) with D_{ik} are equal to zero. The only irreducible tensor of rank two which can be constructed from the components of s_1 and s_2 is the tensor

$$U^2 = [s_1^1 \times s_2^1]^2 . \qquad (5.183)$$

Therefore,

$$[s_1 \cdot s_2 - 3 (s_1 \cdot n) (s_2 \cdot n)] = - \text{const} \sum_m (-1)^m U_m^2 C_{-m}^2 = - \text{const} (U^2 C^2) . \qquad (5.184)$$

To determine the constant in (5.184) it is sufficient to equate the coefficients of the term $s_{10}s_{20}$ in U_0^2 and in the last term of (5.182). From (5.182, 183) we have

$$U_0^2 = \sum_q (11q, -q|1120) s_{1q}s_{2-q}, \quad (1100|1120) = \sqrt{\frac{2}{3}},$$

$$\frac{1}{2} \left(s_{1z}s_{2z} + s_{2z}s_{1z} - \frac{2}{3} s_1 \cdot s_2 \right) = \frac{2}{3} s_{1z}s_{2z} - \frac{2}{3} (s_{1x}s_{2x} + s_{1y}s_{2y}) .$$

Also taking into account (5.181), we obtain

$$\text{const} = 2\sqrt{\frac{3}{2}},$$

$$H_{ss} = -4\sqrt{\frac{3}{2}} \alpha^2 \text{ Ry} \frac{a_0^3}{r_2^3} (U^2 C^2), \qquad (5.185)$$

$$\langle H_{ss} \rangle = -4\sqrt{\frac{3}{2}} \alpha^2 \text{Ry} \left\langle \frac{a_0^3}{r_2^3} \right\rangle \langle l_1' s_1 l_2 s_2 L S J M | U^2 C^2 | l_1' s_1 l_2 s_2 L S J M \rangle . \qquad (5.186)$$

5.5 Multiplet Splitting in LS Coupling

The operator U^2 is a pure spin operator and therefore commutes with the orbital angular momentum L. The operator C^2 commutes with S; therefore for the matrix element in (5.186) we have

$$(-1)^{L+S+J} (s_1 s_2 S \| U^2 \| s_1 s_2 S)\, (l_1' l_2 L \| C^2 \| l_1' l_2 L) \begin{Bmatrix} L & S & J \\ S & L & 2 \end{Bmatrix}. \qquad (5.187)$$

Since in the case under consideration $l_1' = 0$ and $l_2 = L$,

$$(l_1' l_2 L \| C^2 \| l_1' l_2 L) = (L \| C^2 \| L) = -\sqrt{\frac{L(L+1)(2L+1)}{(2L+3)(2L-1)}}. \qquad (5.188)$$

Formula (4.172) can be used in calculating the reduced matrix element of U^2. Taking into account (4.150) for the triplet state $S = 1$, we obtain

$$(s_1 s_2 S \| [s_1^1 \times s_2^1]^2 \| s_1 s_2 S) = \sqrt{\frac{3}{2}} \sqrt{\frac{3}{2}} (2S+1) \sqrt{5} \begin{Bmatrix} \tfrac{1}{2} & \tfrac{1}{2} & 1 \\ \tfrac{1}{2} & \tfrac{1}{2} & 1 \\ 1 & 1 & 2 \end{Bmatrix} = \frac{\sqrt{5}}{2}. \qquad (5.189)$$

The 9j symbol in (5.189) is calculated by means of the formulas of Table 5.23, given in Sect 5.7. Thus,

$$H_{ss} = \alpha^2 \left\langle \frac{a_0^3}{r_2^3} \right\rangle \cdot \frac{1}{2} \cdot \frac{3X(X+1) - 8L(L+1)}{(2L-1)(2L+3)}\, \text{Ry}, \qquad (5.190)$$

$$X = J(J+1) - L(L+1) - 2. \qquad (5.191)$$

The second term in (5.190) is not involved in the splitting and can thus be omitted. Collecting together (5.175, 178, 190), we obtain

$$\langle (H_{so}' + H_{so}'' + H_{ss}) \rangle$$
$$= \alpha^2 \left\langle \frac{a_0^3}{r_2^3} \right\rangle \left[\frac{1}{4}(Z-3)X + \frac{3X(X+1)}{2(2L-1)(2L+3)} \right], \qquad (5.192)$$

where

$$X = \begin{cases} -2(L+1), & J = L-1, \\ -2, & J = L, \\ +2L, & J = L+1. \end{cases} \qquad (5.193)$$

Let us compare (5.192) with (5.175), i.e., with the fine-structure splitting formula in the approximation (5.134). According to (5.175), the He terms must be normal terms obeying the Landé interval rule. Taking into account the spin-other orbit interaction leads to the replacement of Z by $Z-3$. The Landé interval rule is not violated in this case. However, the sign of the splitting constant turns out to depend on Z. $Z-3 = -1$ for He, which corresponds to inverted splitting.

In the case of Li$^+$, $Z-3 = 0$ and taking into account the term H''_{s0} leads to a complete cancellation of the effect. With Be^{++}, $Z-3 = 1$ and, consequently, the normal order of arrangement of the components of the triplet is restored. The factor $Z-3$ is connected, obviously, with the screening of the nuclear charge by the electron $1s$. The greater Z is, the less effectively is the charge of the nucleus screened.

Spin-spin interaction leads to deviations from the Landé interval rule. In order to estimate the part played by this term in the fine-structure splitting of He and Li$^+$, we give the relative magnitude of the splitting of the ^3P terms

$$\frac{\Delta E_{J=2} - \Delta E_{J=1}}{\Delta E_{J=1} - \Delta E_{J=0}} = \xi \tag{5.194}$$

for the three cases: 1) perturbation H'_{s0}, 2) perturbation $H'_{s0} + H''_{s0}$, and 3) perturbation $H'_{s0} + H''_{s0} + H_{ss}$:

$$\text{He} \quad \xi_1 = 2, \quad \xi_2 = -2, \quad \xi_3 = -\frac{2}{35},$$

$$\text{Li}^+ \quad \xi_1 = 2, \quad \text{———}, \quad \xi_3 = -\frac{12}{30}. \tag{5.195}$$

The experimental values of ξ are

$$\xi(2p\ ^3P\ \text{He}) = 0.08,\ \xi(3p\ ^3P\ \text{He}) = 0.08,\ \xi(2p\ ^3P\ \text{Li}^+) = -0.41\ . \tag{5.196}$$

Thus, (5.192) correctly imparts the character of the splitting. For He, the distance between the components $J = 2$ and $J = 1$ is small in comparison with the distance between these components and the component $J = 0$. The incorrect mutual arrangement of the components $J = 1, 2$ must be attributed to simplifications made in the derivation of (5.192). Note that calculations including the exchange term and without neglecting r_1 in comparison with r_2 give the correct sign and somewhat improve the numerical value of ξ_3 [2]. Agreement with experiment is considerably better for Li$^+$.

Formula (5.192) shows that the spin-other orbit and spin-spin interaction are particularly important for light atoms. These interactions are proportional to Z^3 because the factor $\langle r^{-3} \rangle$ common to all three terms in (5.192) is proportional to Z^3, whereas $\langle H'_{s0} \rangle \propto Z^4$.

5.5 Multiplet Splitting in LS Coupling

For a multielectron atom, H_{s0}'' and H_{ss} contain terms of three types: interaction between electrons of filled shells, interaction of electrons of filled shells with electrons of unfilled shells, and interaction between electrons of unfilled shells. Only terms of the last type are involved in the splitting of terms. Thus, the term splitting for the configuration $nsn'l$ of an alkaline-earth atom is approximately described by (5.192), in which it is necessary to replace Z by the effective charge of the atomic core. For a sufficiently high value of this charge, the terms $\langle H_{s0}'''\rangle$ and $\langle H_{ss}'''\rangle$ are small in comparison with H_{s0}'. This circumstance justifies the approximation (5.134) when calculating fine-structure splitting. Let us note in conclusion that deviations from the Landé interval rule are not necessarily determined by the spin-spin interaction only. In cases when $\langle H_{ss}'\rangle \ll \langle H_{s0}'\rangle$, corrections of the second order in H_{s0}' can have a greater effect than $\langle H_{ss}\rangle$.

5.5.7 Spin-Spin and Spin-Other Orbit Interactions

The relative contribution of the interactions H_{s0}'' and H_{ss} to the splitting of terms of other multielectron atoms also decreases as the atomic number increases. This question has been treated in many papers. Calculations are easiest to carry out for l^n configurations since in this case there are no exchange terms and, in addition, one can express the matrix elements of H_{ss} and H_{s0}'' in terms of the reduced matrix elements of the operators V^{1k}.

We shall give the results of a calculation of the fine structure for terms of configurations p^n. For the configuration p^2, taking into account the correction in first-order perturbation theory from $H_{s0}'' + H_{ss}$ and the corrections in first- and second-orders perturbation theory from H_{s0}, one can obtain

$$(^3P_2) - (^3P_1) = (\zeta' - 55M_0) - 12M_0$$
$$-(\zeta' + 10M_0)^2 \frac{1}{2[(^1D) - (^3P)]},$$
$$(^3P_1) - (^3P_0) = \frac{1}{2}(\zeta' - 55M_0) \qquad (5.197)$$
$$+ 30M_0 + 2(\zeta' + 10M_0)^2 \frac{1}{(^1S) - (^3P)},$$
$$\zeta' = \zeta_p - 5M_0,$$

where M_0 is the radial integral in the matrix elements of H_{s0}'' and H_{ss}.

The splitting of terms of the configuration p^4 is determined by the same formulas if one replaces $(\zeta_p - 5M_0)$ by $-(\zeta_p - 25M_0)$. Comparison of (5.197) with experimental magnitudes of splitting enables one to determine the parameters ζ_p and M_0. Results are given in Table 5.21. It is noteworthy that with increasing Z the relative importance of the interactions H_{ss} and H_{s0}'' falls.

The quantities ζ_p and M_0, given in the table, fit with good accuracy the straight lines $\zeta^{1/4} \propto (Z - \sigma)$ and $M_0^{1/3} \propto (Z - \sigma')$, where σ and σ' are screening constants.

Table 5.21 Experimental values of the parameters ζ_p and M_0

$2p^2$	ζ_p [cm^{-1}]	M_0 [cm^{-1}]	$2p^4$	ζ_p [cm^{-1}]	M_0 [cm^{-1}]
C I	32.8	0.079	O I	146.4	0.274
N II	97.0	0.202	F II	320.0	0.471
O III	222.2	0.38	Ne III	606	0.92
F IV	436.0	0.61	Na IV	1039	1.22
Ne V	788.6	1.10	Mg V	1667	1.85
Na VI	1304	1.64	Al VI	2556	2.75
Mg VII	2054	2.57	Si VII	3743	3.57
Al VIII	3080	3.67			
Si IX	4368	3.93			

Additional data on the relative magnitude of the interactions under consideration can be given by the splitting of terms of the configuration p^3. In this case $\langle H'_{s0} \rangle = 0$; therefore, to the same approximation as in (5.197),

$$(^2D_{5/2}) - (^2D_{3/2}) = -\frac{185}{2} M_0 + \frac{5}{4} \frac{\zeta'^2}{(^2P) - (^2D)}, \tag{5.198}$$

$$(^2P_{3/2}) - (^2P_{1/2}) = -\frac{75}{2} M_0 + \zeta'^2 \left\{ \frac{5}{4[(^2P) - (^2D)]} + \frac{1}{(^2P) - (^4S)} \right\}. \tag{5.199}$$

If the first terms in (5.198, 199) are greater than the second terms, the splitting is inverted. But if the second terms play the major part, normal doublets must be observed. The experimental data given in Table 5.22 show that the splitting is inverted only for small values of Z. With increasing Z, the second-order correction from H'_{s0} exceeds $\langle H''_{s0} \rangle$ and $\langle H_{ss} \rangle$. In the configurations $3p^n$ the interactions H_{ss} and H''_{s0} play an even smaller part than in the configurations $2p^n$.

Table 5.22 Splitting of ^2D and ^2P terms of the configuration p^3

p^3		$(^2D_{5/2}) - (^2D_{3/2})$ [cm^{-1}]	$(^2P_{3/2}) - (^2P_{1/2})$ [cm^{-1}]
N	I	-8	0
O	II	-21	-1.5
F	III	-36	0
Ne	IV	-25	10
Na	V	-25	39
Mg	VI	-21	122
Al	VII	60	270
Si	VIII	280	580

A similar situation also occurs for the configuration $3d^n$. The second-order corrections from H'_{s0} lead to greater deviations from the Landé interval rule than the first-order corrections from H_{ss} and H''_{s0}.

5.6 jj Coupling

5.6.1 Wave Functions

In the jj coupling approximation, the electron in a central field is described by the wave function ψ_{nljm} and the system of electrons by the determinant (5.2), in which the set of quantum numbers $nljm$ is denoted by the letter a. For two electrons

$$\Psi = \frac{1}{\sqrt{2}} \left[\psi_{nljm}(\xi_1) \psi_{n'l'j'm'}(\xi_2) - \psi_{nljm}(\xi_2) \psi_{n'l'j'm'}(\xi_1) \right]. \tag{5.200}$$

The wave functions Ψ_{JM} describing the states of a system with assigned values of the total angular momentum J and its z components M can be constructed by the general rule for addition of angular momenta (4.18). Here one can use exactly the same methods as in constructing the functions $\Psi_{LSM_LM_S}$. Thus, for two electrons

$$\Psi_{JM}(j_1 j_2') = \sum_{mm'} C^J_{mm'} \psi_{jm}(\xi_1) \psi_{j'm'}(\xi_2), \tag{5.201}$$

$$\Psi_{JM}(j_2 j_1') = \sum_{mm'} C^J_{mm'} \psi_{jm}(\xi_2) \psi_{j'm'}(\xi_1), \tag{5.202}$$

$$\Psi_{JM} = \frac{1}{\sqrt{2}} \left[\Psi_{JM}(j_1 j_2') - \Psi_{JM}(j_2 j_1') \right]. \tag{5.203}$$

Using the symmetry properties of the Clebsch–Gordan coefficients (4.40)

$$(jj'mm' | jj'JM) = (-1)^{j+j'-J} (j'jm'm | j'jJM), \tag{5.204}$$

we obtain

$$\Psi_{JM} = \frac{1}{\sqrt{2}} \left[\Psi_{JM}(j_1 j_2') - (-1)^{j+j'-J} \Psi_{JM}(j_1' j_2) \right]. \tag{5.205}$$

For equivalent electrons $n = n'$, $l = l'$, and when $j = j'$

$$\Psi_{JM}(j_1 j_2') = \Psi_{JM}(j_1' j_2) = \Psi_{JM}(j_1 j_2);$$

therefore

$$\Psi_{JM} = \frac{1}{2} \left[\Psi_{JM}(j_1 j_2) - (-1)^{2j-J} \Psi_{JM}(j_1 j_2) \right]$$

$$= \frac{1 - (-1)^{2j-J}}{2} \Psi_{JM}(j_1 j_2). \tag{5.206}$$

In (5.206) it is taken into account that when $j = j'$ the normalization factor in (5.205) must be 1/2 and not $1/\sqrt{2}$. From (5.206) it follows that $\Psi_{JM} \neq 0$ for odd values $2j - J$.

Since $2j$ is odd and J is an integer,

$$\Psi_{JM} = \Psi_{JM}(j_1 j_2), \quad J \text{ even}$$
$$\Psi_{JM} = 0, \quad J \text{ odd}. \tag{5.207}$$

Relation (5.207) is in agreement with the table of allowed terms in jj coupling. In the case $n = n'$, $l = l'$, but $j = l \pm 1/2$ the wave function is defined by relation (5.205). This shows that by equivalent electrons in jj coupling one has to understand electrons with the same values of n, l, j.

In the parentage scheme approximation, the wave function of a system of electrons can be presented in a form analogous to (5.33),

$$\Psi_{JM}(J_1, j) = \frac{1}{\sqrt{N}} \sum_{i=1}^{N} (-1)^{N+i} \Psi_{JM}(J_1, j_i), \tag{5.208}$$

where

$$\Psi_{JM}(J_1, j_i) = \sum_{M_1 m} C^J_{M_1 m} \Psi_{J_1 M_1} \psi_{jm}(\xi_i). \tag{5.209}$$

In (5.208, 209), J_1 is the total angular momentum of the parent ion. The wave function of the parent ion $\Psi_{J_1 M_1}$ is antisymmetric with respect to electrons $1, 2, \ldots, i-1, i+1, \ldots, N$; therefore the linear combination (5.208) is antisymmetric with respect to all N electrons of the system.

For equivalent electrons, just as in the case of LS coupling, the parentage characteristic of terms does not make sense even in a first approximation. The wave functions $\Psi_{JM}(j^n)$, can be represented in the form of a linear combination of the functions $\Psi_{JM}(j^{n-1}[J_1]j)$ obtained by the addition of an electron with angular momentum j to the state J_1 of configuration j^{n-1} with the aid of the fractional parentage coefficients

$$(j^{n-1} [J_1] jJ \} j^n J) = G^J_{J_1}, \tag{5.210}$$

$$\Psi_{JM}(j^n) = \sum_{J_1} G^J_{J_1} \Psi_{JM}(j^{n-1} [J_1], j). \tag{5.211}$$

The coefficients $G^J_{J_1}$ are calculated by the same methods as the coefficients $G^{LS}_{L_1 S_1}$. We shall not dwell in detail on this question.

Among the terms of the configuration j^n, as a rule one encounters terms with one and the same values of J. By way of an additional quantum number, enabling one to distinguish identical terms, the seniority quantum number v can be introduced. Classification with respect to v is introduced in exactly the same way as in the case of the *LS* coupling. Identical terms of the configuration j^n are divided into two classes. The states JM of the first class can be obtained from states of the same type of configuration j^{n-2} by the addition of the closed pair $j^2[J=0]$. The states of the second class cannot be obtained by such means and in this sense appear first in the configuration j^n.

The quantum number v shows at what value $n=v$ the term $j^n vJ$ appears first. Thus, for the configuration j^3 the values $v=1$ are possible, for which $(j^2[0]jJ|j^3J) \neq 0$, and $v=3$, for which $(j^2[0]jJ|j^3J) = 0$ (Sect. 5.1).

5.6.2 Spin-Orbit and Electrostatic Interactions

In this case we need to treat first the spin-orbit interaction of electrons and then the electrostatic interaction. As before we shall start from (5.134) for spin-orbit interaction. In this approximation the correction to the energy of the level n_1l_1, n_2l_2 is the sum of the single-electron terms

$$\Delta E_{j_1 j_2 \ldots} = \sum_k \Delta E_{n_k l_k j_k}, \tag{5.212}$$

$$\Delta E_{n_k l_k j_k} = \frac{1}{2}\zeta_{n_k l_k}\left[j_k(j_k+1) - l_k(l_k+1) - \frac{3}{4}\right]. \tag{5.213}$$

Thus the spin-orbit splitting in the *jj* coupling scheme is determined directly by the single-electron parameters ζ_{nl}. The level $j_1 j_2 \ldots$ is degenerate with respect to J. For example, one value of energy corresponds to the states $(1/2\ 3/2)_1$ and $(1/2\ 3/2)_2$.

Degeneracy with respect to J is removed by the electrostatic interaction between the electrons. This splitting is calculated by the same methods as for *LS* coupling. We shall show this in the example of the splitting of the level $nljn'l'j'$ of a two-electron system. By substituting in the matrix element

$$\langle jj'\ JM \left| \frac{e^2}{r_{12}} \right| jj'JM \rangle \tag{5.214}$$

the wave functions (5.205), we obtain

$$\langle jj'JM \left| \frac{e^2}{r_{12}} \right| jj'JM \rangle$$

$$= \langle j_1 j'_2 JM \left| \frac{e^2}{r_{12}} \right| j_1 j'_2 JM \rangle - (-1)^{j+j'-J}\langle j_1 j'_2 JM \left| \frac{e^2}{r_{12}} \right| j'_1 j_2 JM \rangle, \tag{5.215}$$

$$\langle jj'JM \left| \frac{e^2}{r_{12}} \right| jj'JM \rangle = \sum_k (f_k F^k - g_k G^k), \tag{5.216}$$

$$f_k = \langle j_1 j'_2 JM | C_1^k C_2^k | j_1 j'_2 JM \rangle$$

$$= (-1)^{J+j'+J} (slj\|C^k\|slj)(sl'j'\|C^k\|sl'j') \begin{Bmatrix} j & j' & J \\ j' & j & k \end{Bmatrix}, \tag{5.217}$$

$$g_k = (-1)^{J+j'-J} \langle j_1 j'_2 JM | C_1^k C_2^k | j'_1 j_2 JM \rangle$$

$$= (-1)^{2J} (slj\|C^k\|sl'j')^2 \begin{Bmatrix} j & j' & J \\ j & j' & k \end{Bmatrix}. \tag{5.218}$$

The reduced matrix elements of C^k in (5.217, 218) are determined by (4.183, 184). It is evident from these formulas that the coefficients f_k do not depend on l and are uniquely defined by the magnitude jj'. These coefficients, in particular, are identical for the interaction of the electrons $np_{3/2}$; $n'p_{3/2}$ and $np_{3/2}$; $n'd_{3/2}$, $np_{3/2}$; $n'd_{5/2}$ and $nd_{3/2}$; $n'f_{5/2}$, and so on.

Also, l and l' do not enter explicitly into the formula for g_k. However, these coefficients depend indirectly on l, l' since different expressions for the reduced matrix elements of C^k correspond to the two possible cases $j = l \pm 1/2$, $j' = l' \pm 1/2$ and $j = l \pm 1/2$, $j' = l' \mp 1/2$.

For equivalent electrons

$$\langle j^2 JM \left| \frac{e^2}{r_{12}} \right| j^2 JM \rangle = \sum_k f_k F^k, \tag{5.219}$$

$$f_k = (-1)^{2J-J} (slj\|C^k\|slj)^2 \begin{Bmatrix} j & j & J \\ j & j & k \end{Bmatrix}. \tag{5.220}$$

For the case $n = n'$, $l = l'$, but $j \neq j'$ ($j = l \pm 1/2$; $j' = l \mp 1/2$), we have $F^k = G^k$; therefore

$$\langle jj' JM \left| \frac{e^2}{r_{12}} \right| jj' JM \rangle = \sum_k (f_k - g_k) F^k. \tag{5.221}$$

Formulas (5.216–221) enable one to express the electrostatic splitting for any two-electron configuration in terms of the Slater parameters F^k, G^k. In calculating multielectron configurations, the same methods are used as in the case of the LS coupling.

5.7 Intermediate Coupling and Other Types of Coupling

5.7.1 Transformations Between LS and jj Coupling Schemes

The wave functions Ψ_{SLJM} and $\Psi_{jj'JM}$ correspond to the following two schemes of addition of angular momenta

$$s + s' = S, \quad l + l' = L, \tag{5.222}$$

5.7 Intermediate Coupling and Other Types of Coupling

$$S + L = J, \tag{5.223}$$

$$s + l = j, \quad s' + l' = j', \tag{5.224}$$

$$j + j' = J. \tag{5.225}$$

Therefore

$$\Psi_{jj'JM} = \sum_{L,S} (ss'\,[S],\,ll'[L]\,J|sl[j],\,s'l'[j']\,J)\,\Psi_{SLM_SM_L}. \tag{5.226}$$

In the expansion (5.226), all terms are presented for which

$$L + S \geqslant J \geqslant |L - S|.$$

For example, in the case of the configuration $np\,n'p$ the wave function

$$\Psi_{j=3/2;\,j'=1/2;\,J=2;\,M}$$

can be represented by an expansion in terms of the functions

$$\Psi(^1D_2), \quad \Psi(^3D_2), \quad \Psi(^3P_2).$$

The transition from LS coupling to jj coupling is a change in the scheme of addition of four angular momenta. So the transformation coefficients in (5.226) are expressed in terms of $9j$ symbols. From (4.103) it follows that

$$(ss'\,[S];\,ll'[L]\,J|sl[j];\,s'l'[j']\,J)$$
$$= (-1)^{S+L-J+s+l-j+s'+l'-j'}\,(ll'\,[L]\,ss'\,[S]\,J|ls\,[j]\,l's'\,[j']\,J)$$
$$= \sqrt{(2S+1)(2L+1)(2j+1)(2j'+1)} \begin{Bmatrix} l & l' & L \\ j & j' & J \\ \frac{1}{2} & \frac{1}{2} & S \end{Bmatrix}. \tag{5.227}$$

The $9j$ symbols in (5.227) can be calculated in explicit form. The values of the factor

$$\begin{Bmatrix} l & l' & L \\ j & j' & J \\ \frac{1}{2} & \frac{1}{2} & S \end{Bmatrix} = A\,(SLJ;\,jj'J) \tag{5.228}$$

are given in Table 5.23 [17].

Table 5.23

l	l'	$S=0,\quad L=J$
		$A(SLJ;jj'J)$
$j+\tfrac{1}{2}$	$j'+\tfrac{1}{2}$	$\left[\dfrac{(j+j'+J+2)(j+j'-J+1)}{2(2j+1)(2j+2)(2j'+1)(2j'+2)(2J+1)}\right]^{1/2}$
$j+\tfrac{1}{2}$	$j'-\tfrac{1}{2}$	$\left[\dfrac{(j-j'+J+1)(-j+j'+J)}{2(2j+1)(2j+2)(2j')(2j'+1)(2J+1)}\right]^{1/2}$
$j-\tfrac{1}{2}$	$j'+\tfrac{1}{2}$	$(-1)\left[\dfrac{(j-j'+J)(-j+j'+J+1)}{2(2j)(2j+1)(2j'+1)(2j'+2)(2J+1)}\right]^{1/2}$
$j-\tfrac{1}{2}$	$j'-\tfrac{1}{2}$	$\left[\dfrac{(j+j'+J+1)(j+j'-J)}{2(2j)(2j+1)(2j')(2j'+1)(2J+1)}\right]^{1/2}$

l	l'	$S=1,\; L=J+1$
		$A(SLJ;jj'J)$
$j+\tfrac{1}{2}$	$j'+\tfrac{1}{2}$	$\left[\dfrac{(j+j'+J+2)(j+j'+J+3)(j-j'+J+1)(-j+j'+J+1)}{3(2j+1)(2j+2)(2j'+1)(2j'+2)(2J+1)(2J+2)(2J+3)}\right]^{1/2}$
$j+\tfrac{1}{2}$	$j'-\tfrac{1}{2}$	$\left[\dfrac{(j+j'+J+2)(j+j'-J)(j-j'+J+1)(j-j'+J+2)}{3(2j+1)(2j+2)(2j')(2j'+1)(2J+1)(2J+2)(2J+3)}\right]^{1/2}$
$j-\tfrac{1}{2}$	$j'+\tfrac{1}{2}$	$\left[\dfrac{(j+j'+J+2)(j+j'-J)(-j+j'+J+1)(-j+j'+J+2)}{3(2j)(2j+1)(2j'+1)(2j'+2)(2J+1)(2J+2)(2J+3)}\right]^{1/2}$
$j-\tfrac{1}{2}$	$j'-\tfrac{1}{2}$	$(-1)\left[\dfrac{(j+j'-J-1)(j+j'-J)(j-j'+J+1)(-j+j'+J+1)}{3(2j)(2j+1)(2j')(2j'+1)(2J+1)(2J+2)(2J+3)}\right]^{1/2}$

l	l'		$S=1,\quad L=J$
			$A(SLJ;jj'J)$
$j+\tfrac{1}{2}$	$j'+\tfrac{1}{2}$	$(j-j')$	$\left[\dfrac{(j+j'+J+2)(j+j'-J+1)}{6(2j+1)(2j+2)(2j'+1)(2j'+2)J(J+1)(2J+1)}\right]^{1/2}$
$j+\tfrac{1}{2}$	$j'-\tfrac{1}{2}$	$(j+j'+1)$	$\left[\dfrac{(j-j'+J+1)(-j+j'+J)}{6(2j+1)(2j+2)(2j')(2j'+1)(J(J+1)(2J+1)}\right]^{1/2}$
$j-\tfrac{1}{2}$	$j'+\tfrac{1}{2}$	$(j+j'+1)$	$\left[\dfrac{(j-j'+J)(-j+j'+J+1)}{6(2j)(2j+1)(2j'+1)(2j'+2)J(J+1)(2J+1)}\right]^{1/2}$
$j-\tfrac{1}{2}$	$j'-\tfrac{1}{2}$	$(-j+j')$	$\left[\dfrac{(j+j'+J+1)(j+j'-J)}{6(2j)(2j+1)(2j')(2j'+1)J(J+1)(2J+1)}\right]^{1/2}$

l	l'		$S=1,\; L=J-1$
			$A(SLJ;jj'\,J)$
$j+\tfrac{1}{2}$	$j'+\tfrac{1}{2}$	(-1)	$\left[\dfrac{(j+j'-J+1)(j+j'-J+2)(j-j'+J)(-j+j'+J)}{3(2j+1)(2j+2)(2j'+1)(2j'+2)(2J-1)(2J)(2J+1)}\right]^{1/2}$
$j+\tfrac{1}{2}$	$j'-\tfrac{1}{2}$	(-1)	$\left[\dfrac{(j+j'+J+1)(j+j'-J+1)(-j+j'+J-1)(-j+j'+J)}{3(2j+1)(2j+2)(2j')(2j'+1)(2J-1)(2J)(2J+1)}\right]^{1/2}$
$j-\tfrac{1}{2}$	$j'+\tfrac{1}{2}$		$\left[\dfrac{(j+j'+J+1)(j+j'-J+1)(j-j'+J-1)(j-j'+J)}{3(2j)(2j+1)(2j'+1)(2j'+2)(2J-1)(2J)(2J+1)}\right]^{1/2}$
$j-\tfrac{1}{2}$	$j'-\tfrac{1}{2}$		$\left[\dfrac{(j+j'+J)(j+j'+J+1)(j-j'+J)(-j+j'+J)}{3(2j)(2j+1)(2j')(2j'+1)(2J-1)(2J)(2J+1)}\right]^{1/2}$

Transformations between LS and jj coupling schemes in the case of equivalent electrons require special consideration. When $j=j'$

$$\Psi_{JM}(j^2)=\sum_{LS}(l^2SLJ|j^2J)\,\Psi_{JM}(l^2LS)\,. \tag{5.229}$$

When $j \neq j'$ ($j = l \pm 1/2$; $j' = l \mp 1/2$)

$$\Psi_{JM}(jj') = \sum_{LS} (l^2 SLJ|jj'J) \Psi_{JM}(l^2 SL). \tag{5.230}$$

Using the expressions given above for the functions $\Psi_{JM}(j^2)$, $\Psi_{JM}(jj')$, $\Psi_{JM}(l^2 SL)$, and also the symmetry properties of $9j$ symbols, it is not difficult to obtain

$$(l^2 SLJ|j^2 J) = (ss\,[S],\,ll[L]J|sl[j]\,sl[j]\,J), \tag{5.231}$$

$$(l^2 SLJ|jj'J) = \sqrt{2}\,(ss\,[S];\,ll[L]\,J|sl[j]\,sl[j']\,J). \tag{5.232}$$

Formulas (5.227–232) enable one to represent the functions $\Psi_{jj'JM}$ in the form of linear combinations of functions Ψ_{SLJM} for any two-electron configuration.

5.7.2 Intermediate Coupling

If the electrostatic interaction between electrons U and the spin-orbit interaction W are of the same order of magnitude, then both the LS coupling approximation and the jj coupling approximation are inapplicable. Such cases are spoken of as intermediate coupling. The qualitative picture of the level scheme in intermediate coupling can be obtained by comparing the schemes of levels in the two limiting cases of LS and jj coupling. In a quantitative treatment of intermediate coupling, to determine the energy it is necessary to solve a secular equation composed of the matrix elements of the perturbation $U + W$. In conducting such calculations it is convenient to use the fact that one can choose as zeroth approximation functions the central field functions $\Psi_{m\mu m'\mu'}$ or any independent linear combinations of these functions. In particular, one can proceed from the functions Ψ_{SLJM}. In this case the matrix of the electrostatic interaction U is diagonal with respect to $SLJM$, which significantly simplifies calculations. Since the matrix of W is also diagonal with respect to JM (but not with respect to SL), the secular equation corresponding to specific values of JM has the form

$$\begin{vmatrix} \langle L_1 S_1 JM|U + W|L_1 S_1 JM\rangle - \varepsilon; & \langle L_1 S_1 JM|W|L_2 S_2 JM\rangle; & \ldots \\ \langle L_2 S_2 JM|W|L_1 S_1 JM\rangle; & \langle L_2 S_2 JM|U + W|L_2 S_2 JM\rangle - \varepsilon; & \ldots \\ \ldots \ldots \ldots \ldots \ldots \ldots & & \end{vmatrix} = 0.$$

$$\tag{5.233}$$

The roots of the secular equation (5.233), $\varepsilon_1, \varepsilon_2, \ldots, \varepsilon_k, \ldots, \varepsilon_f$, are the required corrections to the energy. Having solved the secular equation, it is possible to find the eigenfunctions $\Psi_{JM}^{(k)}$. Let us consider, as an example, the configuration p^2. The electrostatic splitting in LS coupling is defined by (5.98). Fine splitting is easy to obtain from (5.159) and Table 5.18. Thus

$$
\begin{aligned}
(^1S_0) &= F_0 + 10F_2, \quad F_2 = F^2/25, \\
(^1D_2) &= F_0 + F_2, \\
(^3P_0) &= F_0 - 5F_2 - \zeta_{np}, \\
(^3P_1) &= F_0 - 5F_2 - \tfrac{1}{2}\zeta_{np}, \quad (^3P_2) = F_0 - 5F_2 + \tfrac{1}{2}\zeta_{np}.
\end{aligned}
\tag{5.234}
$$

In the other limiting case, the jj coupling approximation, it follows from (5.212, 213, 219, 220) that

$$
\begin{aligned}
\left(\tfrac{3}{2}\tfrac{3}{2}\right)_0 &= \zeta_{np} + F_0 + 5F_2, \quad \left(\tfrac{3}{2}\tfrac{1}{2}\right)_2 = -\tfrac{1}{2}\zeta_{np} + F_0 - F_2, \\
\left(\tfrac{3}{2}\tfrac{3}{2}\right)_2 &= \zeta_{np} + F_0 - 3F_2, \quad \left(\tfrac{3}{2}\tfrac{1}{2}\right)_1 = -\tfrac{1}{2}\zeta_{np} + F_0 - 5F_2, \\
\left(\tfrac{1}{2}\tfrac{1}{2}\right)_0 &= -2\zeta_{np} + F_0.
\end{aligned}
\tag{5.235}
$$

To obtain the secular equation (5.233), it is necessary to calculate the matrix of the spin-orbit interaction. In the case which interests us, (5.160) gives

$$
\langle p^2 SLJM | \zeta_{np} \sum_i \mathbf{l}_i \cdot \mathbf{s}_i | p^2 S'L'JM \rangle
$$

$$
= (-1)^{S'+L+J}\zeta_{np}\sqrt{6}(p^2SL\|V^{11}\|p^2S'L')\begin{Bmatrix} S & L & J \\ L' & S' & 1 \end{Bmatrix}.
\tag{5.236}
$$

Substituting the values of the reduced matrix elements of V^{11} from Table 5.18, we obtain

	$^1S_0$	$^3P_0$	$^3P_1$	$^3P_2$	$^1D_2$
$^1S_0$		$-\sqrt{2}\,\zeta_{np}$			
$^3P_0$	$-\sqrt{2}\,\zeta_{np}$	$-\zeta_{np}$			
$^3P_1$			$-\tfrac{1}{2}\zeta_{np}$		
$^3P_2$				$\tfrac{1}{2}\zeta_{np}$	$\tfrac{1}{\sqrt{2}}\zeta_{np}$
$^1D_2$				$\tfrac{1}{\sqrt{2}}\zeta_{np}$	

(5.237)

5.7 Intermediate Coupling and Other Types of Coupling

In accordance with (5.237), the secular equation (5.233) has the following form

$$J = 0 \quad \begin{vmatrix} F_0 + 10F_2 - \varepsilon & -\sqrt{2}\,\zeta_{np} \\ -\sqrt{2}\,\zeta_{np} & F_0 - 5F_2 - \zeta_{np} - \varepsilon \end{vmatrix} = 0, \quad (5.238)$$

$$J = 1 \quad F_0 - 5F_2 - \frac{1}{2}\zeta_{np} - \varepsilon = 0. \quad (5.239)$$

$$J = 2 \quad \begin{vmatrix} F_0 - 5F_2 + \frac{1}{2}\zeta_{np} - \varepsilon & \frac{1}{\sqrt{2}}\zeta_{np} \\ \frac{1}{\sqrt{2}}\zeta_{np} & F_0 + F_2 - \varepsilon \end{vmatrix} = 0. \quad (5.240)$$

From (5.238–240) it follows that

$$\left.\begin{matrix}\varepsilon_1 \\ \varepsilon_2\end{matrix}\right\} = \left(F_0 + \frac{5}{2}F_2 - \frac{1}{2}\zeta_{np}\right) \pm \sqrt{\frac{225}{4}F_2^2 + \frac{15}{2}F_2\zeta_{np} + \frac{9}{4}\zeta_{np}^2}, \quad (5.241)$$

$$\varepsilon_3 = F_0 - 5F_2 - \frac{1}{2}\zeta_{np}, \quad (5.242)$$

$$\left.\begin{matrix}\varepsilon_4 \\ \varepsilon_5\end{matrix}\right\} = \left(F_0 - 2F_2 + \frac{1}{2}\zeta_{np}\right) \mp \sqrt{9F_2^2 - \frac{3}{2}F_2\zeta_{np} + \frac{9}{16}\zeta_{np}^2}. \quad (5.243)$$

If $F_2 \gg \zeta_{np}$

$$\left.\begin{matrix}\varepsilon_1 \\ \varepsilon_2\end{matrix}\right\} \approx F_0 + \frac{5}{2}F_2 - \frac{1}{2}\zeta_{np} \pm \frac{15}{2}F_2\sqrt{1 + \frac{2}{15}\frac{\zeta_{np}}{F_2}} \approx \begin{cases} F_0 + 10F_2, \\ F_0 - 5F_2 - \zeta_{np}, \end{cases} \quad (5.244)$$

$$\left.\begin{matrix}\varepsilon_4 \\ \varepsilon_5\end{matrix}\right\} \approx F_0 - 2F_2 + \frac{1}{4}\zeta_{np} \mp 3F_2\sqrt{1 - \frac{1}{6}\frac{\zeta_{np}}{F_2}} \approx \begin{cases} F_0 - 5F_2 + \frac{1}{2}\zeta_{np}, \\ F_0 + F_2. \end{cases} \quad (5.245)$$

Therefore, in the limit of weak spin-orbit interaction, we obtain the *LS* coupling approximation

$$\varepsilon_1 \to (^1S_0), \quad \varepsilon_2 \to (^3P_0), \quad \varepsilon_3 \to (^3P_1), \quad \varepsilon_4 \to (^3P_2), \quad \varepsilon_5 \to (^1D_2). \quad (5.246)$$

The relations (5.246) establish a unique correspondence between the levels ε_k and the levels of *LS* coupling approximation. This enables us to use the *LS* coupling terminology in cases when the *LS* coupling approximation itself loses its meaning. Thus the levels ε_1, ε_2 are often denoted in terms of $^1S_0'$, $^3P_0'$, and so on. The wave functions corresponding to these levels are related to the functions Ψ_{SLJM} as follows:

$$\begin{aligned}
\Psi(^1S_0') &= c_{11}\,\Psi(^1S_0) + c_{12}\,\Psi(^3P_0)\,,\\
\Psi(^3P_0') &= c_{21}\,\Psi(^1S_0) + c_{22}\,\Psi(^3P_0)\,,\\
\Psi(^3P_1') &= \Psi(^3P_1)\,,\\
\Psi(^3P_2') &= b_{11}\,\Psi(^3P_2) + b_{12}\,\Psi(^1D_2)\,,\\
\Psi(^1D_2') &= b_{21}\,\Psi(^3P_2) + b_{22}\,\Psi(^1D_2)\,.
\end{aligned} \qquad (5.247)$$

The coefficients on the right of (5.247) are determined together with the energy corrections. Comparison of (5.246) and (5.247) shows that in the limiting case of small spin-orbit interaction

$$c_{11},\ c_{22} \to 1;\quad c_{12},\ c_{21} \to 0 \quad b_{11}, b_{22} \to 1,\quad b_{12}, b_{21} \to 0.$$

Formulas (5.247) show that in the presence of spin-orbit interaction it is impossible to characterize the states of an atom by specific values of *L* and *S*. Orbital angular momentum and spin are not conserved separately. Thus the state $^1S_0'$ is the superposition of the singlet state with $L = 0$ and the triplet state with $L = 1$. To characterize the relative magnitudes of the electrostatic and spin-orbit interaction it is convenient to introduce the dimensionless parameter $\chi = \tfrac{1}{5}\cdot\zeta_{np}/F_2$. Small deviations from *LS* coupling correspond to values $\chi \ll 1$. When $\chi \gg 1$, there is a transition to *jj* coupling. In fact, expanding the roots (5.241, 243) in powers of $1/\chi$, it is easy to obtain (5.235) and show that

$$\begin{aligned}
(^1S_0') &\to \left(\tfrac{3}{2}\ \tfrac{3}{2}\right)_0,\quad (^3P_2') \to \left(\tfrac{3}{2}\ \tfrac{1}{2}\right)_2,\\
(^3P_0') &\to \left(\tfrac{1}{2}\ \tfrac{1}{2}\right)_0,\quad (^1D_2') \to \left(\tfrac{3}{2}\ \tfrac{3}{2}\right)_2,\\
(^3P_1') &\to \left(\tfrac{3}{2}\ \tfrac{1}{2}\right)_1.
\end{aligned} \qquad (5.248)$$

The full picture of the transition from *LS* coupling to *jj* coupling is shown in Fig. 5.1.

For small deviations from *LS* coupling ($\chi \ll 1$), the coefficients in the wave functions (5.247) can be represented in the form of an expansion in powers of χ

$$\begin{aligned}
c_{11} = c_{22} &= 1 - \tfrac{1}{9}\chi^2 + \tfrac{2}{27}\chi^3 + \ldots,\\
c_{21} = c_{12} &= \tfrac{1}{3}\sqrt{2}\,\chi\left(1 - \tfrac{1}{3}\chi - \tfrac{2}{9}\chi^2 + \ldots\right),
\end{aligned} \qquad (5.249)$$

5.7 Intermediate Coupling and Other Types of Coupling

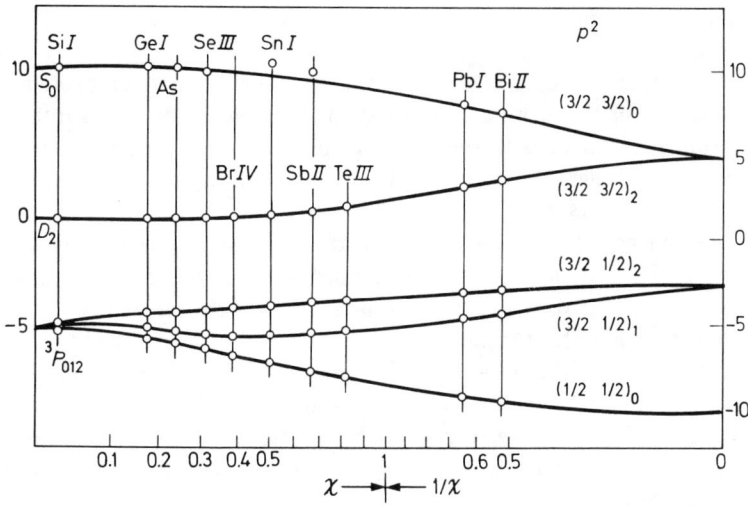

Fig. 5.1. Transition from LS coupling to jj coupling for the configuration p^2

$$b_{11} = b_{22} = 1 - \frac{25}{144}\chi^2 - \frac{125}{864}\chi^3 + \cdots ,$$

$$b_{21} = b_{12} = \frac{5}{12}\sqrt{2}\chi\left(1 + \frac{5}{12}\chi - \frac{25}{72}\chi^2 + \cdots\right).$$

It has been shown above that in the LS coupling approximation one can obtain for the relative distances between terms a series of relations which do not depend on the parameters F^k and G^k. Similarly, for some configurations in intermediate coupling it is possible to exclude the parameters F^k, G^k, and ζ_{np} and express the relative distances between levels in terms of dimensionless parameters describing the relative magnitudes of the electrostatic and spin-orbit interaction. In the case of the configuration p^2 considered above, and also for the configurations p^3 and p^4, such a parameter is χ. Knowing from experiment the relative arrangement of levels of a given atom, one can determine the magnitude of χ and thereby give a quantitative estimate of the deviations from SL coupling (or jj coupling). At the same time it is possible to determine the coefficients in the expansion of the wave functions in intermediate coupling with respect to the functions Ψ_{LSJM}. This is of great importance for a number of applications. The deviation from LS coupling is also described by the magnitude of the off-diagonal matrix elements $\langle S_1L_1JM|W|S_2L_2JM\rangle$ connecting the terms L_1S_1 and L_2S_2.

5.7.3 jl Coupling

jl coupling is realized, as a rule, in cases when the optical electron is on average far from the electrons of the atomic core. In such cases the electrostatic interaction of the optical electron with the electrons of the atomic core is usually small in comparison with the spin-orbit interaction of the electrons of the atomic core. Such a situation occurs in atoms of noble gases (Sect. 3.4).

In the *jl* coupling approximation the levels are described by the quantum numbers $\gamma S_1 L_1 j, l[K]J$. Obviously, such a description makes sense only in the case when the distances between the two components of the level $jlKJ = K \pm 1/2$ is considerably less than the distances between the different K levels. The conditions for this are that both the spin-orbit interaction of the optical electron and also exchange terms in the electrostatic interaction are small. The condition on the magnitude of the exchange terms arises because the exchange interaction depends on the mutual orientation of the angular momentum K and spin of the optical electron.

This circumstance allows one to omit the exchange terms in calculating electrostatic splitting of the levels $S_1 L_1 jlK$ and $S_1 L_1 jlK'$. Therefore, for the two-electron configuration ll'

$$\langle slj, l'K \left| \frac{e^2}{r_{12}} \right| slj, l'K \rangle \approx \sum_k f_k F^k , \qquad (5.250)$$

$$f_k = \langle s_1 l_1 j_1, l'_2 K | C_1^k C_2^k | s_1 l_1 j_1, l'_2 K \rangle$$
$$= (-1)^{J+l'+K} (slj\|C^k\|slj)(l'\|C^k\|l') \begin{Bmatrix} j & l' & K \\ l' & j & k \end{Bmatrix} . \qquad (5.251)$$

Formula (5.251) includes the very important case of the noble gases. In this case the interaction of the l' electron with the almost filled shell p^5 has the form

$$\langle W(p^5 l') \rangle = F_0 + f_2 F^2 , \qquad (5.252)$$

where, in accordance with the general rule established in Sect. 5.4,

$$f_2(p^5 l') = - f_2(pl') . \qquad (5.253)$$

In calculating the spin-orbit splitting of the level $\gamma S_1 L_1 jlK$, one can proceed from (5.161) for the spin-orbit interaction operator. In the given case the average of the first term in (5.161) over the state $\gamma S_1 L_1 jlK$

$$A_1 \frac{1}{2}[j(j+1) - L_1(L_1+1) - S_1(S_1+1)]$$

does not depend on the orientation of the angular momentum j with respect to the angular momenta l, K, s, and this term can therefore be omitted. Thus

$$\langle \gamma_1 S_1 L_1 jl\, [K]\, sJM | W | \gamma_1 S_1 L_1 jl\, [K]\, sJM \rangle$$
$$= \zeta_{nl} \langle \gamma jl_N\, [K]\, s_N JM | l_N s_N | \gamma jl_N [K]\, s_N JM \rangle$$
$$= \zeta_{nl} (-1)^{K+S+J} (jlK\|l\|jlK)(s\|s\|s) \begin{Bmatrix} K & s & J \\ s & K & 1 \end{Bmatrix}. \tag{5.254}$$

Using (4.178), we obtain

$$\langle \gamma jl\, [K]\, sJM | W | \gamma jl\, [K]\, sJM \rangle$$
$$= \zeta_{nl} \frac{K(K+1) + l(l+1) - j(j+1)}{2K(K+1)} \frac{1}{2} \left[J(J+1) - K(K+1) - \frac{3}{4} \right]. \tag{5.255}$$

Generally speaking, the spin-orbit interaction has to be taken into account together with the exchange part of the electrostatic interaction. We do not need such calculations anywhere below, however, and therefore we shall not discuss this question in any more detail. Let us consider in conclusion the transition from LS and jj coupling to jl coupling. The transformation from jj coupling to jl coupling is particularly simple because in this transformation it is sufficient to change the order of addition of the three angular momenta. In the transition from LS coupling to jl coupling it is necessary to change the order of addition of the three angular momenta twice. Using the general formulas of Sect. 4.2, we obtain

$$(j, l'sj'J|jl'[K]\,sJ) = (-1)^{j'-s-l'} (j, sl'J|jl'[K]\,sJ)$$
$$= (-1)^{s+l'+j+J} \sqrt{(2j'+1)(2K+1)} \begin{Bmatrix} s & l' & j' \\ j & J & K \end{Bmatrix}, \tag{5.256}$$

$$(S_1 s\, [S]\, L_1 l'\, [L]\, J | S_1 L_1 jl'\, [K]\,sJ)$$
$$= (-1)^{L+S-J+L_1+S_1-j} (L_1 l'\, [L]\, S_1 s[S]\, J | L_1 S_1 jl'\, [K]\, sJ)$$
$$= (-1)^{S_1+s+S+L_1+l'+L+2K} \sqrt{(2j+1)(2L+1)(2S+1)(2K+1)}$$
$$\times \begin{Bmatrix} s & L_1 & j \\ l' & K & L \end{Bmatrix} \begin{Bmatrix} S & L & J \\ K & s & S_1 \end{Bmatrix}. \tag{5.257}$$

5.7.4 Experimental Data

A qualitative suggestion of how well a system of levels corresponds to the LS coupling approximation can be obtained by comparing the magnitude of the fine-structure splitting of terms with the energy differences between the terms. Such a comparison is possible, of course, only in a case when deviations from LS coupling are small. To obtain any quantitative measure of the type of coupling, it is necessary to solve the secular equation for the electrostatic and spin-orbit interactions. Such calculation has been carried out above for the configuration p^2. In this case the relative magnitude of the electrostatic and spin-orbit interaction is described by one dimensionless parameter $\chi = \zeta_{nl}/5F_2$. For LS coupling $\chi \to 0$; for jj coupling $\chi \to \infty$. Comparison of experimental values of energy levels with calculated values enables one to define the parameter χ and thereby to give a quantitative description of the type of coupling.

The problem has been investigated in detail for a series of atoms and ions with the ground state configurations p^2, p^3, and p^4 [18]. Calculated values of the splitting of levels as a function of the parameter χ together with experimental values are given in Fig. 5.1. These data show that the parameter χ increases monotonically with increase of Z. The LS coupling scheme provides a sufficiently good approximation for an atoms at the beginning of the periodic system. For heavy atoms, such as Pb and Bi, deviations from LS coupling are so great that the classification of levels in terms of LS coupling becomes questionable. The same type of regularity is also observed for atoms of other isoelectronic sequences [1]. The larger Z is the more LS coupling is violated. A similar situation occurs for elements of the transition groups.

The LS coupling approximation is good enough for atoms of the iron group. For atoms of the palladium group, deviations from LS coupling increase but not so much as to make this approximation inapplicable. For atoms of the platinum group, the spin-orbit interaction is so large that intermediate coupling occurs.

Atoms of the noble gases and rare earths occupy a special place. In the first case we have jl coupling. As already noted above, this type of coupling is also characteristic of highly excited states of a number of other atoms. In the case of rare earths, the situation is more complicated. A number of cases are known in which levels of the configurations $f^n l$ and $f^n ll'$ fit well into the scheme of $J_\mathrm{I} j$ and $J_\mathrm{I} J_\mathrm{II}$ couplings, where J_I is the total angular momentum of the group f^n and J_II the total angular momentum of the group ll'.

5.7.5 Other Types of Coupling

A number of other types of coupling are possible besides the types LS, jj, jl, and Jj considered above. Let us consider as an example electron configurations containing one strongly excited electron $n'l'$. The distance of this electron from the electrons of the atomic core is on average much greater than the interelectron distance in the atomic core. Let LS coupling occur for the atomic core. We shall denote the total spin and the total orbital angular momentum of the atomic core by S_0 and L_0. The character of the coupling of the excited electron to the core in this case is determined by the relative magnitude of the spin-orbit interaction of the electrons of the core W^0; the Coulomb and exchange interactions of electron l'

5.7 Intermediate Coupling and Other Types of Coupling

with the core, H' and H'_{exch}; and spin-orbit interaction $W_{l'}$ for electron l'. The following types of couplings are possible

LS: $S_0 s\,[S]\,L_0 l'\,[L]\,J$, $H'; H'_{\text{exch}} \gg W^0, W_{l'}$

LS_0: $S_0 L_0 l'\,[L]\,KsJ$, $H' \gg W^0 \gg H'_{\text{exch}}, W_{l'}$,

Jl: $S_0 L_0\,[J_0]\,l'\,[K]\,sJ$, $W^0 \gg H' \gg H'_{\text{exch}}, W_{l'}$

Jj: $S_0 L_0\,[J_0]\,l's\,[j']\,J$, $W^0 \gg H', W_{l'} \gg H', H'_{\text{exch}}$.

If jj coupling occurs for the core, then two types of coupling of electron l' to the core are possible:

$J_0 l$: $J_0 l'\,[K]\,sJ$, $H' \gg W_{l'}; H'_{\text{exch}}$.

jj: $J_0 l's'\,[j']\,J$, $W_{l'} \gg H', H'_{\text{exch}}$.

In addition to the types of coupling considered above, LS_0 and $J_0 l$ coupling can be observed in a number of spectra. Thus, for example, levels of the configuration $2s2p4f$ C II fit well into the LS_0 coupling scheme. Couplings of the type LS_0, Jl, Jj, $J_0 l$, and so on, are often called nonhomogeneous couplings.

Chapter 6 Hyperfine Structure of Spectral Lines

In this chapter the theory of hyperfine splitting caused by nuclear magnetic dipole and electric quadrupole moments is considered. Isotopic shift of spectral lines is also discussed.

6.1 Nuclear Magnetic Dipole and Electric Quadrupole Moments

6.1.1 Magnetic Moments

Nuclear magnetic moments are usually expressed in nuclear magnetons, i.e., in the units

$$\frac{e\hbar}{2m_p c} = \left(\frac{m}{m_p}\right)\mu_0 , \qquad (6.1)$$

where m_p is the mass of proton, μ_0 is the Bohr magneton. In these units we have for the magnetic moment of a proton

$$\mu = g_l l + g_s s: \quad g_l = 1 , \quad g_s = 5.58 ; \qquad (6.2)$$

and for a neutron

$$\mu = g_l l + g_s s; \quad g_l = 0, \quad g_s = -3.82 . \qquad (6.3)$$

Here the different factors g are called the gyromagnetic ratios. The magnetic moment of the nucleus is also given by the gyromagnetic ratio

$$\mu = g_I I , \qquad (6.4)$$

where I is the nuclear spin. By magnetic moment of the nucleus is usually understood the maximum projection of the magnetic moment on the direction of a field

$$\mu = g_I I . \qquad (6.5)$$

This is the quantity which is usually tabulated.

6.1.2 Quadrupole Moments

The electric quadrupole moments $Q_{\alpha\beta}$ are the second important characteristics of the structure of a nucleus. Usually the quadrupole moment tensor is defined by the relation

$$Q_{\alpha\beta} = \int \rho \left(3 r_\alpha r_\beta - \delta_{\alpha\beta} r^2\right) dr. \tag{6.6}$$

According to this definition the operator of the proton quadrupole moment (neutrons obviously do not contribute to electric moments) has the form

$$Q_{\alpha\beta} = e \left(3 r_\alpha r_\beta - \delta_{\alpha\beta} r^2\right). \tag{6.7}$$

In nuclear physics, however, it is standard to omit the charge, and to measure the quadrupole moments in barns (10^{-24} cm^2). Thus for a nucleus

$$Q_{\alpha\beta} = \sum \left(3 r_\alpha r_\beta - \delta_{\alpha\beta} r^2\right). \tag{6.8}$$

Summation in (6.8) is carried out over all protons of the nucleus. It is conventional to describe the magnitude of the quadrupole moment by the mean value of the component Q_{zz} in the state I, $M = I$. This quantity is denoted by

$$Q = \langle \gamma I M | Q_{zz} | \gamma I M \rangle_{M=I}. \tag{6.9}$$

Calculations of (6.9) are significantly simplified in spherical coordinates if the quadrupole moment is defined by the relation

$$Q_{2q} = r^2 C_q^2(\theta, \varphi). \tag{6.10}$$

Taking into account that $Q_{zz} = 2 Q_{20}$, we obtain

$$Q = 2 \langle \gamma II | Q_{20} | \gamma II \rangle = 2 (\gamma I \| Q \| \gamma I) \begin{pmatrix} I & 2 & I \\ -I & 0 & I \end{pmatrix}$$

$$= 2 (\gamma I \| Q_2 \| \gamma I) \sqrt{\frac{I(2I-1)}{(2I+3)(2I+1)(I+1)}}. \tag{6.11}$$

Thus the quadrupole moment is equal to zero in the states $I = 0$ and $I = 1/2$. It also follows from (4.120, 6.11) that in the state $M \neq I$

$$\langle \gamma IM | Q_{zz} | \gamma IM \rangle = 2 (\gamma I \| Q_2 \| \gamma I) (-1)^{I-M} \begin{pmatrix} I & 2 & I \\ -M & 0 & M \end{pmatrix}$$

$$= Q \frac{3M^2 - I(I+1)}{I(2I-1)}. \tag{6.12}$$

6. Hyperfine Structure of Spectral Lines

We determine now the quadrupole moment of a charged particle in a spherically symmetric field in a state with angular momentum l. Assuming $I = l$, it is easy to obtain, see (4.144),

$$(l\|Q_2\|l) = \langle r^2 \rangle (l\|C^2\|l) = - \langle r^2 \rangle \sqrt{\frac{l(l+1)(2l+1)}{(2l-1)(2l+3)}}, \tag{6.13}$$

whence

$$Q = - \langle r^2 \rangle \frac{2l}{2l+3}. \tag{6.14}$$

For a particle with spin in a spherically symmetric field in the state slj, we obtain, in a similar way, from (4.186),

$$(\gamma j | Q_2 | \gamma j) = \langle r^2 \rangle (slj\|C^2\|slj)$$
$$= - \langle r^2 \rangle \frac{1}{4} \sqrt{\frac{(2j+1)(2j-1)(2j+3)}{j(j+1)}}, \tag{6.15}$$

$$Q = - \langle r^2 \rangle \frac{2j-1}{2j+2}. \tag{6.16}$$

Typical values of Q are given in Table 6.1. The maximum values of Q correspond to the nonspherical nucleus. For nonspherical nuclei, the values of Q prove to be considerably larger than would follow from (6.16).

Table 6.1 Spins and quadrupole moments of various nuclei

Nucleus	I	$Q\ [10^{-24}\mathrm{cm}^2]$	Nucleus	I	$Q\ [10^{-24}\mathrm{cm}^2]$
Be^9	3/2	+0.02	As^{75}	3/2	+0.3
B^{11}	3/2	+0.0355	Kr^{83}	9/2	+0.15
O^{17}	5/2	−0.004	Kr^{85}	9/2	+0.25
O^{16}	0	0	In^{113}	9/2	+0.750
Al^{27}	5/2	+0.149	In^{115}	9/2	+0.761
S^{33}	3/2	−0.0064	Eu^{151}	5/2	+1.2
S^{35}	3/2	+0.038	Eu^{153}	5/2	+2.5
Cl^{35}	3/2	−0.0789	Ta^{181}	7/2	+5.9
Cu^{65}	3/2	−0.15	Re^{185}	5/2	+2.8
Ga^{69}	3/2	+0.178	Hg^{201}	3/2	0.65
Ga^{71}	3/2	+0.112	U^{238}	0	+11

For various applications it is useful to express the operator of a quadrupole moment in a state with a given value I in terms of the components of $\mathbf{I}$. The tensor $Q_{\alpha\beta}$ is symmetric and has a trace equal to zero. The only tensor of this type which can be built from components of the vector $\mathbf{I}$ is the tensor

$$D_{\alpha\beta} = I_\alpha I_\beta + I_\beta I_\alpha - \frac{2}{3} I^2 \delta_{\alpha\beta}. \tag{6.17}$$

Assuming $Q_{\alpha\beta} = AD_{\alpha\beta}$, it is possible to determine the constant A by comparing the matrix elements $Q_{\alpha\beta}$ and $D_{\alpha\beta}$. According to (6.12)

$$\langle \gamma IM | Q_{zz} | \gamma IM \rangle = A \left\langle \gamma IM \left| 2I_z^2 - \frac{2}{3} I^2 \right| \gamma IM \right\rangle$$

$$= \frac{2}{3} A \left[3M^2 - I(I+1) \right], \tag{6.18}$$

whence

$$A = \frac{3}{2} \frac{Q}{I(2I-1)}, \tag{6.19}$$

$$Q_{\alpha\beta} = \frac{3}{2} \frac{Q}{I(2I-1)} \left(I_\alpha I_\beta + I_\beta I_\alpha - \frac{2}{3} I^2 \delta_{\alpha\beta} \right). \tag{6.20}$$

In the case of (6.14)

$$Q_{\alpha\beta} = -\langle r^2 \rangle \frac{3}{(2I-1)(2I+3)} \left(l_\alpha l_\beta + l_\beta l_\alpha - \frac{2}{3} l^2 \delta_{\alpha\beta} \right). \tag{6.21}$$

6.2 Hyperfine Splitting

6.2.1 General Character of the Splitting

For nuclei with nonzero moments μ and Q, an additional interaction with the electron shell takes place, namely

$$W = W_\mu + W_Q = -\boldsymbol{\mu} \cdot \boldsymbol{H}(0) + \frac{1}{6} e \sum_{\alpha\beta} Q_{\alpha\beta} \frac{\partial^2 \varphi}{\partial x_\alpha \partial x_\beta}. \tag{6.22}$$

Here $\boldsymbol{H}$ and φ are, respectively, the magnetic field strength and the electrostatic potential created by the electrons at the nucleus. The interaction (6.22) leads to a splitting of a level with angular momentum $\boldsymbol{J}$ into a number of components, each of which corresponds to a definite value of the total angular momentum of the atom

$$\boldsymbol{F} = \boldsymbol{I} + \boldsymbol{J}.$$

This splitting is called hyperfine splitting. The physical meaning of hyperfine splitting is obvious. As a result of the interaction (6.22), neither of the angular momenta I and J is conserved separately. Only the total angular momentum of the atom F is conserved. The interaction (6.22) is always very small, so the splitting of each level can be considered independently of the splitting of all the others. To determine the energy of splitting in this approximation, it is necessary to average (6.22) over the state $JIFM_F$. The situation is fully analogous to the one we met above in considering spin-orbit interaction in LS coupling.

We shall begin by considering the first term in (6.22). The magnetic moment of a nucleus with spin I is directed along I and equals $g_I I$. The mean value of H in a state with a given value of J is directed along J and so

$$W_\mu = -g_I a \mathbf{I} \cdot \mathbf{J} = A \mathbf{I} \cdot \mathbf{J} = \frac{1}{2} A (F^2 - J^2 - I^2)$$

$$\langle \gamma JIFM | W_\mu | \gamma JIFM \rangle \tag{6.23}$$

$$= \frac{1}{2} A [F(F+1) - J(J+1) - I(I+1)].$$

With the replacements $J \to L$, $I \to S$, and $F \to J$, (6.23) coincides with (5.137) for the spin-orbit splitting of a term. Thus the level J as a result of the interaction of the magnetic moment of the nucleus with the electron shell splits into a number of components

$$F = J+I, \ J+I-1, \ldots |J-I|.$$

The number of hyperfine structure components is $2I+1$ when $J \geqslant I$, and $2J+1$ when $J < I$. Hyperfine splitting obeys the Landé interval rule

$$\Delta E_F - \Delta E_{F-1} = AF. \tag{6.24}$$

This rule is analogous to the Landé interval rule for multiplet splitting. Just as in the case of fine-structure splitting, the center of gravity of the hyperfine structure of a level is not shifted:

$$\sum_F (2F+1) \Delta E_F = 0.$$

We shall consider now the quadrupole interaction. It is convenient to write the second term in (6.22) in a somewhat different form. Let us consider the interaction of two charges distributed in space with densities $\rho(r)$ and $\rho'(r')$, these densities being nonvanishing in the region $r' < r$. In this case

$$W = \int \frac{\rho(r) \rho'(r')}{|\mathbf{r} - \mathbf{r}'|} dr\, dr' = \int dr\, dr'\, \rho(r) \rho'(r') \sum_k \frac{r'^k}{r^{k+1}} P_k(\cos \alpha)$$

$$= \int dr\, dr'\, \rho(r) \rho'(r') \sum_k \frac{r'^k}{r^{k+1}} [C^k(\theta, \varphi) \cdot C^k(\theta', \varphi')]. \tag{6.25}$$

6.2 Hyperfine Splitting

The term $k = 2$ in the sum (6.25) corresponds to the quadrupole interaction. We shall define the quadrupole moment Q_{2m} by (6.10)

$$e' Q_{2m} = \int \rho'(\mathbf{r}') \, r'^2 \, C_m^2(\theta', \varphi') \, d\mathbf{r}' \tag{6.26}$$

and introduce the notation

$$e \eta_{2m} = \int \rho(\mathbf{r}) \frac{1}{r^3} C_m^2(\theta, \varphi) \, d\mathbf{r} \,. \tag{6.27}$$

After this the term $k = 2$ in (6.25) takes on the form

$$ee' \sum_m Q_{2m} \eta_{2m}^* \,. \tag{6.28}$$

According to (6.28), the interaction of the quadrupole moment of the nucleus with the electron shell can be written

$$W_Q = -e^2 \sum_m Q_{2m} \, \eta_{2m}^* \,, \tag{6.29}$$

$$\eta_{2m} = \sum_i \frac{1}{r^3} C_m^2(\theta_i, \varphi_i) \,. \tag{6.30}$$

Expression (6.29) is the scalar product of irreducible tensor operators of rank two, where Q_{2m} does not contain electronic variables nor η_{2m} nuclear ones. Thus, using (4.169), we obtain

$$\langle \gamma IJFM | W_Q | \gamma JIFM \rangle$$
$$= -e^2 (-1)^{I+J+F} (\gamma I \| Q_2 \| \gamma I)(\gamma J \| \eta_2 \| \gamma J) \begin{Bmatrix} J & I & F \\ I & J & 2 \end{Bmatrix}$$
$$= \Delta + BC(C+1), \tag{6.31}$$

$$C = F(F+1) - J(J+1) - I(I+1), \tag{6.32}$$

where the splitting constant B and the shift Δ, which does not depend on F, are determined by the expressions

$$B = -\frac{3}{2} \frac{e^2 (\gamma I \| Q_2 \| \gamma I)(\gamma J \| \eta_2 \| \gamma J)}{\sqrt{J(J+1)(2J-1)(2J+1)(2J+3) \, I(I+1)(2I-1)(2I+1)(2I+3)}}, \tag{6.33}$$

$$A = 2 \frac{e^2(\gamma I \|Q_2\| \gamma I)(\gamma J \|\eta_2\| \gamma J) J(J+1) I(I+1)}{\sqrt{J(J+1)(2J-1)(2J+1)(2J+3) I(I+1)(2I-1)(2I+2)(2I+3)}}. \tag{6.34}$$

Using (6.11) it is also easy to obtain

$$B = -\frac{3}{4} \frac{e^2 Q}{I(2I-1)} \frac{(\gamma J \|\eta_2\| \gamma J)}{\sqrt{J(J+1)(2J-1)(2J+1)(2J+3)}}, \tag{6.35}$$

$$A = \frac{e^2 Q(I+1)}{(2I-1)} \frac{(\gamma J \|\eta_2\| \gamma J) J(J+1)}{\sqrt{J(J+1)(2J-1)(2J+1)(2J+3)}}, \tag{6.36}$$

Therefore, the resulting expression for the hyperfine splitting of a level has the form (the term not dependent on F is omitted)

$$\Delta E_F = \frac{1}{2} AC + BC(C+1),$$
$$C = F(F+1) - J(J+1) - I(I+1). \tag{6.37}[1]$$

The splitting of a level defined by (6.37) is considerably more complex than the purely magnetic splitting (6.23). In particular, when $B \neq 0$, the Landé interval rule is not fulfilled.

6.2.2 Calculation of the Hyperfine Splitting Constant A

Having determined experimentally the hyperfine splitting constants A and B, one can find the values of the nuclear moments μ and Q. For this, however, it is necessary to know how the constants A and B are connected with μ and Q. The problem of calculating the hyperfine splitting constants consists of finding this connection. From (6.22) it follows that to calculate the constants A and B it is necessary to know the values of the magnetic field and of the second derivatives of the electrostatic potential at the position of the nucleus, i.e., at the origin.

The magnetic field $\boldsymbol{H}(0)$ created by the electrons at the position of the nucleus can be represented in the form of the sum

$$\boldsymbol{H}_l(0) + \boldsymbol{H}_s(0).$$

The first term is due to the orbital motion of the electrons and the second to the spins of the electrons. The magnetic field produced by a charged particle in steady motion is determined by the well-known expression

---

[1] Sometimes the quadrupole splitting constant is determined slightly differently, utilizing the term $BC(C+1)$ in the form $3/8\, B' \frac{C(C+1)}{I(2I-1) J(2J-1)}$; $B' = 8/3\, I(2I-1) J(2J-1) B$.

6.2 Hyperfine Splitting

$$H = e[\mathbf{r}, \mathbf{v}]/cr^3 .$$

Substitution of $[\mathbf{r},\mathbf{v}]m = \hbar \mathbf{l}$ in this expression gives

$$H_l(0) = -\frac{e\hbar}{mc}\frac{1}{r^3}\mathbf{l} = -\frac{2}{r^3}\mathbf{l}\mu_0 . \tag{6.38}$$

The corresponding term in the energy of interaction has the form

$$W_{Il} = -\boldsymbol{\mu} \cdot \mathbf{H}_l(0) = g_I\mu_0^2 \left(\frac{m}{m_p}\right)\frac{2}{r^3}\mathbf{I}\cdot\mathbf{l} = a_l \mathbf{I}\cdot \mathbf{l} , \tag{6.39}$$

where

$$a_l = \frac{2}{r^3} g_I\mu_0^2 \left(\frac{m}{m_p}\right) = g_I\alpha^2 a_0^3 \left(\frac{m}{m_p}\right)\frac{1}{r^3}\, \text{Ry} , \tag{6.40}$$

$\alpha = e^2/\hbar c$ is the fine-structure constant, and $a_0 = \hbar^2/me^2$. The magnetic field produced by the spin of magnetic moment of the electron $\boldsymbol{\mu}_s = -2\mu_0\mathbf{s}$ is defined by the expression

$$\mathbf{H}_s(0) = \frac{2\mu_0}{r^3}[\mathbf{s} - 3(\mathbf{s}\cdot\mathbf{n})\mathbf{n}] , \tag{6.41}$$

where $\mathbf{n}$ is the unit vector directed along $\mathbf{r}$. From (6.41) it follows that

$$W_{Is} = -a_l [\mathbf{I}\cdot\mathbf{s} - 3(\mathbf{s}\cdot\mathbf{n})(\mathbf{I}\cdot\mathbf{n})] . \tag{6.42}$$

Thus the complete expression for the interaction energy of the magnetic moment of the nucleus with an atomic electron is

$$W = a_l \mathbf{l}\cdot\mathbf{I} - a_l[\mathbf{s} - 3(\mathbf{s}\cdot\mathbf{n})\mathbf{n}]\cdot \mathbf{I} . \tag{6.43}$$

We shall begin by considering the single-electron problem. In this case, to calculate the energy of splitting it is necessary to average (6.43) over the state $ljIF$. Using the results of Sect. 4.3, we shall write [see (4.158, 167)] the second term of (6.43) in the form

$$-a_l \sqrt{10} \sum_q (-1)^q [C^2 \times s^1]_q^1 I_{-q} . \tag{6.44}$$

Therefore,

$$\langle sljIFM| W |sljIFM\rangle = \langle a_l\rangle (-1)^{j+I+F}(I||I||I)(slj||l||slj)$$
$$- \sqrt{10}\, (slj||[C^2 \times s^1]^1||slj)\begin{Bmatrix} j & I & F \\ I & j & 1 \end{Bmatrix} . \tag{6.45}$$

The reduced matrix elements contained on the right-hand side of (6.45) are determined by (4.148, 173, 177), where the 9j symbol in (4.173) can be calculated from the equations given in Sect. 5.7.1. Comparing (6.45) with (6.23), we obtain

$$A = \langle a_l \rangle \frac{l(l+1)}{j(j+1)} = \alpha^2 g_I \left(\frac{m}{m_p}\right) \left(\frac{a_0^3}{r^3}\right) \frac{l(l+1)}{j(j+1)} \text{Ry} . \tag{6.46}$$

Expression (6.46) is not applicable when $l = 0$. In this special case the interaction of an electron with the magnetic moment of a nucleus is [3]

$$W = a_s \mathbf{I} \cdot \mathbf{s} , \tag{6.47}$$

$$a_s = \frac{8\pi}{3} \alpha^2 g_I \left(\frac{m}{m_p}\right) a_0^3 |\psi_s(0)|^2 \text{ Ry} , \tag{6.48}$$

$$\langle W \rangle = a_s (-1)^{j+I+F} (I\|I\|I) (slj\|s\|slj) \begin{Bmatrix} j & I & F \\ I & j & I \end{Bmatrix} . \tag{6.49}$$

Therefore, in the general case,

$$A = \langle a_l \rangle \frac{l(l+1)}{j(j+1)} (1 - \delta_{l0}) + a_s \delta_{l0} . \tag{6.50}$$

For a hydrogenlike atom,

$$\left(\frac{a_0^3}{r^3}\right) = \frac{Z^3}{n^3(l+1)\left(l+\frac{1}{2}\right)l} , \tag{6.51}$$

$$|\psi_s(0)|^2 = \frac{Z^3}{\pi a_0^3 n^3} \tag{6.52}$$

and

$$l \neq 0 , \quad A_l = \frac{\alpha^2 g_I Z^3}{n^3 \left(l+\frac{1}{2}\right) j(j+1)} \left(\frac{m}{m_p}\right) \text{Ry} , \tag{6.53}$$

$$l = 0, \quad A_s = a_s = \frac{8\alpha^2 g_I Z^3}{3n^3} \left(\frac{m}{m_p}\right) \text{Ry} . \tag{6.54}$$

Equation (6.50) is also applicable to the alkali metals. In this case, however, the factors $\langle r^{-3} \rangle$ and $|\psi_s(0)|^2$ cannot be calculated exactly. Therefore, instead of (6.51, 54) one has to use various approximate expressions. The approximation on which (5.151) is based, consisting in the replacement of the factor Z^3/n^3 by $Z_a^2 Z_i/n_*^3$, is usually used. In this case

$$\left(\frac{a_0^3}{r^3}\right) = \frac{Z_a^2 Z_i}{n_*^3 (l+1)(l+1/2)\, l}, \tag{6.55}$$

where the parameter Z_i can be determined from the magnitude of the multiplet splitting. It is possible to also express a_l directly in terms of ζ_l

$$l \neq 0, \quad A_l = \zeta_l \frac{g_I l(l+1)}{Z_i j(j+1)} \left(\frac{m}{m_p}\right). \tag{6.56}$$

In the same approximation it is also not difficult to determine $|\psi_s(0)|^2$. For this it is sufficient to replace Z^3/n^3 by $Z_a^2 Z_i/n_*^3$ in (6.52) and put $Z_i = Z$. Better results, however, are obtained if it is assumed that

$$Z_i = Z\left(1 + \left|\frac{\partial \Delta}{\partial n}\right|\right), \tag{6.57}$$

where Δ is the quantum defect for the terms $ns\ ^2S_{1/2}$. In this case [19]

$$|\psi_s(0)|^2 = \frac{Z_a^2 Z}{\pi a_0^3 n_*^3}\left(1 + \left|\frac{\partial \Delta}{\partial n}\right|\right), \tag{6.58}$$

$$A_s = \frac{8}{3}\frac{\alpha^2 g_I Z_a^2 Z}{n_*^3}\left(\frac{m}{m_p}\right)\left(1 + \left|\frac{\partial \Delta}{\partial n}\right|\right) \text{Ry}. \tag{6.59}$$

Expression (6.59) is called the Fermi-Segré formula. The dependence of the quantum defect Δ on n is small and $(1 + |\partial \Delta/\partial n|)$ is close to 1, because Δ is almost constant for a given series (Sect. 3.3). Nevertheless, taking into consideration the term $\partial \Delta/\partial n$ proves to be important in a number of cases.

The expressions given above for the hyperfine splitting constant A have been obtained in a nonrelativistic approximation. For hydrogen and hydrogenlike ions with a small value of Z, the relativistic corrections are not important. For large values of Z the results of relativistic calculations differ greatly from those given above. It is usual to introduce into the nonrelativistic formulas the correction factor $F_r(jZ_i)$ – the so-called relativistic correction. It is also necessary to introduce the correction $H_r(lZ_i)$ into (6.56) at the same time as $F_r(jZ_i)$. To refine even further the expressions for the constant A, certain additional effects are taken into account and these lead to additional corrections. Most important is the correction for the finite size of the nucleus which is usually written as $(1-\delta)$. We give the final formulas:

$$l = 0,\ A_s = \frac{8}{3}\frac{\alpha^2 g_I Z_a^2 Z}{n_*^3}\left(1 + \left|\frac{\partial \Delta}{\partial n}\right|\right)(1 - \delta)\, F_r\!\left(\frac{1}{2}Z\right)\!\left(\frac{m}{m_p}\right)\text{Ry}, \tag{6.60}$$

$$l \neq 0,\ A_l = \frac{\alpha^2 g_I Z_a^2 Z_i F_r(jZ_i)}{n_*^3 \left(l + \frac{1}{2}\right) j(j+1)}\left(\frac{m}{m_p}\right)(1 - \delta)\,\text{Ry},$$

$$= \zeta_{l} \frac{g_{I}l(l+1) F_{r}(jZ_{i})}{j(j+1) H_{r}(lZ_{i})} (1-\delta) \left(\frac{m}{m_{p}}\right) \text{Ry}. \tag{6.61}$$

The dependence of F_r and H_r on Z_i when $l = 1$ is shown in Fig. 6.1. It is easy to see from the figure that relativistic corrections are necessary when Z_i is greater than about 20–30, since F_r and H_r rapidly increase with increasing Z_i.

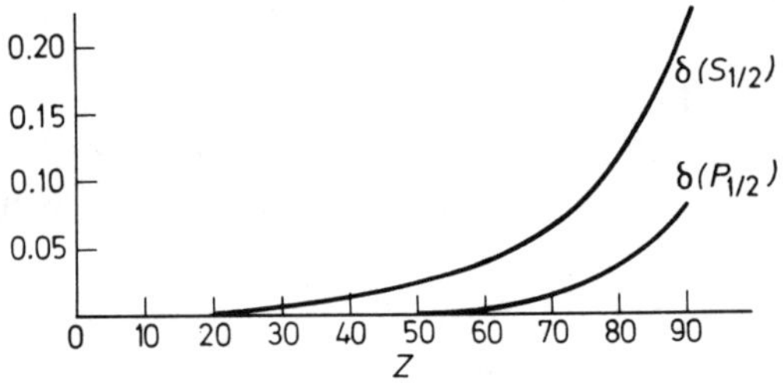

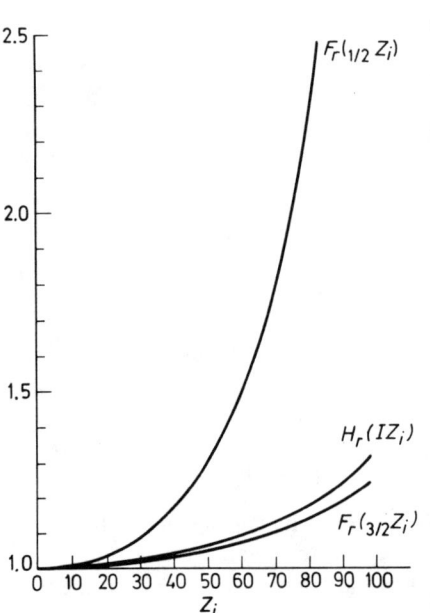

Fig. 6.2. Dependence of the factor δ on Z

Fig. 6.1. Relativistic corrections F_r and H_r when $l = 1$

The factor $(1-\delta)$ also becomes significant for large Z. δ increases monotonically with increasing Z (Fig. 6.2). For not very large values of Z, $\delta \lesssim 0.1$, and is not very important; δ reaches values of $0.15-0.20$ only when $Z \simeq 80-90$.

Table 6.2 Values of the correction factors $(1 + |\partial A/\partial n|)$, $F_r(1/2Z)$ and $(1-\delta)$

Element	Z	Level	Z_a	n_*^3	$1 + \left\|\frac{\partial A}{\partial n}\right\|$	$F_r\left(\frac{1}{2}Z\right)$	$(1-\delta)$
Li I	3	$2s\ ^2S_{1/2}$	1	4.02	1.00	1.00	1.00
Na I	11	$3s\ ^2S_{1/2}$	1	4.31	1.03	1.01	1.00
K I	19	$4s\ ^2S_{1/2}$	1	5.55	1.06	1.04	1.00
Sc III	21	$4s\ ^2S_{1/2}$	3	13.8	—	1.04	0.998
Rb I	37	$5s\ ^2S_{1/2}$	1	5.88	1.08	1.15	0.995
In III	49	$5s\ ^2S_{1/2}$	3	9.127	1.124	1.29	0.973
Cs I	55	$6s\ ^2S_{1/2}$	1	6.532	1.101	1.39	0.96
La III	57	$6s\ S_{1/2}$	3	18.53	1.08	1.43	0.955
Hg II	80	$6s\ ^2S_{1/2}$	2	4.943	1.943	2.26	0.88
Tl III	81	$7s\ ^2S_{1/2}$	3	30.40	1.100	2.32	0.88
Bi V	83	$6s\ ^2S_{1/2}$	5	12.98	1.14	2.46	0.86

Values of $(1-\delta)$, $F_r(1/2Z)$, and $(1+|\partial\Delta/\partial n|)$ for a number of atoms are given in Table 6.2.

We shall now consider multielectron atoms. Let us first of all consider an atom, one of whose valence electrons is an s electron. In this case, as a rule, the hyperfine splitting is determined mainly by the interaction of the magnetic moment of the nucleus with this s electron. Thus one can assume approximately

$$W = a_s \mathbf{I} \cdot \mathbf{s}. \tag{6.62}$$

Averaging the operator (6.62) can be carried out in several stages. First we shall average (6.62) over the state with a given value of the spin of the atom $\mathbf{S}$. This gives

$$\langle s \rangle = \frac{\langle \mathbf{s} \cdot \mathbf{S} \rangle}{S(S+1)} \mathbf{S} = \frac{S(S+1) + s(s+1) - S_1(S_1+1)}{2S(S+1)} \mathbf{S}, \tag{6.63}$$

where S_1 is the spin of the parent ion ($\mathbf{S} = \mathbf{S}_1 + \mathbf{s}$). Then we average $\mathbf{S}$ over the state with a given value of $\mathbf{J}$

$$\langle \mathbf{S} \rangle = \frac{J(J+1) + S(S+1) - L(L+1)}{2J(J+1)} \mathbf{J}. \tag{6.64}$$

Using (6.63, 64) it is easy to obtain

$$A = a_s \frac{S(S+1) + s(s+1) - S_1(S_1+1)}{2S(S+1)} \frac{J(J+1) + S(S+1) - L(L+1)}{2J(J+1)}. \tag{6.65}$$

We shall now consider the configuration l^n. In this case

$$W = a_l \mathbf{L} \cdot \mathbf{I} - a_l \sum_i [\mathbf{s}_i - 3(\mathbf{s}_i \cdot \mathbf{n}_i) \mathbf{n}_i] \cdot \mathbf{I}. \tag{6.66}$$

Averaging the first term obviously presents no difficulty. In averaging the second term, it is again convenient to use (6.44). This enables us to express the mean value of the operator in terms of the reduced matrix element of the operator V^{12} (Sects. 5.4 and 5.5). We give the final result:

$$A = \langle a_l \rangle \left[(2-g) - \sigma \frac{6\Gamma(2-g) - 2(g-1)L(L+1)}{(2L-1)(2L+3)} \right], \tag{6.67}$$

$$\Gamma = \frac{1}{2}[J(J+1) - L(L+1) - S(S+1)], \tag{6.68}$$

$$g = 1 + \frac{J(J+1) + S(S+1) - L(L+1)}{2J(J+1)}, \tag{6.69}$$

$$\sigma = -(l|C^2|l) \sqrt{\frac{(2L-1)(2L+3)}{L(L+1)(2L+1)S(S+1)(2S+1)}}$$
$$(l^n\gamma SL|V^{12}|l^n\gamma SL). \tag{6.70}$$

This equation is the generalization of (6.46) to the case of several equivalent electrons. For $n=1$, $L=l$, $S=1/2$ $(l^n\gamma SL\|V^{12}\|l^n\gamma SL) = (3/2)^{1/2}$, $\sigma = -1$ and (6.67) turns into (6.46). With the aid of the tables of the reduced matrix elements of V^{12} (Sect. 5.4), it is easy to calculate the value of A for any of the configurations p^n. For configurations d^n, one can use (5.122). In the case of more complex configurations the constant A contains several different parameters $\langle a_l \rangle$ which increase the inaccuracy of numerical estimates.

6.2.3 Calculation of the Hyperfine Splitting Constant B

For an atom with one valence electron, the calculation of the quadrupole splitting constant B, according to (6.27, 6.35), reduces to a calculation of the reduced matrix element

$$(lsj\|\eta_2\|lsj) = \left\langle \frac{1}{r^3} \right\rangle (lsj\|C^2\|lsj). \tag{6.71}$$

Using (4.186), we obtain

$$(lsj\|\eta_2\|lsj) = -\frac{1}{4} \left\langle \frac{1}{r^3} \right\rangle \sqrt{\frac{(2j+3)(2j+1)(2j-1)}{j(j+1)}}, \tag{6.72}$$

$$B = \frac{3e^2Q}{16I(2I-1)j(j+1)} \left\langle \frac{1}{r^3} \right\rangle. \tag{6.73}$$

In a nonrelativistic approximation (light nuclei) the factor $\langle 1/r^3 \rangle$ can be calculated by means of the approximation (6.55). Taking into account relativistic effects leads to the appearance of the correction factor R_r. Thus

$$B = \frac{3e^2Q}{16I(2I-1)j(j+1)} \frac{Z_a^2 Z_i R_r}{n_*^3 a_0^3 (l+1)(l+1/2)l}. \tag{6.74}$$

The relativisitc correction R_r, just like F_r becomes particularly important for heavy nuclei. The dependences of F_r, H_r and R_r on Z_i is given in Table 6.3.

The constant B can also be expressed in terms of the multiplet splitting constant ζ_l. Using (5.152) and remembering that $e^2/a_0 = 2\text{Ry}$, we obtain

$$B = \frac{3e^2Q}{16I(2I-1)j(j+1)} \frac{\zeta_l}{\alpha^2 a_0^3 Z_i \text{Ry}} \frac{R_r}{H_r}.$$

6.2 Hyperfine Splitting

Table 6.3 Dependence of the values F_r, H_r, and R_r on Z_i

Z_i	$F_r\left(\frac{1}{2}Z_i\right)$	$F_r\left(\frac{3}{2}Z_i\right)$	$H_r(1Z_i)$	$R_r(1Z_i)_j = \frac{3}{2}$
1	1.0001	1.0000	1.0000	1.0000
5	1.0024	1.0005	1.0006	1.0010
10	1.0098	1.0021	1.0023	1.0042
15	1.0224	1.0047	1.0053	1.0095
20	1.0404	1.0084	1.0094	1.0171
25	1.0643	1.0132	1.0148	1.0268
30	1.0948	1.0191	1.0216	1.0389
35	1.1328	1.0261	1.0296	1.0535
40	1.1795	1.0343	1.0391	1.0706
45	1.2365	1.0438	1.0502	1.0905
50	1.3058	1.0545	1.0629	1.1133
55	1.3904	1.0666	1.0775	1.1392
60	1.4941	1.0801	1.0940	1.1686
65	1.6226	1.0951	1.1128	1.2016
70	1.7837	1.1116	1.1340	1.2387
75	1.9892	1.1299	1.1581	1.2803
80	2.2573	1.1500	1.1853	1.3268
85	2.6174	1.1721	1.2164	1.3790
90	3.1205	1.1963	1.2518	1.4373

$$= \frac{3}{8} \frac{Q}{I(2I-1)j(j+1)} \frac{\zeta_l}{\alpha^2 a_0^2 Z_i} \frac{R_r}{H_r} . \tag{6.75}$$

We shall also consider how the constant B is calculated for the group of equivalent electrons l^n. In this case

$$(l^n\gamma SLJ||\eta_2||l^n\gamma SLJ) = \left\langle \frac{1}{r^3} \right\rangle (l||C^2||l)(l^n\gamma SLJ|| \sum_i u_i^2||l^n\gamma SLJ) , \tag{6.76}$$

where u_i^2 is the unit tensor of rank two introduced in Sects. 5.3 and 5.4 and defined by the relation

$$(l||u^2||l) = 1 . \tag{6.77}$$

The operator

$$U^2 = \sum_i u_i^2 \tag{6.78}$$

does not contain spin variables. Therefore,

$$(l^n\gamma SLJ||U^2||l^n\gamma SLJ) = (-1)^{S+2+L+J} (l^n\gamma SL||U^2||l^n\gamma SL)$$
$$\times (2J+1) \begin{Bmatrix} L & J & S \\ J & L & 2 \end{Bmatrix}, \tag{6.79}$$

$(l^n \gamma SLJ \| \eta_2 \| l^n \gamma SLJ)$

$$= -\left\langle \frac{1}{r^3} \right\rangle \sqrt{\frac{l(l+1)(2l+1)}{(2l-1)(2l+3)}} (-1)^{S+L+J}$$

$$\times (2J+1)(l^n \gamma SL \| U^2 \| l^n \gamma SL) \begin{Bmatrix} L & J & S \\ J & L & 2 \end{Bmatrix} \quad (6.80)$$

and

$$B = \frac{3}{4} \frac{e^2 Q}{I(2I-1)} \left\langle \frac{1}{r^3} \right\rangle \sqrt{\frac{l(l+1)(2l+1)}{(2l-1)(2l+3)}} (l^n \gamma SL \| U^2 \| l^n \gamma SL)$$

$$\times (-1)^{S+L+J} \frac{(2J+1)}{\sqrt{J(J+1)(2J-1)(2J+1)(2J+3)}} \begin{Bmatrix} L & J & S \\ J & L & 2 \end{Bmatrix}. \quad (6.81)$$

Values of the reduced matrix elements of U^2 for configurations p^n are given in Table 5.18.

6.2.4 Radiative Transitions Between Hyperfine-Structure Components

Electric dipole transitions between the components of the hyperfine structure of two different levels γJ and $\gamma' J'$ (it is assumed that transitions between these levels are allowed) obey the additional selection rules

$$\Delta F = 0, \pm 1; \quad F + F' \geqslant 1. \quad (6.82)$$

For the relative intensities of the transitions it is possible to formulate the following sum rules.

The sum of the intensities of all lines of the hyperfine structure of the transition $\gamma J \to \gamma' J'$ originating from the component F of the level γJ is proportional to the statistical weight of this component $2F + 1$.

The sum of the intensities of all lines of the hyperfine structure of the transition $\gamma J \to \gamma' J'$ ending on the component F' of the level $\gamma' J'$ is proportional to the statistical weight of this component $2F' + 1$.

Electric dipole transitions between components of the hyperfine structure of the same level are forbidden by the parity selection rule. Only magnetic-dipole transitions and electric-quadrupole transitions are allowed. In the first case the selection rules (6.82) apply, and in the second

$$\Delta F = 0, \pm 1, \pm 2; \quad F + F' \geqslant 2. \quad (6.83)$$

6.2.5 Isotope Shift of the Atomic Levels

The energy levels of two isotopes of any element are shifted relative to each other. The simplest example of this isotope shift is the difference between the terms of hydrogen and deuterium. In this case

$$E_n = -\frac{1}{2}\frac{\mu e^4}{\hbar^2 n^2} = -\frac{1}{2}\frac{me^4}{\hbar^2 n^2}\cdot\frac{M}{m+M} \approx -\frac{\text{Ry}}{n^2}\left(1-\frac{m}{M}\right) = E_n^0\left(1-\frac{m}{M}\right),$$
(6.84)

where E_n^0 is the zeroth approximation corresponding to a nucleus with $M \to \infty$. For hydrogen $M = m_p$ and for deuterium $M = 2m_p$; therefore the deuterium levels are shifted downwards relative to the hydrogen levels by an amount $\text{Ry}/n^2 \cdot \frac{m}{2m_p}$. Thus the lines of the deuterium spectrum are shifted towards higher frequencies or shorter wavelengths.

The isotope shift (6.84) is caused by the motion of the nucleus relative to the center of mass of the atom. In the case of the complex atoms, in addition to this effect of the finite mass there is the effect of the finite size of the nucleus. The field inside the nucleus is not a Coulomb field, and this is naturally reflected in the energy levels. The addition of one or a pair of neutrons to the nucleus leads to a change of the nuclear radius r_0 and, consequently, to a displacement of the levels. The binding energy of electrons in an atom is less for an isotope of larger mass ($M' > M$; $r_0' > r_0$). The levels of this isotope are accordingly shifted upwards. Thus the volume effect is opposite in sign to the mass effect. It is usual to regard the isotope shift as positive if the spectral line corresponding to the heavier isotope is shifted towards higher frequencies [as in the case of (6.84)]. Thus the volume effect gives a negative shift.

The nuclei of isotopes can differ not only in mass and radius but also in other properties. The nuclei can be nonspherical; this also leads to an isotope shift. We shall include all these phenomena within the volume effect. Investigation of the volume effect enables one to obtain valuable information on nuclear structure and thus it is of the greatest interest. To obtain the volume effect, it is necessary to subtract from the observed shift the shift caused by the difference of masses of the isotopes. For light elements the volume effect is negligibly small in comparison with the mass effect. For heavy elements ($Z \geqslant 60$), the volume effect becomes dominant.

It is necessary to take into account the possible existence of hyperfine splitting when analyzing the isotope effect. The isotope shift is determined by the distance between the center of gravity of the hyperfine structures. In a gas mixture of different isotopes, the isotope shift leads to a further splitting of the observed spectral lines.

In the case of a multielectron atom, the generalization of (6.84) can be obtained in the following way. The kinetic energy of the nucleus is approximately m/M times less than the kinetic energy of the electrons. This permits one to consider the motion of the nucleus by means of perturbation theory, assuming $\Delta E = \langle P^2/2M \rangle$, where $\boldsymbol{P}$ is the nuclear momentum. The sum of the nuclear momentum $\boldsymbol{P}$ and the electron momenta $\sum_i \boldsymbol{P}_i$ is conserved

$$\boldsymbol{P} + \sum_i \boldsymbol{P}_i = 0.$$

Thus,

$$\Delta E = \Delta E_n + \Delta E_s = \frac{m}{M}\left(\sum_i \frac{P_i^2}{2m}\right) + \frac{m}{M}\left(\sum_{i \neq k} \frac{P_i \cdot P_k}{2m}\right). \tag{6.85}$$

The first term in (6.85) is called the normal shift and the second the specific shift. According to the virial theorem

$$\Delta E_n = -\frac{m}{M} E^0, \tag{6.86}$$

which coincides with (6.84). The second term in (6.85) is a symmetrical two-electron operator. Thus in calculating ΔE_s it is possible to use the results previously obtained in Sects. 5.2–5.4. The term ΔE_s in (6.85) can have either sign. Therefore, the sum $\Delta E_n + \Delta E_s$ can be either positive or negative. In the simplest case of the two-electron configuration ll', the term ΔE_s is nonzero only when $l = l' \pm 1$, i.e., for configurations $nsn'p$, $npn'd$, and so on. Of greatest practical interest are the configurations containing s electrons.

The isotope shift of s electron energy levels caused by the difference in radii δr_0 of the nuclei is usually given by the equation of Racah, Rosenthal, and Breit

$$\delta E = \frac{4\pi a_0^3}{Z} |\psi_s(0)|^2 \frac{\gamma + 1}{[\Gamma(2\gamma + 1)]^2} B(\gamma) \left(\frac{2Zr_0}{a_0}\right)^{2\gamma} \frac{\delta r_0}{r_0} \text{Ry}, \tag{6.87}$$

where $\gamma = \sqrt{1 - \alpha^2 Z^2}$, $\alpha = e^2/\hbar c$; the factor $B(\gamma)$ depends on the distribution of proton charge in the nucleus. For a uniform distribution of proton charge over the volume of the nucleus and for the charge distributed uniformly over the surface of the nucleus, we have correspondingly

$$B(\gamma) = 3\left[(2\gamma + 1)(2\gamma + 3)\right]^{-1}; \quad B(\gamma) = (2\gamma + 1)^{-1}. \tag{6.88}$$

The factor $|\psi_s(0)|^2$ can be determined from (6.58). As was stated above, for light nuclei $\delta E < \Delta E_n + \Delta E_s$ and for heavy nuclei $\delta E > \Delta E_n + \Delta E_s$.

Chapter 7 The Atom in an External Electric Field

The influence of electric fields (especially, plasma microfields) on atomic levels and transition probabilities is very important for various applications of atomic spectroscopy. For this reason, not only quadratic and linear Stark effects in the homogeneous and static electric field, but also the effects of inhomogeneous and variable fields are considered in this chapter.

7.1 Quadratic Stark Effect

The Stark effect is the splitting and shifting of atomic levels under the action of an external electric field.

The energy of an atom in a homogeneous electric field is equal to the scalar product of the strength of the electric field $\mathscr{E}$ and the electric dipole moment of the atom D taken with negative sign

$$H' = - \mathscr{E} \cdot D = e\mathscr{E} \cdot \sum_i r_i . \tag{7.1}$$

The matrix elements of D connecting states of the same parity, including the diagonal matrix elements, are zero. Therefore the interaction (7.1) does not lead to any change in the energy of an atom in first-order perturbation theory. The splitting of levels is determined by second-order corrections. Let us choose the z axis to be along the direction of the field $\mathscr{E}$. Then $H' = -\mathscr{E} D_z$, and for the correction to the energy of the state γJM we obtain

$$\Delta E_{\gamma JM} = \mathscr{E}^2 \sum_{\gamma' J'} \frac{|\langle \gamma JM | D_z | \gamma' J' M \rangle|^2}{E_{\gamma J} - E_{\gamma' J'}} . \tag{7.2}$$

The dependence of the matrix elements of D_z on M can be calculated in explicit form (Sect. 9.2)

$$\langle \gamma JM | D_z | \gamma' J' M \rangle \propto \begin{cases} \sqrt{J^2 - M^2} & J' = J-1 , \\ M , & J' = J , \\ \sqrt{(J+1)^2 - M^2} & J' = J+1 , \end{cases} \tag{7.3}$$

whence it follows

$$\Delta E_{\gamma JM} = \mathscr{E}^2 (A_{\gamma J} + B_{\gamma J} M^2) , \tag{7.4}$$

where the constants $A_{\gamma J}$ and $B_{\gamma J}$ are discussed below [see (7.6–8)]. Thus the homogeneous electric field splits the level γJ into the components

$$|M| = J, J-1 \ldots, \tag{7.5}$$

the magnitude of the splitting being proportional to the square of $\mathscr{E}$. All levels, with the exception of the level $M = 0$, are twofold degenerate with respect to the sign of the projection of the angular momentum. The levels $J = 0$ and $J = 1/2$ obviously do not split and undergo a shift only. The asymmetry of the splitting is a special feature of (7.4).

The general features of the splitting are practically completely described by the statements above. Further investigation of (7.2) requires one to specify the particular case being considered. The case of LS coupling is of the greatest interest for applications. If the multiplet structure of the perturbing terms is neglected and it is assumed that $E_{\gamma'J''} \simeq E_{\gamma'}$, it is then possible to calculate the dependence of $A_{\gamma J}$ and $B_{\gamma J}$ on J in explicit form. We give the results

$$A_{\gamma J} = \alpha_\gamma + \frac{\beta_\gamma C_2}{2}, \quad B_{\gamma J} = \beta_\gamma C_1, \tag{7.6}$$

$$C_1 = \frac{3\langle \mathbf{L}\cdot\mathbf{J}\rangle(2\langle \mathbf{L}\cdot\mathbf{J}\rangle - 1) - 2J(J+1)L(L+1)}{J(J+1)(2J-1)(2J+3)}, \tag{7.7}$$

$$C_2 = 2\frac{L(L+1)[2J(J+1)-1] - \langle \mathbf{L}\cdot\mathbf{J}\rangle(2\langle \mathbf{L}\cdot\mathbf{J}\rangle - 1)}{(2J-1)(2J+3)}. \tag{7.8}$$

Here α_γ, β_γ are new constants; γ is the set of quantum numbers describing the term and

$$2\langle \mathbf{L}\cdot\mathbf{J}\rangle = J(J+1) + L(L+1) - S(S+1). \tag{7.9}$$

Equation (7.2) can be simplified in the case of the ground state and in the case of a strong interaction with the nearest level, when the main contribution gives only one of the terms of the sum. In the case of the ground state, the energy differences $E_{\gamma J} - E_{\gamma' J'}$ in (7.2) for the levels of the discrete spectrum are more than E_r but less than E_i. Since $E_i - E_r \lesssim E_i, E_r$ (remember that for hydrogen $E_r = \frac{3}{4}E_i$), the sum (7.2) can be written approximately in the form:

$$\Delta E_{\gamma JM} = \mathscr{E}^2 I^{-1} \sum_{\gamma'J'}{}' |\langle \gamma JM|D_z|\gamma'J'M\rangle|^2 = \mathscr{E}^2 I^{-1} \langle \gamma JM|D_z^2|\gamma JM\rangle,$$

where I has the order of magnitude E_r, E_i. Taking $I = E_r$ or $I = E_i$, it is possible to give an approximate estimate of the sum (7.2). For excited states such estimates are not possible because the differences $E_{\gamma J} - E_{\gamma'J'}$ can be either positive or negative.

7.1 Quadratic Stark Effect

The second case usually occurs if one of the differences $E_{\gamma J} - E_{\gamma' J'}$ is much less than all the others. For two such strongly interacting levels one can assume approximately,

$$\Delta E_{\gamma JM} = \mathscr{E}^2 (E_{\gamma J} - E_{\gamma' J'})^{-1} |\langle \gamma JM | D_z | \gamma' J' M \rangle|^2, \tag{7.10}$$

$$\Delta E_{\gamma' J' M'} = \mathscr{E}^2 (E_{\gamma' J'} - E_{\gamma J})^{-1} |\langle \gamma' J' M | D_z | \gamma JM \rangle|^2 = - \Delta E_{\gamma JM}. \tag{7.11}$$

In (7.10, 11) the square of the matrix element of D_z can be replaced by the oscillator strength of the transition $f(\gamma J; \gamma' J')$, (Sect. 9.2)

$$\Delta E_{\gamma JM} = - \Delta E_{\gamma' J'M} = \frac{3e^2}{2m\omega^2}(2J+1)f(\gamma J; \gamma' J') \begin{pmatrix} J & 1 & J' \\ -M & 0 & M \end{pmatrix}^2 \mathscr{E}^2. \tag{7.12}$$

Formula (7.12) is suitable only for rough estimates. The contribution of the large number of small terms omitted in (7.12) can be of the same order as (7.12). In a number of cases of closely spaced interacting levels, experiment shows a symmetry of splitting ($\Delta E_{\gamma JM} = - \Delta E_{\gamma' J' M'}$) which is characteristic in the two-level approximation.

Equations (7.10, 12) are valid as long as the corrections to the energy are small in comparison with the initial splitting $E_{\gamma J} - E_{\gamma' J'}$. In the general case it is necessary to treat, at the same time, the interaction with the field H' and the interatomic interaction H'' responsible for the splitting of the levels γJ and $\gamma' J'$. The latter is the sum of three parts: the central potential, the electrostatic interaction between the electrons, and spin-orbit interaction. The matrices of all these interactions are diagonal in the quantum numbers J and M. We shall define H'' in such a way that the matrix of H'' is also diagonal in the quantum numbers γ, whence

$$E_{\gamma J} = E_0 + \langle \gamma JM | H'' | \gamma JM \rangle = E_0 + \frac{\Delta}{2},$$
$$E_{\gamma' J'} = E_0 + \langle \gamma' J' M | H'' | \gamma' J' M \rangle = E_0 - \frac{\Delta}{2}, \tag{7.13}$$

$$E_0 = \frac{1}{2}(E_{\gamma J} + E_{\gamma' J'}), \quad \Delta = E_{\gamma J} - E_{\gamma' J'}. \tag{7.14}$$

It is easy to see that such a choice of H'' is really possible, because in the absence of the field, the first-order corrections from the perturbation H'' give the correct values for the energy of the states γJ and $\gamma' J'$. Taking into account the interactions H' and H'', one can find the energy levels from the secular equation:

$$\begin{vmatrix} \Delta/2 - \Delta E & \langle \gamma JM | H' | \gamma' J' M \rangle \\ \langle \gamma' J' M | H' | \gamma JM \rangle & -\Delta/2 - \Delta E \end{vmatrix} = 0. \tag{7.15}$$

Substituting $H' = -\mathscr{E} D_z$ in (7.15), we find

$$\Delta E_{1,2} = \pm \sqrt{(\Delta/2)^2 + |\langle \gamma J M | D_z | \gamma' J' M \rangle|^2 \, \mathscr{E}^2}. \tag{7.16}$$

In the absence of the field, as it should be,

$$\Delta E_{1,2} = \pm \Delta/2. \tag{7.17}$$

If $\Delta/2 \gg |\langle \gamma J M | D_z | \gamma' J' M \rangle|^2 \mathscr{E}^2$, the expansion of the square root in (7.16) in powers of $\mathscr{E}$ gives the formulas for the quadratic Stark effect (7.10, 11):

$$\Delta E_1 = -\Delta E_2 = \frac{\Delta}{2} + \frac{|\langle \gamma J M | D_z | \gamma' J' M \rangle|^2}{\Delta} \mathscr{E}^2. \tag{7.18}$$

But if the interaction with the field is so great that the second term under the root in (7.16) becomes considerably greater than the first, then

$$\Delta E_{1,2} = \pm |\langle \gamma J M | D_z | \gamma' J' M \rangle| \, \mathscr{E}. \tag{7.19}$$

Thus with increase of $\mathscr{E}$ there occurs a transition of the quadratic effect into a linear effect. The dependence of the splitting on the field strength is shown in Fig. 7.1. This dependence, of course, is typical only in the two-level approximation. With increase of $\mathscr{E}$, the terms omitted from the sum (7.2) which are quadratic in $\mathscr{E}$ become more and more important. As a result of this, the linear dependence on $\mathscr{E}$ is replaced by a more complex one. The validity of the general formula (7.2) for the quadratic Stark effect is also limited by the condition for the smallness of $\Delta E_{\gamma J M}$ in comparison with the differences $E_{\gamma J} - E_{\gamma' J'}$. If the shift $\Delta E_{\gamma J M}$ becomes comparable with one of these differences, the quadratic dependence of the splitting on $\mathscr{E}$ is violated. A special situation arises when the levels γJ, $\gamma' J'$ are strictly degenerate: the splitting then depends linearly on $\mathscr{E}$ for indefinitely small values of $\mathscr{E}$. An example of this is hydrogen, the levels of which are degenerate with respect to l. This case will be examined in the next subsection.

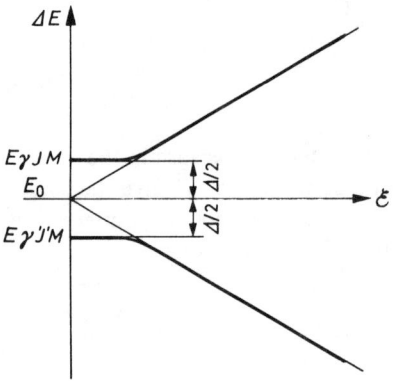

Fig. 7.1. Transition of the quadratic effect into a linear effect

Let us discuss now the splitting of spectral lines. This splitting and also the polarization of radiation depend on the direction of observation. In the case of observation along the z axis (in the direction of the field $\mathscr{E}$), the radiation is polarized in the plane x, y and is connected with the transitions $M \to M \pm 1$. The corresponding components of a line are called σ components. In a direction perpendicular to the z axis, besides the σ components, there are also observed π components, polarized along the z axis and arising from the transitions $M \to M$. The frequencies of the π and σ components are given by the obvious relations

$$\omega_\pi(M) = \omega_0 + [(A - A') + (B - B') M^2] \mathscr{E}^2,$$
$$\omega_\sigma(M) = \omega_0 + [(A - A') + BM^2 - B'(M \pm 1)^2] \mathscr{E}^2.$$

The strengths of the electric fields which are used in experiments do not considerably exceed 10^5 V cm^{-1} (0.33×10^3 cgse units). By substituting this magnitude into (7.12), we find that when $f \simeq 1$ and $(E_{\gamma J} - E_{\gamma' J'})/2\pi\hbar c \sim 10^3$ cm^{-1}, the splitting has an order of magnitude of 1 cm^{-1}. The magnitude of the splitting rapidly decreases with increasing $(E_{\gamma J} - E_{\gamma' J'})$, so as a rule the observed splitting of a line is determined entirely by the splitting of the upper level. In this case

$$\omega_\pi(M) = \omega_0 + (A + BM^2) \mathscr{E}^2,$$
$$\omega_\sigma(M) = \omega_0 + (A + BM^2) \mathscr{E}^2.$$
(7.20)

We also give in Table 7.1 the relative intensities of the π and σ components of a line for transverse observation (the calculations are made in Sect. 9.2).

Table 7.1 Relative intensities of the π and σ components of a line observed in the transverse direction

Transition	I_π	I_σ
$\gamma J \to \gamma' J$	$2M^2$	$\alpha J (J + 1) - M^2$
$\gamma J \to \gamma' J - 1$	$2(\alpha J^2 - M^2)$	$\alpha J (J - 1) + M^2$
$\gamma J \to \gamma' J + 1$	$2[\alpha(J + 1)^2 - M^2]$	$\alpha(J + 1)(J + 2) + M^2$
$\alpha = 1$ when $M \neq 0$ and $1/2$ when $M = 0$		

7.2 Hydrogenlike Levels. Linear Stark Effect

As already noted, the energy level of hydrogen undergo splitting proportional to $\mathscr{E}$ as a result of degeneracy with respect to l. This linear Stark effect is due to the mutual perturbation of the states with the same value of n and different l. For low levels (small n), calculation is comparatively simple, especially in the case when fine splitting is neglected. Let us consider the level $n = 2$. Four states correspond to this level: $l = 0$, $m = 0$; $l = 1$, $m = 0$, ± 1, for which only the matrix element $\langle 00|D_z|10\rangle$ is nonzero.

Thus the general secular equation of fourth order splits into two equations of first order for $m = \pm 1$,

$$\Delta E_1 = \Delta E_{-1} = 0, \tag{7.21}$$

and an equation of second order for $m = 0$,

$$\begin{vmatrix} \Delta E_0 & \langle 00|D_z|10\rangle\,\mathscr{E} \\ \langle 10|D_z|00\rangle\,\mathscr{E} & \Delta E_0 \end{vmatrix} = 0, \tag{7.22}$$

$$\Delta E_0^{(1)} = +\langle 00|D_z|10\rangle\,\mathscr{E}, \qquad \Delta E_0^{(2)} = -\langle 00|D_z|10\rangle\,\mathscr{E}. \tag{7.23}$$

Consequently, the level $n = 2$ splits into three sublevels, one of which is twofold degenerate. This splitting is symmetrical.

Let us also consider the splitting of the level $n = 3$. The states $l = 0$, $m = 0$; $l = 1$, $m = 0, \pm 1$; $l = 2$, $m = 0, \pm 1, \pm 2$ correspond to this level. The corrections to the energy are determined by the equations

$m = \pm 2$
$$\Delta E_2 = \Delta E_{-2} = 0, \tag{7.24}$$

$m = \pm 1$
$$\begin{vmatrix} \Delta E_m & \langle 1m|D_z|2m\rangle\,\mathscr{E} \\ \langle 2m|D_z|1m\rangle\,\mathscr{E} & \Delta E_m \end{vmatrix} = 0, \tag{7.25}$$

$$\Delta E_1^{(1)} = \Delta E_{-1}^{(1)} = \langle 1m|D_z|2m\rangle\,\mathscr{E}, \tag{7.26}$$

$$\Delta E_1^{(2)} = \Delta E_{-1}^{(2)} = -\langle 1m|D_z|2m\rangle\,\mathscr{E}, \tag{7.27}$$

$m = 0$
$$\begin{vmatrix} \Delta E_0 & \langle 00|D_z|10\rangle\,\mathscr{E} & 0 \\ \langle 10|D_z|00\rangle\,\mathscr{E} & \Delta E_0 & \langle 10|D_z|20\rangle\,\mathscr{E} \\ 0 & \langle 20|D_z|10\rangle\,\mathscr{E} & \Delta E_0 \end{vmatrix} = 0, \tag{7.28}$$

$$\Delta E_0 = 0 \tag{7.29}$$

$$\Delta E_0^{(1)} = \sqrt{|\langle 00|D_z|10\rangle|^2 + |\langle 10|D_z|20\rangle|^2}\cdot\mathscr{E}, \tag{7.30}$$

$$\Delta E_0^{(2)} = -\sqrt{|\langle 00|D_z|10\rangle|^2 + |\langle 10|D_z|20\rangle|^2}\cdot\mathscr{E}. \tag{7.31}$$

Thus the level $n = 3$ splits into 5 components, the splitting being symmetrical and linear in $\mathscr{E}$. The scheme of splitting of the levels $n = 2, n = 3$ (an arbitrary scale) and also the possible radiative transitions are shown in Fig. 7.2. As is seen from Fig. 7.2, the line H_α splits into 15 components (8 π components and 7 σ components).

7.2 Hydrogenlike Levels. Linear Stark Effect

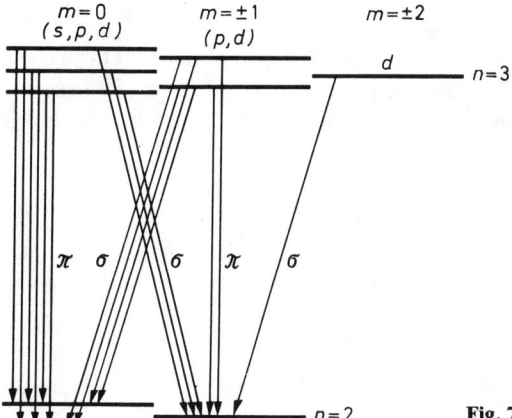

Fig. 7.2. Splitting of the levels $n = 2,3$ of hydrogen in an electri field

It is not reasonable to extend these calculations to other excited levels because then a secular equation of high order has to be solved. It is more convenient to use the fact that the transition from cartesian to parabolic coordinates

$$\xi = \sqrt{x^2 + y^2 + z^2} + z = r(1 + \cos\theta), \tag{7.32}$$

$$\eta = \sqrt{x^2 + y^2 + z^2} - z = r(1 - \cos\theta), \tag{7.33}$$

$$\varphi = \arctan(y/x) \tag{7.34}$$

diagonalizes the matrix $D_z = 1/2 \, (D_\xi - D_\eta)$.[1] In parabolic coordinates the stationary state of a discrete spectrum is defined by the "parabolic" quantum numbers n_1, n_2 and the magnetic quantum number m. The principal quantum number n is connected with n_1 and n_2 by the relation

$$n = n_1 + n_2 + |m| + 1. \tag{7.35}$$

For a given n the number $|m|$ can take n different values $0, 1, 2 \ldots (n-1)$. For each $|m|$ the number n_1 takes the values, $0, 1 \ldots n - |m| - 1$. The first-order correction of perturbation theory to the energy levels has the form

$$\Delta E^{(1)} = \frac{3}{2} n(n_1 - n_2) e \, \mathscr{E} \, a_0 \,. \tag{7.36}$$

---

[1] For a discussion of the Schrödinger equation for hydrogen in parabolic coordinates with and without taking the electric field into account, and also for a derivation of the formulas given below, see [1,3,20].

For a given n the difference $n_1 - n_2$ can take the values $n - 1, n - 2, n - 3, \ldots -(n - 1)$.

Thus the level n splits into $2(n - 1) + 1 = 2n - 1$ components. This agrees with the results for $n = 2$ and $n = 3$ obtained above. The splitting of a spectral line corresponding to the transition $n - n'$ is characterized by the possible values of the difference

$$\Delta = n(n_1 - n_2) - n'(n_1' - n_2'). \tag{7.37}$$

The selection rules with respect to the magnetic quantum number m remain the same as without an external field [see (1.15)]:

$$\begin{aligned} \Delta m &= 0, \quad \pi \text{ components} \\ \Delta m &= \pm 1, \quad \sigma \text{ components}. \end{aligned} \tag{7.38}$$

In parabolic coordinates it is possible to also obtain a simple expression for the second-order correction to the energy:

$$\Delta E^{(2)} = -\frac{n^4}{16}[17n^2 - 3(n_1 - n_2)^2 - 9m^2 + 19]\, \mathscr{E}^2 a_0^3. \tag{7.39}$$

In contrast to (7.36), the quadratic effect depends on $|m|$. Thus for large values of $\mathscr{E}$ there is no degeneracy with respect to $|m|$. Comparison of (7.36) and (7.39) shows that the violation of the linear dependence of the splitting on $\mathscr{E}$ begins with fields on the order of

$$\mathscr{E} \sim 0.1\, e/n^4 a_0^2 \sim n^{-4} 10^6 \text{ cgse units} \sim n^{-4} 3 \cdot 10^8 \text{ V cm}^{-1}. \tag{7.40}$$

A Stark effect of the same type as in the case of hydrogen is also characteristic for highly excited hydrogenlike levels of other atoms.

It follows from the selection rules for radiation that radiative transitions $2s$—$1s$ are forbidden. It is not difficult to show that even in an extremely weak electric field, such transitions become allowed.

The level $n = 2$ in an electric field splits into the components $n_1 = 1, n_2 = 0, m = 0$; $n_1 = 0, n_2 = 1, m = 0$ and $n_1 = 0, n_2 = 0, m = \pm 1$. Consequently, the eigenfunctions of the Hamiltonian $H_0 - \mathscr{E}D_z$ are the parabolic functions $\psi_{n_1 n_2 m}$. These functions can be represented in the form of a linear combination of the spherical functions ψ_{lm}. The expansion coefficients are easily defined by the general formulas of perturbation theory

$$\psi_{001} = \psi_{p,1}; \quad \psi_{00,-1} = \psi_{p,-1}; \quad \psi_{100} = \frac{1}{\sqrt{2}}(\psi_{s0} - \psi_{p0});$$

$$\psi_{010} = \frac{1}{\sqrt{2}}(\psi_{s0} + \psi_{p0}). \tag{7.41}$$

Let us assume that as a result of some excitation process the atom at the initial time $t = 0$ is in the state ψ_{s0}. For $t > 0$ the time-dependent wave function of the atom in an electric field $\Psi(t)$ can be written in the form of a linear combination of the wave functions of the stationary states $n_1 = 1, n_2 = 0, m = 0$ and $n_1 = 0, n_2 = 1, m = 0$

$$\Psi(t) = A\psi_{100}\, e^{-i(E_0+\Delta)t/\hbar} + B\psi_{010}\, e^{-i(E_0-\Delta)t/\hbar}, \tag{7.42}$$

where, in accordance with (7.36), $\Delta = 3ea_0\, \mathscr{E}$.

The coefficients A and B are found from the initial condition: when $t = 0$, $\Psi(0) = \psi_{s0}$. Substituting (7.41) in (7.42), we have

$$\begin{aligned}\Psi(t) &= \frac{1}{\sqrt{2}} \psi_{100}\, e^{-i(E_0+\Delta)t/\hbar} + \frac{1}{\sqrt{2}} \psi_{010}\, e^{-i(E_0-\Delta)t/\hbar} \\ &= \left[\psi_{s0}\cos\frac{\Delta}{\hbar}t + \psi_{p0} i\sin\frac{\Delta}{\hbar}t\right] e^{-iE_0t/\hbar}. \end{aligned} \tag{7.43}$$

It follows from (7.43) that the orbital angular momentum of an electron is not conserved in an electric field. The atom oscillates between the states ψ_{s0} and ψ_{p0} with a period $T = \pi\hbar/\Delta$. Let us estimate the magnitude of this period. When $\mathscr{E} \sim 300$ V cm^{-1}, $\Delta/\hbar \sim 7.5 \times 10^9$. Consequently, even in such a weak field the period of oscillation between the states $2s0$ and $2p0$ is of the same order as the time τ necessary for the radiative transition $2p$—$1s$.

Thus an electric field applied to the atom in the state $2s$ can induce a radiative transition to the state $1s$. The probability of this transition for $\mathscr{E} \sim 300$ V cm^{-1} is approximately equal to the probability of the allowed transition $2p$—$1s$.

In a strong electric field, when $T \ll \tau$, during the whole time of decay the states $2s0$ and $2p0$ are populated roughly equally (independent of which state the atom was in at the initial moment $t = 0$). Therefore, the probabilities of radiative transitions $2s$—$1s$, $2p$—$1s$ in the presence of a strong electric field are the same and equal $(2\tau)^{-1}$. It is obvious that an electric field also violates the restrictions on other ns—$n's$ transitions.

7.3 Inhomogeneous Field. Quadrupole Splitting

In the case of an inhomogeneous electric field it is necessary to add to the dipole interaction (7.1) terms which take into account higher multipole moments of the atom. If the variation of the field in a distance of the order of the dimensions of the atom is small, then among these terms the quadrupole interaction plays the principal role. Of greatest interest are fields produced by charged particles—electrons and ions. In this case the energy of the quadrupole interaction can be written in the form (6.28).

Let us locate the origin of coordinates at the center of the atom and direct the z axis towards the charge e' producing the field. Then

$$H' = -\frac{ee'}{R^3} Q_{20}, \tag{7.44}$$

where R is the distance to the charge e', Q_{20} is the component $q = 0$ of the quadrupole moment of the atom

$$Q_{2q} = \sum_i r_i^2 C_q^2(\theta_i, \varphi_i). \tag{7.45}$$

From (6.11, 12) it follows

$$\langle \gamma JM|H'|\gamma JM\rangle = -\frac{ee'}{R^3} \frac{1}{2} Q \frac{3M^2 - J(J+1)}{J(2J-1)}, \tag{7.46}$$

$$Q = 2(\gamma J\|Q_2\|\gamma J)\sqrt{\frac{J(2J-1)}{(2J+3)(2J+1)(J+1)}}. \tag{7.47}$$

Thus for levels $J \neq 0, 1/2$ there occurs a quadrupole splitting which is linear in the field.

For single-electron atoms (one electron outside closed shells)

$$Q = -\langle r^2\rangle \frac{2j-1}{2j+2} \tag{7.48}$$

[see (6.16)]; therefore,

$$\langle \gamma jm|H'|\gamma jm\rangle = -\frac{ee'}{R^3}\langle r^2\rangle \frac{j(j+1) - 3m^2}{4j(j+1)}. \tag{7.49}$$

For $j = 3/2$ the splitting is symmetrical

$$\langle \gamma\, 3/2\, 3/2|H'|\gamma\, 3/2\, 3/2\rangle = -\langle \gamma\, 3/2\, 1/2|H'|\gamma\, 3/2\, 1/2\rangle.$$

For all other values of j the splitting is asymmetrical. We shall also find the dependence of the splitting on J in the general case of LS coupling. From (7.47) we have

$$Q = 2(\gamma SL\|Q_2\|\gamma SL)(-1)^{S+L+J}\sqrt{\frac{J(2J-1)(2J+1)}{(2J+3)(J+1)}} \begin{Bmatrix} L & J & S \\ J & L & 2 \end{Bmatrix}. \tag{7.50}$$

The reduced matrix element $(\gamma SL\|Q_2\|\gamma SL)$ can be calculated by using the general formulas of Sects. 4.3 and 5.2. We shall consider the two simplest examples — the configurations ll' and l^n. In the first case it is easy to obtain

$$(ll'SL\|Q_2\|ll'SL) = (l_1 l_2' SL\|Q_2(1)\|l_1 l_2' SL) + (l_1 l_2' SL\|Q_2(2)\|l_1 l_2' SL)$$

$$= \langle r_1^2 \rangle (l\|C^2\|l)(-1)^{l'+l+L}(2L+1)\begin{Bmatrix} l & L & l' \\ L & l & 2 \end{Bmatrix}$$

$$+ \langle r_2^2 \rangle (l'\|C^2\|l')(-1)^{l+l'+L}(2L+1)\begin{Bmatrix} l' & L & l \\ L & l' & 2 \end{Bmatrix}. \tag{7.51}$$

Thus

$$Q(ll'SLJ) = (-1)^{S+J+l-l'} 2 \left[\langle r_1^2 \rangle (l\|C^2\|l) \begin{Bmatrix} l & L & l' \\ L & l & 2 \end{Bmatrix} \right.$$

$$\left. + \langle r_2^2 \rangle (l'\|C^2\|l') \begin{Bmatrix} l' & L & l \\ L & l' & 2 \end{Bmatrix} \right] (2L+1) \sqrt{\frac{J(2J-1)(2J+1)}{(2J+3)(J+1)}} \begin{Bmatrix} L & J & S \\ J & L & 2 \end{Bmatrix}. \tag{7.52}$$

In the case of configuration l^n,

$$(l^n \gamma SL\|Q_2\|l^n \gamma SL)$$

$$= n \sum_{\gamma' S'L'}' |G^{\gamma SL}_{\gamma' S'L'}|^2 (\gamma' S'L'l_n SL\|Q_2(n)\|\gamma' S'L'l_n SL)$$

$$= n \sum_{\gamma' S'L'}' |G^{\gamma SL}_{\gamma' S'L'}|^2 \langle r^2 \rangle (l\|C^2\|l)(-1)^{L'+l+L}(2L+1) \begin{Bmatrix} l & L & L' \\ L & l & 2 \end{Bmatrix}, \tag{7.53}$$

$$Q(l^n \gamma SL) = \langle r^2 \rangle (l\|C^2\|l)(2L+1)$$

$$\times (-1)^{L+l+L} n \sum_{\gamma' S'L'}' |G^{\gamma SL}_{\gamma' S'L'}|^2 (-1)^{L'} \begin{Bmatrix} l & L & L' \\ L & l & 2 \end{Bmatrix}.$$

We have considered above the special case of a field of charge e'. It is easy to generalize all the results to the case of an arbitrary inhomogeneous field having axial symmetry replacing e'/R^3 by $(1/2)\partial^2\varphi/\partial Z^2$, where φ is the electrostatic potential.

7.4 Time-Dependent Field

7.4.1 Amplitude Modulation

We shall begin the study of the Stark effect in a time-dependent field by considering the general case of a pertubation $V(t)$ depending explicitly on time. Let us assume that for $t < t_0$, the atom is in the state n. We shall expand the wave function $\varphi_n(t)$ in wave functions of the unperturbed atom

$$\psi_k^{(0)}(t) = \psi_k \, e^{-iE_k t/\hbar} \tag{7.54}$$

184   7. The Atom in an External Electric Field

$$\psi_n(t) = \sum_k a_{kn} \psi_k \, e^{-iE_k t/\hbar} \tag{7.55}$$

The coefficients of this expansion $a_{kn}(t)$ are determined by the well-known equations of perturbation theory

$$i\hbar \dot{a}_{kn} = \sum_s V_{ks} a_{sn} e^{i\omega_{ks} t}, \tag{7.56}$$

$$\hbar \omega_{ks} = E_k - E_s \tag{7.57}$$

and satisfy the initial conditions

$$a_{kn}(t_0) = \delta_{kn}. \tag{7.58}$$

In what follows it is convenient to make the substitution

$$a_{nn} = e^{-i\alpha_n} \tag{7.59}$$

and put $t_0 = 0$. After this, we obtain the system of equations

$$\hbar \dot{\alpha}_n = V_{nn} + {\sum_s}' e^{i\alpha_n} V_{ns} a_{sn} e^{i\omega_{ns} t}, \tag{7.60}$$

$$k \neq n \quad i\hbar \dot{a}_{kn} = e^{-i\alpha_n} V_{kn} e^{i\omega_{kn} t} + {\sum_s}' V_{ks} a_{sn} e^{i\omega_{ks} t}$$

with the initial conditions

$$a_{nk}(0) = \delta_{nk}, \quad \alpha_n(0) = 0. \tag{7.61}$$

By integrating the system of (7.60) in the framework of perturbation theory, it is possible in the second equation to omit the sum over s, containing the small quantities a_{ns} and V_{ns}, and assume $\exp(-i\alpha_n) \simeq 1$. After this,

$$a_{kn} = -\frac{i}{\hbar} \int_0^t V_{kn}(t') \, e^{i\omega_{kn} t'} \, dt'. \tag{7.62}$$

Substituting this expression in the first of (7.60) in which it is also assumed that $\exp(i\alpha_n) \simeq 1$, we obtain in a second approximation of perturbation theory

$$\alpha_n(t) = \frac{1}{\hbar} \int_0^t V_{nn}(t') \, dt' - \frac{i}{\hbar^2} {\sum_s}' \int_0^t V_{ns}(t') e^{i\omega_{ns} t'} \, dt' \int_0^t V_{ns}^*(t'') e^{-i\omega_{ns} t''} \, dt''. \tag{7.63}$$

In the general case the phase $\alpha_n(t)$ is complex

$$\alpha_n(t) = \eta_n(t) - i\Gamma_n(t). \tag{7.64}$$

We now explain the physical meaning of the quantities η_n and Γ_n. From (7.63) it is not difficult to obtain[2]

$$2\Gamma_n(\infty) = -\operatorname{Im}\{2\alpha_n(\infty)\} = \frac{1}{\hbar^2}\sum_s{}' \left|\int_0^\infty V_{ns}\, e^{i\omega_{ns}t}\, dt\right|^2. \qquad (7.65)$$

The right-hand side of (7.65) coincides with the familiar expression for the total probability of transitions from the level n to all other levels. Thus the imaginary part of the phase α_n describes the decay of the state n caused by the perturbation $V(t)$. It is simplest of all to explain the physical meaning of η_n if a constant or slowly varying perturbation is considered. In this case, on integrating the second term of (7.63) by parts, we obtain

$$\int_0^{t'} V_{ns}^*(t'')\, e^{-i\omega_{ns}t''}\, dt''$$

$$= V_{ns}^*(t'')\frac{e^{-i\omega_{ns}t''}}{-i\omega_{ns}}\bigg|_0^{t'} - \int_0^{t'}\frac{dV^*}{dt''}\frac{e^{-i\omega_{ns}t''}}{-i\omega_{ns}}\, dt'' \approx V_{ns}^*(t')\frac{e^{-i\omega_{ns}t'}}{-i\omega_{ns}}. \qquad (7.66)$$

Therefore,

$$\eta_n(t) \approx \alpha_n(t) \approx \frac{1}{\hbar}\int_0^t\left[V_{nn}(t') + \sum_s{}'\frac{|V_{ns}(t')|^2}{\hbar\omega_{ns}}\right]dt'. \qquad (7.67)$$

The expression in brackets under the integral is the shift of the level n under the action of the perturbation V.

Thus η_n is an increase of phase $\hbar^{-1}\int_0^t \Delta E(t')\,dt'$ caused by the shift of the level n [remember that the phase of the unperturbed wave function ψ_n equals $(E_n/\hbar)\int_{-\infty}^t dt'$ and the shift of the level ΔE_n in a constant field leads to an additional change of phase by an amount $(\Delta E_n/\hbar)\int_{-\infty}^t dt'$].

It follows from (7.66, 67) that a perturbation varying only a little in a time of order ω_{ns}^{-1} does not cause transitions from the state n to other states. The phase α_n is real.

If we assume that in (7.67) $V = -\mathscr{E}D_z$, $V_{nn} = 0$, we obtain the formula for the quadratic Stark effect

---

[2] In the derivation of (7.65) the obvious relation was used
$$\operatorname{Im}\left\{i\int_0^t \Phi(t')\,dt'\int_0^{t'}\Phi^*(t'')\,dt''\right\} = \int_0^t \operatorname{Re}\{\Phi(t')\}\,dt'\int_0^{t'}\operatorname{Re}\{\Phi(t'')\}\,dt''$$
$$+ \int_0^t \operatorname{Im}\{\Phi(t')\}\,dt'\int_0^{t'}\operatorname{Im}\{\Phi(t'')\}\,dt'',$$
and also the fact that for an arbitrary function $f(t)$
$$\int_0^t f(t')\,dt'\int_0^{t'}f(t'')\,dt'' = \frac{1}{2}\left|\int_0^t f(t')\,dt'\right|^2.$$

$$\Delta E(t') = \mathscr{E}^2(t') \sum_s{}' \frac{|(D_z)_{ns}|^2}{\hbar \omega_{ns}}. \tag{7.68}$$

The time-dependent quantity $\mathscr{E}^2(t')$ now enters into this formula. Thus the shift of the level at each given moment of time is determined by the same equation as in the case of a constant field.

A completely different situation arises for a rapidly changing field. Let us assume that the field is applied for the short time Δt, small compared to the periods of motion of the electrons $T_{ns} = 2\pi\omega_{ns}^{-1}$. In this case the factor $\exp(-i\omega_{ns}t'')$ can be carried outside the integral. Thus the phase α_n proves to be purely imaginary and $\eta_n = 0$. This means that a rapidly varying perturbation causes transitions between levels but does not give a shift. We shall consider this effect in more detail in the particular case of a perturbation which is constant in the interval $\Delta t(t_0, t_0 + \Delta t)$. By integrating the second term of (7.63), we obtain the following expression for the increase of the phase η in the time Δt

$$\eta = \frac{1}{\hbar} \sum_s{}' \frac{|V_{ns}|^2}{\hbar \omega_{ns}} \left(\Delta t - \frac{\sin \omega_{ns} \Delta t}{\omega_{ns}} \right) \approx \frac{1}{\hbar} \sum_s{}' \frac{|V_{ns}|^2}{\hbar \omega_{ns}} \frac{\omega_{ns}^2 \Delta t^2}{6} \Delta t. \tag{7.69}$$

Calculating the same quantity by the quadratic Stark effect formula for a constant field we have

$$\eta = \frac{1}{\hbar} \sum_s{}' \frac{|V_{ns}|^2}{\hbar \omega_{ns}} \Delta t. \tag{7.70}$$

Thus the instantaneous shift of a level is considerably less than in a constant field of the same magnitude. The atom does not succeed in following the field. This effect has a simple physical meaning. In the absence of the field, the atom does not have a dipole moment. The latter appears only as a result of the polarization of the atom by the field, i.e., as a result of the deformation of the electron shells. If the field is applied only for a small interval of time $\Delta t < T_{ns}$, then the shells are not deformed because of the inertia of the system.

In the example considered above, the reduction of shift because of the delay effects is determined by the factors $(\Delta t/T_{ns})^2$. For an atom, $T_{ns} \lesssim 10^{14}$ s (with the exception of highly excited states). Thus by time-dependent field it is necessary to understand those fields whose magnitude varies in a time of the order of 10^{-14} s. If, for example, a charged particle with a velocity $v \simeq 10^8$ cm s^{-1} (in the case of an electron, a kinetic energy of the order of 3 eV corresponds to such a velocity) and impact radius of 10^{-7} cm interacts with the atom, a field is applied for a time of the order of 10^{-15} s. In this case it is very important to take into consideration the nonstationary nature of the field.

We shall also consider the periodic perturbation

$$V = -D_z \mathscr{E}_0 \cos \omega t = -D_z \mathscr{E}_0 \frac{1}{2} (e^{i\omega t} + e^{-i\omega t}). \tag{7.71}$$

7.4 Time-Dependent Field

In this case by integrating the second term in (7.63), we obtain the following expression for the time average of the derivative of the phase

$$\bar{\alpha}_n = \frac{1}{4}\frac{\mathscr{E}_0^2}{\hbar^2}\sum_s{}'\left(\frac{1}{\omega_{ns}+\omega}+\frac{1}{\omega_{ns}-\omega}\right)|(D_z)_{ns}|^2. \qquad (7.72)$$

Thus the mean shift of the level $\overline{\Delta E_n}$ is related to the mean value of the square of the field $\overline{\mathscr{E}^2} = 1/2\,\overline{\mathscr{E}_0^2}$ by

$$\begin{aligned}\overline{\Delta E_n} &= \frac{1}{2\hbar}\sum_s{}'|(D_z)_{ns}|^2\left(\frac{1}{\omega_{ns}+\omega}+\frac{1}{\omega_{ns}-\omega}\right)\overline{\mathscr{E}^2}\\ &= \frac{1}{\hbar}\sum_s{}'\frac{\omega_{ns}|(D_z)_{ns}|^2}{\omega_{ns}^2-\omega^2}\overline{\mathscr{E}^2}\,.\end{aligned} \qquad (7.73)$$

In the limiting case of a static field $\omega \to 0$, (7.73) becomes the usual quadratic Stark effect formula. For high frequencies $\omega \gg \omega_{ns}$ the corresponding terms of the sums (7.68) and (7.73) differ approximately by the factor $(\omega_{ns}/\omega)^2$.

7.4.2 The Hydrogen Atom in a Rotating Electric Field[3]

We shall consider the excited hydrogen atom placed in electric field perpendicular to the z axis and rotating in the plane xy with angular velocity Ω. The amplitude of the field $\mathscr{E}$ is supposed to be constant. We shall introduce the rotating system of coordinates $x'y'z'(z'=z)$, the x' axis being directed along field $\mathscr{E}$. The wave functions $\varphi'(t)$ in the rotating system of coordinates are related to the wave functions $\varphi(t)$ in the initial system by

$$\varphi'(t) = e^{iL_z\Omega t}\varphi(t)\,, \qquad (7.74)$$

where L is the angular momentum operator. Substituting (7.74) into the Schrödinger equation, we obtain

$$i\hbar\frac{\partial\varphi'}{\partial t} = (H_0 + d_x\mathscr{E} + \hbar L_z\Omega)\,\varphi' \equiv (\hat{H}_0 + \hat{V})\,\varphi'\,. \qquad (7.75)$$

It is easy to see that in the rotating system, besides the interaction with electric field $d_x\mathscr{E}$, there is an additional interaction $L_z\Omega$ of the magnetic type due to the rotation. As a result of this interaction the energy level n splits into further components so that resulting number of these components exceeds the number of states n^2 corresponding to the level. The picture of the corresponding splitting of the spectral line is determined by the magnitude of the parameter $x = \Omega\,[(B/e)\cdot\mathscr{E}]^{-1}$, where

---

[3] We follow [21] here.

$B = (3/2)ne^2 a_0/\hbar$. As an example, the splitting of the line L_α for different values of Ω (of parameter x) is shown in Fig. 7.3. In a constant field the line L_α splits into three components ($x = 0$). In a rotating field ($x \neq 0$) the above-mentioned central component splits into three new components and each of the left and right components splits into two new components.

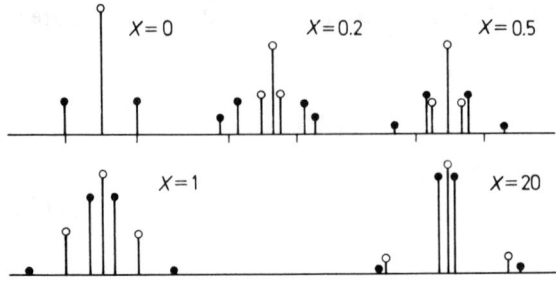

Fig. 7.3. Splitting of the L_α line in a rotating electric field

Chapter 8 The Atom in an External Magnetic Field

In this chapter the level splitting in weak and strong magnetic fields (including the splitting of hyperfine structure components) is considered.

8.1 Zeeman Effect

A magnetic field, in contrast to an electric field, completely removes the degeneracy of levels with respect to M. The interaction of an atom with a magnetic field has the form

$$W = -\boldsymbol{\mu}\cdot\boldsymbol{H}, \tag{8.1}$$

where $\boldsymbol{\mu}$ is the magnetic moment of the atom. This moment, generally speaking, is composed of two parts – electronic and nuclear. The latter, however, is at least three orders of magnitude less than the former. Therefore, for the magnetic moment of an atom in the state γJ it can be assumed that

$$\boldsymbol{\mu} = -\mu_0 g \boldsymbol{J}. \tag{8.2}$$

Here $\mu_0 = e\hbar(2mc)^{-1}$ is the Bohr magneton, $\boldsymbol{J}$ is the total electronic angular momentum, and g is the gyromagnetic ratio, which is usually called the g factor (Sect. 6.1). With the z axis along the direction of $\boldsymbol{H}$, we obtain

$$\langle W \rangle = g\mu_0 H M. \tag{8.3}$$

Thus the level γJ in a magnetic field splits into $(2J+1)$ components $M=0$, $\pm, \pm 2, \ldots \pm J$. This splitting is linear in H and is symmetrical. The absolute magnitude of the splitting is determined by the magnitude of H and the g factor. As g is of the order 1, the absolute magnitude of the splitting in cm^{-1} is $eH/4\pi mc^2 \approx 4.7\cdot 10^{-5} H$. When H is of the order 10^4 Oe, the splitting reaches 0,5 cm^{-1}. The value of the g factor depends on the type of coupling. In the case of LS coupling, calculation of the g factor is very simple. The operator of the magnetic moment of an electron is given by the expression

$$\boldsymbol{\mu} = -\mu_0(g_l \boldsymbol{l} + g_s \boldsymbol{s}), \tag{8.4}$$

where $g_l = 1$, $g_s = 2$; therefore

$$g\boldsymbol{J} = \langle g_l \sum_i \boldsymbol{l}_i + g_s \sum_i \boldsymbol{s}_i \rangle = \langle \boldsymbol{L} + 2\boldsymbol{S} \rangle. \tag{8.5}$$

Averaging in (8.5) is done over states with a given value of the total angular momentum. Using the equality

$$L + 2S = J + S$$

and calculating the mean value of S with the aid of (4.180)

$$\langle S \rangle = \frac{\langle S \cdot J \rangle}{J(J+1)} J, \qquad (8.6)$$

we obtain

$$g = 1 + \frac{J(J+1) - L(L+1) + S(S+1)}{2J(J+1)}. \qquad (8.7)$$

This is the so-called Landé factor. When $S = 0$, $g = 1$; when $L = 0$, $g = 2$; and when $L = S$, $g = 3/2$. In the general case for the components of the fine structure of terms with $L \geqslant S$

$$\frac{L + 2S}{L + S} \geqslant g \geqslant \frac{L - 2S + 1}{L - S + 1},$$

and with $L < S$

$$\frac{L + 2S}{L + S} \leqslant g \leqslant \frac{2S + 2 - L}{S - L + 1}.$$

For one electron outside closed shells

$$g = 1 + \frac{j(j+1) - l(l+1) + 3/4}{2j(j+1)}. \qquad (8.8)$$

For some levels (for example, $^4D_{1/2}$, $^5F_1$) the Landé factor is zero. This means that in first-order perturbation theory these levels do not split.

In the case of jj coupling, calculation of g factors is a more complex problem. Simple general formulas can be obtained only for the configuration jj' and j^n. In the first case

$$gJ = \langle g(j) j + g(j') j' \rangle,$$

$$g(J) = g(j) \frac{J(J+1) - j'(j'+1) + j(j+1)}{2J(J+1)}$$

$$+ g(j') \frac{J(J+1) - j(j+1) + j'(j'+1)}{2J(J+1)}, \qquad (8.9)$$

each of the g factors on the right-hand side of (8.9) being determined by (8.8). In the second case

$$\sum_i g(j)\, j_i = g(j)\, J;$$

therefore

$$g(j^n J) = g(j).\tag{8.10}$$

In the case of an intermediate coupling the g factor for the level αJ can be expressed in terms of the g factors of the LS coupling. The eigenfunctions $\Psi_{\alpha J}$ are defined by the expansion

$$\Psi_{\alpha J} = \sum_{\gamma SL} (\gamma SLJ|\alpha J)\, \Psi_{\gamma SLJ}.\tag{8.11}$$

Therefore

$$g(\alpha J) = \sum_{\gamma SL} |(\gamma SLJ|\alpha J)|^2 g(\gamma SLJ).\tag{8.12}$$

Summing over γSL means summing over all terms of the given configuration for which $L + S \geqslant J \geqslant |L - S|$. From the unitary property of the transformation coefficients $(\gamma SLJ|\alpha J)$

$$\sum_\alpha (\gamma SLJ|\alpha J)(\alpha J|\gamma' S'L'J) = \delta_{\gamma\gamma'}\delta_{ss'}\delta_{LL'}$$

there follows the important sum rule

$$\sum_\alpha g(\alpha J) = \sum_{\gamma SL} g(\gamma SLJ).\tag{8.13}$$

Thus the sum of the g factors for all levels of a given configuration having one and the same value of J does not depend on the type of coupling. In particular, this sum is the same in the two limiting cases of LS coupling and jj coupling. We shall consider as an example the levels $J = 1$ of the configuration $npn'p$. In the LS coupling approximation

$$g(^1P_1) = 1,\quad g(^3S_1) = 2,\quad g(^3P_1) = 3/2,\quad g(^3D_1) = 1/2,\quad \sum g = 5.$$

In the jj coupling approximation

$$g\left(\tfrac{1}{2}\tfrac{1}{2}\right) = \tfrac{2}{3},\quad g\left(\tfrac{1}{2}\tfrac{3}{2}\right) = g\left(\tfrac{3}{2}\tfrac{1}{2}\right) = \tfrac{3}{2},\quad g\left(\tfrac{3}{2}\tfrac{3}{2}\right) = \tfrac{4}{3},\quad \sum g = 5.$$

In these cases when there is a strong interaction between any two configurations, the summation in (8.13) has to be extended to the terms of both configurations.

Let us consider the splitting of spectral lines in a magnetic field. Just as in the case of the Stark effect, σ components ($\Delta M = \pm 1$) are observed in the direction of the z axis and σ and π components ($\Delta M = 0$) in a direction perpendicular to the z axis. From (8.9) it follows that

$$\omega_\pi = \omega_0 + \frac{1}{\hbar} \mu_0 H (g - g') M ,$$
$$\omega_\sigma = \omega_0 + \frac{1}{\hbar} \mu_0 H [gM - g' (M \pm 1)] . \tag{8.14}$$

If $g = g'$,

$$\omega_\pi = \omega_0 , \quad \omega_\sigma = \omega_0 \pm \frac{1}{\hbar} \mu_0 H g . \tag{8.15}$$

Consequently in this case a doublet is observed along the field, the components of the doublet being on either side of ω_0 at equal distances $\mu_0 H/\hbar$. In a direction perpendicular to the field a triplet is observed — an unshifted π component added to the σ components. Splitting of this type is usually called the normal Zeeman effect. The general case (8.14) is called the anomalous effect. The splitting (8.14) had not found any theoretical explanation before the discovery of electron spin, whereas (8.15) followed from classical electron theory. When $S = 0, g = g' = 1$.

In the general case of (8.14) the splitting has a considerably more complex form. As an example, the splitting of spectral lines corresponding to various transitions between doublet terms is shown in Fig. 8.1. The usual notation for the π and σ components (π components above and σ components below the line) is adopted in this figure. The relative intensities of the π and σ components of a line are calculated in Sect. 9.2. Results are collected in Table 8.1. It follows from Table 8.1 that the intensities of those π components (and σ components) which are symmetrical about ω_0 are the same. In the transverse direction the intensity of the σ components is less than along the z axis by a factor of two. This is explained by the fact that D_x and D_y components of the dipole moment contribute to the intensity of radiation in the direction of the z axis, but only one of them contributes in the direction of the x axis or y axis.

From the formulas given there follow a number of general regularities in the distribution of intensity for the π and σ components of a line. Thus for the transition $\gamma J \to \gamma' J$, the intensity of the π components increases with increase of shift (increase of M) and decreases for transitions $\gamma J \to \gamma' J \pm 1$. In Fig. 8.1 the intensity of each of the components is given by the height of the corresponding line.

Levels with $J = 0$ do not split in a magnetic field. These levels, however, undergo a shift in second-order of perturbation theory, because the correction to the energy

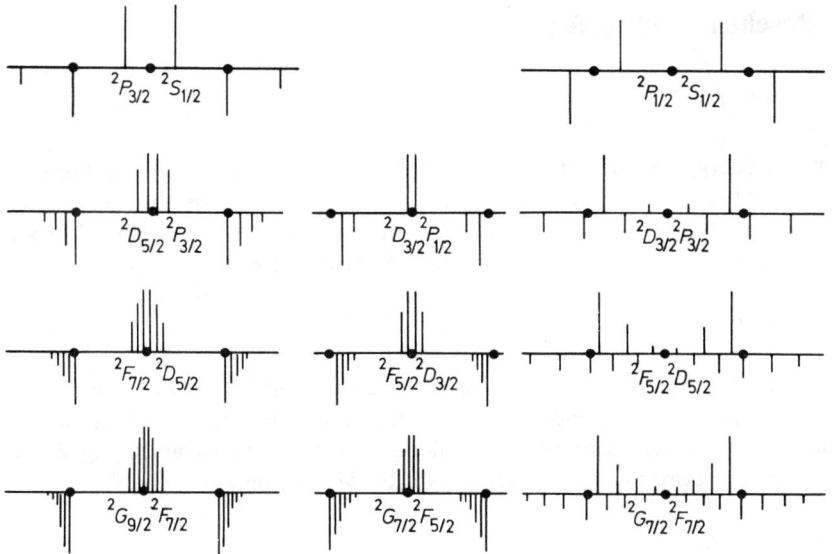

Fig. 8.1. Zeeman splitting of the doublet lines

Table 8.1 Relative intensities of Zeeman components

Transverse observation			
Transition	I_π	$I_\sigma (M \to M-1)$	$I_\sigma (M \to M+1)$
$\gamma J \to \gamma' J$	M^2	$\frac{1}{4}(J+M)(J+1-M)$	$\frac{1}{4}(J-M)(J+1+M)$
$\gamma J \to \gamma'(J-1)$	$J^2 - M^2$	$\frac{1}{4}(J+M)(J-1+M)$	$\frac{1}{4}(J-M)(J-1-M)$
$\gamma J \to \gamma'(J+1)$	$(J+1)^2 - M^2$	$\frac{1}{4}(J+1-M)(J-M+2)$	$\frac{1}{4}(J+1+M)(J+M+2)$
Longitudinal observation			
Transition	I_π	$I_\sigma(M \to M-1)$	$I_\sigma(M \to M+1)$
$\gamma J \to \gamma' J$	0	$\frac{1}{2}(J+M)(J+1-M)$	$\frac{1}{2}(J-M)(J+1+M)$
$\gamma J \to \gamma'(J-1)$	0	$\frac{1}{2}(J+M)(J-1+M)$	$\frac{1}{2}(J-M)(J-1-M)$
$\gamma J \to \gamma'(J+1)$	0	$\frac{1}{2}(J+1-M)(J-M+2)$	$\frac{1}{2}(J+1+M)(J+M+2)$

$$\Delta E''_{JM} = \sum_{\gamma'J'} \frac{|\langle \gamma JM|W|\gamma'J'M\rangle|^2}{E_{\gamma J} - E_{\gamma'J'}} \tag{8.16}$$

is not zero when $J = M = 0$. The matrix elements of W between the components of the fine structure of a term do not vanish. Thus in cases when the fine splitting is small, the corrections (8.16) can play an important role also for levels with $J \neq 0$.

8.2 Paschen–Back Effect

8.2.1 Strong Field

When the energy of interaction of an atom with a magnetic field W becomes greater than the spin-orbit interaction, the character of the splitting alters. We shall consider the splitting of the term γSL in the limiting case $W \gg AL \cdot S$ when the spin-orbit interaction can be neglected. From (8.1, 5) we have

$$W = \mu_0 H(L_z + 2S_z) \ . \tag{8.17}$$

We now need to find the mean value of W for a state with given angular momenta L and S because in the absence of the spin-orbit interaction each of these angular momenta is separately conserved. By taking into account that averaging simple means the replacement of L_z by M_L and S_z by M_S, we obtain

$$\langle W \rangle = \mu_0 H(M_L + 2M_S) \ . \tag{8.18}$$

According to (8.18), the term γSL splits into a number of components, each of which corresponds to definite values of the sum $(M_L + 2M_S)$. Some of these components are degenerate since the same value of $M_L + 2M_S$ can be obtained from different combinations of M_L and M_S.

The correction to the energy of the state $SLM_S M_L$ due to the spin-orbit interaction is

$$\langle AL \cdot S \rangle = AM_L M_S \ ; \tag{8.19}$$

therefore, the energy levels are given by

$$\Delta E_{SLM_S M_L} = \mu_0 H(M_L + 2M_S) + AM_S M_L \ . \tag{8.20}$$

Radiative transitions between the components of the splitting of the two terms satisfy the selection rules

$$\Delta M_S = 0, \quad \Delta M_L = 0, \ \pm 1; \tag{8.21}$$

therefore,

$$\begin{aligned} \hbar \omega_\pi &= \hbar \omega_0 + (A - A') M_S M_L \ , \\ \hbar \omega_\sigma &= \hbar \omega_0 \pm \mu_0 H + A M_S M_L - A' M_S (M_L \mp 1) \ . \end{aligned} \tag{8.22}$$

Thus the splitting of the line is roughly the same as in the normal Zeeman effect. In this instance, however, each of the π and σ components has a multiplet structure. If multiplet splitting is neglected, (8.22) coincides with the formula of the normal Zeeman splitting (8.15). Splitting of lines of the type being considered is called the Paschen–Back effect. It must be noted that the Paschen–Back effect is very seldom observed in pure form. Even in cases when multiplet splitting is comparatively small, this effect should only appear in fields of $H \sim 2.10^5$ Oe. As a rule, in fields of $H \sim 10^4$–10^5 Oe an intermediate case is observed: the deviations from Zeeman splitting are substantial but still not very great.

In the general case of $W \sim AL \cdot S$, both interactions must be taken into account at the same time. In this case either the functions $\Psi_{M_S M_L}$ or any independent linear combinations of these functions can be chosen as zero-order functions. The most convenient in a number of cases is to proceed from the functions Ψ_{JM} because the matrix of the spin-orbit interaction in the representation JM is diagonal. The matrix of $(L_z + 2S_z)$ in this representation is diagonal in M but off-diagonal in J. Therefore the corrections to the energies of M states are determined by the roots of the secular equation

$$\begin{vmatrix} \langle \gamma JM|W+AL\cdot S|\gamma JM\rangle - \Delta E & \langle \gamma JM|W|\gamma J'M\rangle & \ldots \\ \langle \gamma J'M|W|\gamma JM\rangle & \langle \gamma J'M|W+AL\cdot S|\gamma J'M\rangle - \Delta E & \ldots \\ \ldots\ldots\ldots\ldots\ldots\ldots\ldots\ldots\ldots\ldots\ldots\ldots\ldots\ldots\ldots\ldots\ldots \end{vmatrix} = 0. \quad (8.23)$$

A secular equation of the type (8.23) must be written for each possible value of M. For $M = L + S$, the order of this equation equals unity ($J = L + S$), for $M = L + S - 1$ it equals two ($J = L + S, L + S - 1$), for $M = L + S - 2$ it equals three ($J = L + S; L + S - 1; L + S - 2$), and so on.

Let us consider how the off-diagonal matrix elements of W entering into (8.23) are calculated. The matrix J_z is diagonal in J; therefore,

$$\langle \gamma SLJM|L_z + 2S_z|\gamma SLJ'M\rangle = \langle \gamma SLJM|S_z|\gamma SLJ'M\rangle. \quad (8.24)$$

Using, further, the general formulas of Sect. 4.3, it is not difficult to obtain

$\langle \gamma SLJM|S_z|\gamma SLJ'M\rangle$

$$= (-1)^{J-M} (\gamma SLJ\|S\|\gamma SLJ') \begin{pmatrix} J & 1 & J' \\ -M & 0 & M \end{pmatrix}, \quad (8.25)$$

$(\gamma SLJ\|S\|\gamma SLJ')$

$$= (-1)^{L+1+S+J} \sqrt{S(S+1)(2S+1)(2J+1)(2J'+1)} \begin{Bmatrix} S & J & L \\ J' & S & 1 \end{Bmatrix}. \quad (8.26)$$

We shall examine as an example the splitting of the term ^2P. In this case $J = 3/2, 1/2$; $M = \pm 3/2, \pm 1/2$. The matrix elements $AL \cdot S$ do not depend on M and are given by the expression

$$\langle {}^2P_{3/2}|AL\cdot S|{}^2P_{3/2}\rangle = \tfrac{1}{2}A, \ \langle {}^2P_{1/2}|AL\cdot S|{}^2P_{1/2}\rangle = -A, \tag{8.27}$$

where A is the fine-structure splitting constant of the given term. The diagonal matrix elements of W are equal to

$$\langle {}^2P_{3/2}M|W|{}^2P_{3/2}M\rangle = \mu_0 \, Hg({}^2P_{3/2}) \, M = \tfrac{4}{3}\mu_0 HM,$$
$$\langle {}^2P_{1/2}M|W|{}^2P_{1/2}M\rangle = \mu_0 \, Hg({}^2P_{1/2}) \, M = \tfrac{2}{3}\mu_0 HM, \tag{8.28}$$

where $g({}^2P_{3/2})$, $g({}^2P_{1/2})$ are the Landé factors. The off-diagonal matrix elements of W are calculated by means of (8.25, 26). For $M = \pm 1/2$

$$\langle {}^2P_{3/2}M|W|{}^2P_{1/2}M\rangle = \langle {}^2P_{1/2}M|W|{}^2P_{3/2}M\rangle = \mu_0 H \frac{\sqrt{2}}{3}. \tag{8.29}$$

Therefore, for $M = \pm 3/2$,

$$\tfrac{4}{3}\mu_0 HM + \tfrac{1}{2}A - \Delta E = 0 \tag{8.30}$$

and for $M = \pm 1/2$,

$$\begin{vmatrix} \tfrac{4}{3}\mu_0 HM + \tfrac{1}{2}A - \Delta E & \mu_0 H \tfrac{\sqrt{2}}{3} \\ \mu_0 H \tfrac{\sqrt{2}}{3} & \tfrac{2}{3}\mu_0 HM - A - \Delta E \end{vmatrix} = 0. \tag{8.31}$$

Thus the corrections to energy levels have the form

$$\Delta E_{M=\pm 3/2} = \tfrac{4}{3}\mu_0 HM + \tfrac{1}{2}A, \tag{8.32}$$

$$\Delta E^{(1)}_{M=\pm 1/2} = \left(\mu_0 HM - \tfrac{1}{4}A\right)$$
$$+ \sqrt{\left(\mu_0 HM - \tfrac{1}{4}A\right)^2 + \mu_0^2 H^2 \tfrac{2}{9} - \left(\tfrac{4}{3}\mu_0 HM + \tfrac{1}{2}A\right)\left(\tfrac{2}{3}\mu_0 HM - A\right)}, \tag{8.33}$$

$$\Delta E^{(2)}_{M=\pm 1/2} = \left(\mu_0 HM - \tfrac{1}{4}A\right)$$

$$-\sqrt{\left(\mu_0 HM - \frac{1}{4}A\right)^2 + \mu_0^2 H^2 \frac{2}{9}} - \left(\frac{4}{3}\mu_0 HM + \frac{1}{2}A\right)\left(\frac{2}{3}\mu_0 HM - A\right).$$
(8.34)

In the limiting case of a weak field, formulas for the Zeeman effect follow from (8.33, 34):

$$\Delta E_{M=\pm 3/2} = \frac{1}{2}A + \frac{4}{3}\mu_0 HM = \frac{1}{2}A + g\,(^2P_{3/2})\,\mu_0 HM,$$

$$\Delta E^{(1)}_{M=\pm 1/2} = \frac{1}{2}A + \frac{4}{3}\mu_0 HM = \frac{1}{2}A + g\,(^2P_{3/2})\,\mu_0 HM, \quad (8.35)$$

$$\Delta E^{(2)}_{M=\pm 1/2} = -A + \frac{2}{3}\mu_0 HM = -A + g\,(^2P_{1/2})\,\mu_0 HM.$$

In the case of a strong field one can assume $A = 0$, in which case

$$\Delta E_{M=\pm 3/2} = \frac{4}{3}\mu_0 HM,$$

$$\Delta E^{(1)}_{M=\pm 1/2} = 2\mu_0 HM, \qquad (8.36)$$

$$\Delta E^{(2)}_{M=\pm 1/2} = 0.$$

It is easily verified that (8.36) coincide with (8.18). When $M = 3/2$, we have $M_L = \pm 1$, $M_S = \pm 1/2$; therefore $M_L + 2M_S = 4/3M$. When $M = \pm 1/2$ there are two possibilities: $M_L = 0$, $M_S = \pm 1/2$ and $M_L = \pm 1$, $M_S = \mp 1/2$. In the first case $M_L + 2M_S = 2M$, and in the second $M_L + 2M_S = 0$.

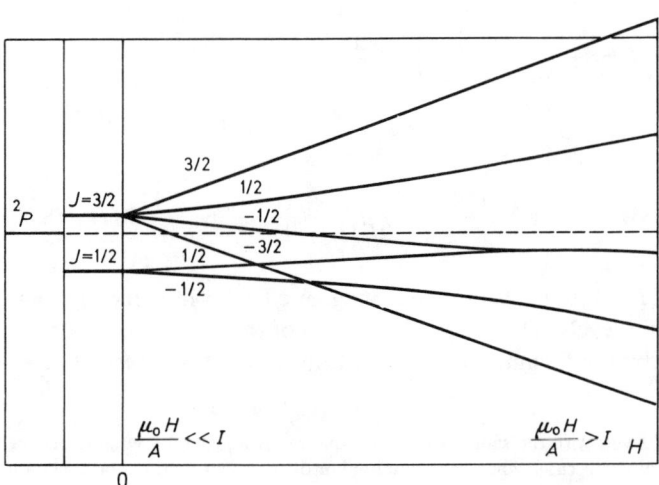

Fig. 8.2. Splitting of the ^2P terms in a weak and strong magnetic field

In Fig. 8.2 the splitting of the ^2P term is plotted against the strength of the magnetic field. A qualitative picture of the splitting in the range of intermediate values of H can be obtained by comparing the two limiting cases of a weak field and a strong field. On increasing H, the Zeeman splitting passes continuously into Paschen–Back splitting. This transition always occurs in such a way that levels with the same value of M do not intersect.[1]

Deviations from Zeeman splitting in the range of intermediate values of H can also be treated by introduction of second-order corrections. Of particular interest is the mutual perturbation of any two levels γJM and $\gamma J'M$. It follows from (8.16) and (8.24–26) that

$$\Delta E_{\gamma JM} = \frac{(J_m - L + S)(J_m + L - S)(L + S + 1 + J_m)(L + S + 1 - J_m)}{4J_m^2(2J_m - 1)(2J_m + 1)}$$

$$\times (J_m^2 - M^2) \frac{\mu_0^2 H^2}{E_{\gamma J} - E_{\gamma J'}}, \qquad (8.37)$$

where J_m is the larger of the numbers J, J'. Due to the perturbation (8.37), a difference in observed values of g factors corresponding to different M sublevels can arise. This difference must increase with increasing H.

8.2.2 Splitting of Hyperfine Structure Components in a Magnetic Field

The splitting of hyperfine structure components in a weak field (the splitting is small as compared with the hyperfine splitting) is determined by the mean value of the operator (8.2) for the state $JIFM$.

The mean value of $\mathbf{J}$ for a state with a given value of F is

$$\langle \mathbf{J} \rangle = \frac{\langle \mathbf{J} \cdot \mathbf{F} \rangle}{F(F+1)} \mathbf{F} = \frac{F(F+1) + J(J+1) - I(I+1)}{2F(F+1)} \mathbf{F};$$

therefore,

$$\langle \gamma JIFM | W | \gamma JIFM \rangle = \mu_0 g \frac{F(F+1) + J(J+1) - I(I+1)}{2F(F+1)} MH. \quad (8.38)$$

Thus the splitting of F components in a weak magnetic field is in every way similar to the splitting of J levels. The relative intensities of the π and σ components are also determined by the formulas of Table 8.1, in which it is necessary to re-

---

[1] The nonintersection of levels with the same M is a consequence of a general theorem defining the behavior of eigenvalues in cases when the Hamiltonian of a system depends on a certain parameter [3]. Note that in the second order of perturbation theory a repulsion of levels with the same value of M occurs.

place J by F. The scale of the splitting is determined by the factor g_F, which is connected with the Landé factor g_J by

$$g_F = g_J \frac{F(F+1) + J(J+1) - I(I+1)}{2F(F+1)}. \tag{8.39}$$

Because of the smallness of the splitting of the hyperfine components (8.38) is valid only for comparatively low values of H. In the limiting case of a strong field, this splitting is superimposed as a small effect on the ordinary Zeeman splitting of a J level. The situation is fully analogous to that of the Paschen-Back effect. The level γJ splits into a number of components, each of which is characterized by definite values of the quantum numbers M_J, M_I

$$\Delta E_{M_J M_I} = \mu_0 g_J H M_J + A M_J M_I, \tag{8.40}$$

where A is the hyperfine splitting constant. Since radiative transitions obey the selection rule $\Delta M_I = 0$, it follows from (8.40) that each of the Zeeman components is in turn split into $(2I+1)$ components.

Chapter 9 Radiative Transitions

This is the most important chapter for those interested in applications of spectroscopic methods. It contains the detailed treatment of radiative phenomena including multipole transitions, bremsstrahlung, photorecombination and photoionization. Approximation methods for calculating transition probabilities and cross sections are discussed. Approximate formulas and tables containing the results of numerical calculations are given.

9.1 Electromagnetic Radiation

9.1.1 Quantization of the Radiation Field

An arbitrary radiation field of the angular frequency ω in a volume V, free from electric charges, can be represented in the form of an expansion in plane waves $\exp[i(\mathbf{k}\cdot\mathbf{r} - \omega t)]$. We shall begin with this expansion for the vector potential of the field

$$\mathbf{A}(\mathbf{r}, t) = \sum_{k} \sum_{\rho=1,2} \mathbf{e}_{k\rho}(a_{k\rho}e^{i\mathbf{k}\cdot\mathbf{r}} + a^*_{k\rho}e^{-i\mathbf{k}\cdot\mathbf{r}}); \quad a_{k\rho} \propto e^{-i\omega t}, \tag{9.1}$$

where $\mathbf{k}$ is the wave vector, $\mathbf{e}_{k\rho}$ is the unit vector of the polarization; $k = \omega/c$, $\mathbf{e}_{k\rho} \perp \mathbf{k}$.

Using the well-known relation between $\mathbf{A}$ and the strength of the electric field $\mathbf{E}$, we obtain

$$\mathbf{E} = \sum_{k} \sum_{\rho=1,2} i\mathbf{k}\cdot\mathbf{e}_{k\rho}(a_{k\rho}e^{i\mathbf{k}\cdot\mathbf{r}} - a^*_{k\rho}e^{-i\mathbf{k}\cdot\mathbf{r}}). \tag{9.2}$$

With the aid of (9.2) it is not difficult to calculate the energy of the field in the volume V

$$\mathscr{E} = \frac{1}{4\pi}\int_V E^2\, dv = \sum_{k\rho} \mathscr{E}_{k\rho}; \quad \mathscr{E}_{k\rho} = \frac{Vk^2}{2\pi} a_{k\rho}a^*_{k\rho}. \tag{9.3}$$

We shall introduce the new "canonically conjugate" variables

$$Q_{k\rho} = \sqrt{\frac{V}{4\pi c^2}}(a_{k\rho} + a^*_{k\rho}); \quad P_{k\rho} = -i\omega\sqrt{\frac{V}{4\pi c^2}}(a_{k\rho} - a^*_{k\rho}). \tag{9.4}$$

The energy of the field $\mathscr{E}_{k\rho}$ and the Hamilton function $H_{k\rho}$ expressed in these variables have the same form as the Hamilton function of the linear harmonic oscillator

$$H_{kp} = \frac{1}{2}(P_{kp} + \omega^2 Q_{kp}); \quad P_{kp} = \dot{Q}_{kp}. \tag{9.5}$$

Thus the expansion obtained above is usually called an expansion in oscillators.

Let us pass now to a quantum description of the field. For this it is necessary to replace the classical variables Q and P by the corresponding operators obeying the commutation relations $PQ - QP = -i\hbar$. The result of such a quantization applied to the harmonic oscillator is well known:

$$\mathscr{E}_{kp} = \hbar\omega \left(n_{kp} + \frac{1}{2} \right), \tag{9.6}$$

where n_{kp} are integers defining the number of quanta in the radiation field, i.e., the number of photons with the wave vector k and polarization e_{kp}.

The matrix $Q_{nn'}$ has the form

$$Q_{n,n+1} = Q^*_{n+1,n} = \sqrt{\frac{\hbar(n+1)}{2\omega}}; \quad Q_{nn'} = 0, \quad n' \neq n \pm 1. \tag{9.7}$$

By using (9.4) it is also possible to obtain

$$(a_{kp})_{n,n+1} = \sqrt{\frac{2\pi c^2 \hbar(n+1)}{\omega V}}, \quad (a^*_{kp})_{n+1,n} = \sqrt{\frac{2\pi c^2 \hbar(n+1)}{\omega V}}. \tag{9.8}$$

All the remaining matrix elements of a_{kp} and a^*_{kp} are zero.

We shall assume, for simplicity, that the volume V is a cube with an edge L, In this case $k_x = 2\pi/L \cdot n_x$, $k_y = 2\pi/L \cdot n_y$, $k_z = 2\pi/L \cdot n_z$, where n_x, n_y, n_z are integers. Thus the number of field oscillators for which k_x, k_y, k_z are included in the intervals $\Delta k_x, \Delta k_y, \Delta k_z$ is equal to $\Delta n = \Delta n_x \Delta n_y \Delta n_z = L^3 \Delta k/(2\pi)^3$. This expression defines also the number of oscillators for which the absolute magnitude of the wave vector is included in the inverval $k, k + dk$ and the direction in the solid-angle element dO,

$$dn = \frac{V}{(2\pi)^3} dk = \frac{V}{(2\pi)^3} k^2 \, dk \, dO. \tag{9.9}$$

Since $dn \propto V$, $dk/(2\pi)^3$ is the number of oscillators with wave vectors in the interval $k, k + dk$ per unit volume.

9.1.2 Radiative Transition Probabilities

Interaction of an atom with each of the plane waves in the expansion (9.1) yields the matrix element

9. Radiative Transitions

$$H' = -\frac{e}{mc} \boldsymbol{p} \cdot \boldsymbol{e}_{k\rho} (a_{k\rho} e^{i\boldsymbol{k}\cdot\boldsymbol{r}} + a_{k\rho}^* e^{-i\boldsymbol{k}\cdot\boldsymbol{r}}), \tag{9.10}$$

where $\boldsymbol{p}$ is the electron momentum. In the case of more than one electron, $\boldsymbol{p}$ must be replaced by $\sum_i \boldsymbol{p}_i$. According to (9.8), H' is nonvanishing only for such transitions in which the number of photons decreases or increases by unity. For such transitions

$$(H')_{an,bn+1} = -\frac{e}{mc}\sqrt{\frac{2\pi c^2(n+1)\hbar}{\omega V}} \, \boldsymbol{e}_{k\rho} \langle a|\boldsymbol{p} e^{i\boldsymbol{k}\cdot\boldsymbol{r}}|b\rangle, \tag{9.11}$$

$$(H')_{bn,an-1} = -\frac{e}{mc}\sqrt{\frac{2\pi c^2 n\hbar}{\omega V}} \, \boldsymbol{e}_{k\rho} \langle b|\boldsymbol{p} e^{-i\boldsymbol{k}\cdot\boldsymbol{r}}|a\rangle. \tag{9.12}$$

The probability of the transition $a \to b$ in which a photon with a wave vector $\boldsymbol{k}$ in the interval $\boldsymbol{k}, \boldsymbol{k}+d\boldsymbol{k}$ and polarization $\boldsymbol{e}_{k\rho}$ is emitted according to the general formula of perturbation theory [3] is

$$dW_{ab} = \frac{2\pi}{\hbar} |H'_{an_{k\rho},bn_{k\rho}+1}|^2 \delta(E_b - E_a + \hbar\omega) \frac{V d\boldsymbol{k}}{(2\pi)^3}. \tag{9.13}$$

Denoting $\hbar^{-1}(E_a - E_b) = \omega_{ab}$, where ω_{ab} is the frequency of the atomic transition, and taking into account that $\delta(\hbar\omega - \hbar\omega_{ab}) = \hbar^{-1}\delta(\omega - \omega_{ab})$, it is not difficult to obtain

$$dW_\rho = \frac{e^2\omega}{2\pi\hbar c^3 m^2} |\boldsymbol{e}_{k\rho} \langle a|\boldsymbol{p} e^{i\boldsymbol{k}\cdot\boldsymbol{r}}|b\rangle|^2 (\bar{n}_{k\rho} + 1) \, dO. \tag{9.14}$$

Here dW_ρ is the probability of radiation of a photon polarized along $\boldsymbol{e}_{k\rho}$ in the element of solid angle dO, $\bar{n}_{k\rho}$ is the mean number of quanta of given polarization in the interval of wave vectors $\boldsymbol{k}, \boldsymbol{k}+d\boldsymbol{k}$, $\omega_{ab} = \omega$. Similarly, for the probability of absorption (transition $b \to a$) we have

$$dW_\rho = \frac{e^2\omega}{2\pi\hbar c^3 m^2} |\boldsymbol{e}_{k\rho} \langle b|\boldsymbol{p} e^{-i\boldsymbol{k}\cdot\boldsymbol{r}}|a\rangle|^2 \bar{n}_{k\rho} \, dO. \tag{9.15}$$

9.1.3 Correspondence Principle for Spontaneous Emission

In classical theory the intensity of radiation is given by the following equation:

$$dI = \frac{\omega^2}{8\pi c^3} \left| \boldsymbol{e}_{k\rho} \int \boldsymbol{j} e^{i\boldsymbol{k}\cdot\boldsymbol{r}} \, d\boldsymbol{r} \right|^2 dO, \tag{9.16}$$

where $\boldsymbol{j}$ is the density of the current. Multiplying (9.14) by $\hbar\omega$, we obtain the corresponding quantum expression. This expression consists of two parts. One of these parts does not contain $\bar{n}_{k\rho}$, i.e., it does not depend on the intensity of the radiation which existed before the transition $a \to b$, and is called the intensity of spontaneous emission. The equation for the intensity of spontaneous emission with the replacement

$$|e_{k\rho} \langle a|eve^{ik\cdot r}|b\rangle|^2 \to \frac{1}{4}\left|e_{k\rho}\int je^{ik\cdot r}\,dv\right|^2 \tag{9.17}$$

coincides with the classical formula (9.16). This is a particular example of the general relation between quantum-mechanical and classical quantities following from the so-called correspondence principle.

9.1.4 Dipole Radiation

Let us assume that the wavelength $\lambda = 2\pi c/\omega$ is much larger than the dimension of the system a, $\lambda \gg a$. Then in (9.14), $k\cdot r \ll 1$ and $\exp(ik\cdot r) \simeq 1$. Therefore, $\langle a|p\exp(ik\cdot r)|b\rangle \simeq m\langle a|v|b\rangle = -i\omega m\langle a|r|b\rangle$. Since $er = D$ is the electric dipole moment, we obtain

$$dW_\rho = \frac{\omega^3}{2\pi c^3 \hbar}|e_{k\rho}\cdot D_{ab}|^2\,dO;\quad dI_\rho = \frac{\omega^4}{2\pi c^3}|e_{k\rho}\cdot D_{ab}|^2\,dO\,. \tag{9.18}$$

We shall direct the z axis along the vector D. The polarization vectors can be chosen so that $\cos(e_{k1}\cdot D) = \sin\theta$ and $e_{k2}\perp D$. Then

$$\sum_{\rho=1,2}\int|e_{k\rho}\cdot D_{ab}|^2\sin\theta\,d\theta\,d\varphi = \frac{8\pi}{3}|D_{ab}|^2,$$

$$W_{ab} = \frac{4\omega^3}{3\hbar c^3}|D_{ab}|^2;\quad I = \frac{4\omega^4}{3c^3}|D_{ab}|^2\,. \tag{9.19}$$

Further expansion of $\exp(ik\cdot r)$ in powers of $k\cdot r$ gives the radiation of the higher electric and magnetic multipole moments. The corresponding formulas for the intensity of radiation will be given in Sect. 9.3.

9.1.5 Stimulated Emission and Absorption

If $\bar{n}_{k\rho}\neq 0$, then a term proportional to $\bar{n}_{k\rho}$ is added to the intensity of the spontaneous emission. This additional emission is called stimulated or induced emission. We shall introduce the spectral intensity $I_{k\rho}$ of radiation with polarization $e_{k\rho}$, having defined this quantity so that $I_{k\rho}\,d\omega dO$ gives the energy in the frequency interval ω, $\omega + d\omega$ incident from the solid-angle element dO on 1 cm² in 1s. This quantity can be obtained by multiplying the number of field oscillators $dk/(2\pi)^3 = \omega^2 d\omega dO/(2\pi c)^3$ by $\bar{n}_{k\rho}$, by the energy of the quanta $\hbar\omega$ and by the velocity of light c. Therefore,

$$\bar{n}_{k\rho} = \frac{8\pi^3 c^2}{\hbar\omega^3} I_{k\rho}. \tag{9.20}$$

Substituting (9.20) in (9.14) and (9.15), we obtain the following relations between the probabilities of spontaneous emission dW_ρ^{sp}, stimulated emission dW_ρ^{st}, and absorption dW_ρ^a:

$$dW_\rho^a(b;a) = dW_\rho^{st}(a,b) = dW_\rho^{sp}(a,b)\frac{8\pi^3 c^2}{\hbar\omega^3} I_{k\rho}. \tag{9.21}$$

If incident radiation is isotropic and naturally polarized, $I_{k1} = I_{k2} = 1/2 I_\omega$, $I_\omega = c U_\omega/4\pi$, where U_ω is the energy density then integrating (9.21) over all angles and summing with respect to ρ, we obtain the following relations for the total probabilities of transition between the states a, b:

$$W^a(b,a) = W^{st}(a,b) = W^{sp}(a,b)\frac{4\pi^3 c^2}{\hbar\omega^3} I_\omega = W^{sp}(a,b)\frac{\pi^2 c^3}{\hbar\omega^3} U_\omega. \tag{9.22}$$

We shall generalize these formulas to transitions between the levels γ and γ'. Let the statistical weights of the levels γ, γ' be g, g' respectively. We shall assume that an atom can be in any of the states a belonging to the level γ with the same probability equal to g^{-1}. Then the total probability of the transition $W(\gamma, \gamma')$ can be obtained by averaging $W(a, b)$ with respect to all initial states a and summing with respect to all final states b

$$W_{\gamma\gamma'} = \frac{1}{g}\sum_{a,b} W(a,b); \quad W_{\gamma'\gamma} = \frac{1}{g'}\sum_{a,b} W(b,a). \tag{9.23}$$

We shall write the probabilities of radiative transitions between the levels γ and γ' in the form

$$W^{sp} = A_{\gamma\gamma'}; \quad W^{st} = B_{\gamma\gamma'} U_\omega; \quad W^a = B_{\gamma'\gamma} U_\omega. \tag{9.24}$$

The quantities $A_{\gamma\gamma'}$, $B_{\gamma\gamma'}$ and $B_{\gamma'\gamma}$ are the Einstein coefficients.

Averaging (9.18) for given directions of k and $e_{k\rho}$ with respect to all possible orientations of the vector D_{ab}, it is not difficult to obtain

$$dW_\rho^{sp}(\gamma,\gamma') = \frac{1}{2} A_{\gamma\gamma'} \frac{dO}{4\pi}$$

$$\frac{g'}{g} dW_\rho^a(\gamma'\gamma) = \frac{1}{2} A_{\gamma\gamma'} \frac{8\pi^3 c^2}{\hbar\omega^3} I_{k\rho} \frac{dO}{4\pi}. \tag{9.25}$$

Integrating over all angles and summing with respect to ρ gives again $W^{sp} = A_{\gamma\gamma'}$. Radiation averaged with respect to all possible orientations of the vector D_{ab} in space is isotropic and nonpolarized.

9.1.6 Effective Cross Sections of Absorption and Stimulated Emission

We shall define the effective absorption cross section σ^a as the ratio of the absorbed energy $dW_p^a\hbar\omega$ to the energy-flow density $I_{k\rho}d\omega dO$. In calculating $dW_p^a\hbar\omega$, it is necessary to take into account the fact that the spectral lines always have a nonvanishing width. An atom is capable of absorbing and emitting not a strictly monochromatic frequency ω but a whole range of frequencies around ω. The transition probabilities which we considered above are integral characteristics. Thus the probability of spontaneous transition A can be written in the form

$$A = \int a_\omega d\omega, \tag{9.26}$$

where $a_\omega d\omega$ is the probability of spontaneous emission in the frequency interval $d\omega$. The quantity a_ω is dimensionless since A has the dimensions of s^{-1}. Multiplying dW_p^a in (9.25) by $\hbar\omega$ and substituting in this equation $a_\omega d\omega$ instead of $A_{\gamma\gamma'}$, we obtain the energy absorbed in the interval ω, $\omega + d\omega$. The ratio of this quantity to $I_{k\rho} d\omega\, dO$ gives

$$\sigma_\omega^a = \frac{g}{g'} a_\omega \frac{\pi^2 c^2}{\omega^2} = \frac{1}{4} \lambda^2 \frac{g}{g'} a_\omega. \tag{9.27}$$

The index ρ is omitted here because the same expression is also valid for the absorption of radiation, polarized in an arbitrary way, in particular, naturally polarized. This follows from the fact that in the general case both the absorbed energy and the density of energy flow are proportional to $(I_{k1} + I_{k2})$. In a similar way for the cross section of stimulated emission we have

$$\sigma_\omega^{st} = \frac{1}{4} \lambda^2 a_\omega \tag{9.28}$$

$$g'\sigma_\omega^a = g\sigma_\omega^{st}. \tag{9.29}$$

9.2 Electric Dipole Radiation

9.2.1 Selection Rules, Polarization, and Angular Distribution

In the case of an electric dipole transition between the states γJM and $\gamma'J'M'$, the probability of spontaneous emission is given by [see (9.18)]

$$dW_\rho(\gamma JM; \gamma'J'M') = \frac{\omega^3}{2\pi\hbar c^3} |e_{k\rho} \langle \gamma JM | D | \gamma'J'M' \rangle|^2 \, dO. \tag{9.30}$$

9. Radiative Transitions

In order to simplify the notation we shall henceforth omit the index k from $e_{k\rho}$.

We shall transform (9.30) by using the addition theorem for spherical harmonics (4.9)

$$e_\rho \cdot D = D \cos \theta_{eD} = D \sum_q C^*_{1q}(\theta_e, \varphi_e) C_{1q}(\theta_D, \varphi_D) = \sum_q e^*_q D_q , \qquad (9.31)$$

$$e_\rho \langle \gamma JM|D|\gamma'J'M'\rangle = \sum_q C^*_{1q}(\theta_e, \varphi_e) \langle \gamma JM|D_q|\gamma'J'M'\rangle$$
$$= \sum_q e^*_q \langle \gamma JM|D_q|\gamma'J'M'\rangle . \qquad (9.32)$$

Here e_q and D_q are the spherical components of the vectors e_ρ and D. According to the general formula (4.120),

$$\langle \gamma JM|D_q|\gamma'J'M'\rangle = (-1)^{J-M} (\gamma J\|D\|\gamma'J') \begin{pmatrix} J & 1 & J' \\ -M & q & M' \end{pmatrix}. \qquad (9.33)$$

From the properties of the 3j symbols (4.33, 34) it follows that the matrix elements (9.33) are nonzero if

$$\Delta J = J - J' = 0, \pm 1; J + J' \geqslant 1 , \qquad (9.34)$$

$$\Delta M = M - M' = 0, \pm 1 . \qquad (9.35)$$

To these selection rules it is necessary to add the selection rule with respect to parity. The components of the electric dipole moment D, like the components of any polar vector, change sign under inversion. Thus, electric dipole transitions are possible only between states of different parity; i.e.,

$$\text{even state} \rightleftarrows \text{odd state} . \qquad (9.36)$$

For each of the three possible transitions $\Delta M = 0, \pm 1$, only one term is nonzero in the sum (9.32). Thus for $\Delta M = 0$,

$$\sum_q = e_0 \langle \gamma JM|D_0|\gamma'J'M\rangle = e_z \langle \gamma JM|D_z|\gamma'J'M\rangle ; \qquad (9.37)$$

for $\Delta M = +1$,

$$\sum_q = e^*_1 \langle \gamma JM|D_1|\gamma'J'M-1\rangle = \frac{1}{2}(e_x - ie_y)\langle \gamma JM|D_x + iD_y|\gamma'J'M-1\rangle, \qquad (9.38)$$

9.2 Electric Dipole Radiation

and for $\Delta M = -1$,

$$\sum_q = e^*_{-1} \langle \gamma JM | D_{-1} | \gamma' J' M + 1 \rangle$$

$$= \frac{1}{2}(e_x + ie_y) \langle \gamma JM | D_x - iD_y | \gamma' J' M + 1 \rangle. \tag{9.39}$$

The angular distribution of radiation for each of the transitions $\Delta M = 0, \pm 1$ is determined by the factor $|C^*_{1q}(\theta_e, \varphi_e)|^2$; $q = 0 \pm 1$, in which the angles θ_e and φ_e describing the direction of the vector $\boldsymbol{e}_\rho$ have to be expressed in terms of $\theta_k = \theta$ and $\varphi_k = \varphi$. Let us consider as an example the transition $\Delta M = 0$. In this case the polarization vectors can be chosen so that

$$\cos \theta_{e1} = \sin \theta, \quad \cos \theta_{e2} = 0.$$

Therefore,

$$dW_1(\gamma JM; \gamma' J' M) = \frac{\omega^3}{2\pi \hbar c^3} |\langle \gamma JM | D_z | \gamma' J' M \rangle|^2 \sin^2 \theta \, dO,$$

$$dW_2(\gamma JM; \gamma' J' M) = 0. \tag{9.40}$$

Summing with respect to $\rho = 1, 2$ and integrating over all angles, we obtain

$$W(\gamma JM; \gamma' J' M) = \frac{4\omega^3}{3\hbar c^3} |\langle \gamma JM | D_z | \gamma' J' M \rangle|^2. \tag{9.41}$$

If all directions in space are equivalent, then the atom can be in any of the M states with equal probability. Thus the probability of a transition $\gamma J \to \gamma' J'$ can be obtained by summing (9.30) with respect to M' and averaging over M

$$dW_\rho(\gamma J, \gamma' J') = \frac{\omega^3}{2\pi \hbar c^3} \frac{1}{2J + 1} \sum_{MM'} |e_\rho \langle \gamma JM | \boldsymbol{D} | \gamma' J' M' \rangle|^2 \, dO. \tag{9.42}$$

Substituting (9.32, 33) into the sum and using (4.42), we have

$$\sum_{MM'} |e_\rho \langle \gamma JM | \boldsymbol{D} | \gamma' J' M' \rangle|^2$$

$$= |(\gamma J \| D \| \gamma' J')|^2 \sum_{MM'} \sum_{qq'} C^*_{1q}(\theta_e, \varphi_e) C_{1q'}(\theta_e, \varphi_e) \begin{pmatrix} J & 1 & J' \\ -M & q & M' \end{pmatrix} \begin{pmatrix} J & 1 & J' \\ -M & q' & M' \end{pmatrix}$$

$$= \frac{1}{3} |(\gamma J \| D \| \gamma' J')|^2 \sum_q C^*_{1q}(\theta_e, \varphi_e) C_{1q}(\theta_e, \varphi_e) = \frac{1}{3} |(\gamma J \| D \| \gamma' J')|^2. \tag{9.43}$$

Expression (9.43) does not depend on the choice of $\boldsymbol{e}_\rho$, i.e., it is valid for any component of the vector $\boldsymbol{D}$. Therefore,

$$\sum_{MM'}|\langle\gamma JM|D_x|\gamma'J'M'\rangle|^2 = \sum_{MM'}|\langle\gamma JM|D_y|\gamma'J'M'\rangle|^2$$
$$= \sum_{MM'}|\langle\gamma JM|D_z|\gamma'J'M'\rangle|^2 = \frac{1}{3}|(\gamma J\|D\|\gamma'J')|^2, \tag{9.44}$$
$$\sum_{MM'}|\langle\gamma JM|D|\gamma'J'M'\rangle|^2 = |(\gamma J\|D\|\gamma'J')|^2$$

and

$$dW_\rho(\gamma J;\gamma'J') = \frac{\omega^3}{6\pi\hbar c^3}\frac{1}{2J+1}\sum_{MM'}|\langle\gamma JM|D|\gamma'J'M'\rangle|^2\, dO$$
$$= \frac{\omega^3}{6\pi\hbar c^3}\frac{1}{2J+1}|(\gamma J\|D\|\gamma'J')|^2\, dO. \tag{9.45}$$

The reduced matrix element does not depend on angular variables. This enables one to integrate (9.45) over all angles and to sum with respect to two independent directions of polarization. As a result we have

$$W_\rho(\gamma J;\gamma'J') = \frac{2\omega^3}{3\hbar c^3}\frac{1}{2J+1}|(\gamma J\|D\|\gamma'J')|^2,$$
$$dW_\rho(\gamma J;\gamma'J') = W_\rho(\gamma J;\gamma'J')\frac{dO}{4\pi}, \tag{9.46}$$

$$W(\gamma J;\gamma'J') = \frac{4\omega^3}{3\hbar c^3}\frac{1}{2J+1}|(\gamma J\|D\|\gamma'J')|^2,$$
$$dW(\gamma J;\gamma'J') = W(\gamma J;\gamma'J')\frac{dO}{4\pi}. \tag{9.47}$$

Thus the radiation of the atom corresponding to the transition $\gamma J \to \gamma'J'$ is isotropic and unpolarized. This result has a simple physical meaning. Until an external field is applied to the atom, all directions in space are equivalent.

9.2.2 Oscillator Strengths and Line Strengths

We shall introduce the oscillator strength $f(\gamma J;\gamma'J')$ for the transition $\gamma J \to \gamma'J'$ by defining this dimensionless quantity by the relation

$$-f(\gamma J;\gamma'J') = \frac{2m}{3\hbar e^2}\frac{\omega_{\gamma J,\gamma'J'}}{2J+1}\sum_{MM'}|\langle\gamma JM|D|\gamma'J'M'\rangle|^2$$
$$= \frac{2m}{3\hbar e^2}\frac{\omega_{\gamma J,\gamma'J'}}{2J+1}|(\gamma J\|D\|\gamma'J')|^2, \tag{9.48}$$

where

$$\omega_{\gamma J, \gamma' J'} = \frac{1}{\hbar}(E_{\gamma J} - E_{\gamma' J'}) .$$

The physical meaning of this quantity is easy to explain by comparing the quantum expression for the polarizability of an atom, averaged over all M states of the level γJ

$$\alpha(\gamma J) = -\frac{2}{3\hbar} \sum_{\gamma' J'} \frac{\omega_{\gamma J, \gamma' J'} |(\gamma J||D||\gamma' J')|^2}{(2J+1)(\omega^2_{\gamma J, \gamma' J'} - \omega^2)} , \qquad (9.49)$$

with the classical formula for the polarizability of a harmonic oscillator of frequency ω_0

$$\alpha = \frac{e^2}{m} \frac{1}{\omega_0^2 - \omega^2} . \qquad (9.50)$$

Using (9.48) it is possible to rewrite (9.49) in the form

$$\alpha(\gamma J) = \frac{e^2}{m} \sum_{\gamma' J'} \frac{f(\gamma J; \gamma' J')}{\omega_0^2 - \omega^2} ; \quad \omega_0 = \omega_{\gamma J, \gamma' J'} . \qquad (9.51)$$

Thus the polarizability of an atom is equal to the sum of the polarizabilities of the "atomic oscillators", in which each oscillator is represented with a weighting factor (strength) $f(\gamma J, \gamma' J')$.

According to (9.48), the probability $W(\gamma J; \gamma' J')$ is expressed in terms of the oscillator strength

$$W(\gamma J; \gamma' J') = \frac{2\omega^2 e^2}{mc^3} |f(\gamma J; \gamma' J')| \qquad (9.52)$$

(we omit the indices γJ, $\gamma' J'$ on ω). It follows from (9.48) that

$$(2J+1)f(\gamma J; \gamma J') = -(2J'+1)f(\gamma' J'; \gamma J) . \qquad (9.53)$$

Oscillator strengths satisfy important sum rules (Sect. 9.4). The sum of the squares of the matrix elements in (9.45) and (9.48) is called the transition-line strength and is denoted by

$$S(\gamma J; \gamma' J') = S(\gamma' J'; \gamma J) = \sum_{MM'} |\langle \gamma JM|D|\gamma' J'M'\rangle|^2$$
$$= |(\gamma J||D||\gamma' J')|^2 . \qquad (9.54)$$

Line strengths are symmetrical with respect to the initial and final states, and are connected with transition probabilities and oscillator strength by

$$W(\gamma J;\,\gamma'J') = \frac{4\omega^3}{3\hbar c^3}\frac{1}{2J+1} S(\gamma J;\,\gamma'J'),\qquad(9.55)$$

$$-f(\gamma J;\,\gamma'J') = \frac{2m}{3\hbar e^2}\frac{\omega_{\gamma J,\gamma'J'}}{2J+1} S(\gamma J;\,\gamma'J').\qquad(9.56)$$

The line strength and oscillator strength can be defined for a transition between any levels γ, γ' with statistical weights g, g'

$$S(\gamma\gamma') = S(\gamma'\gamma) = \sum_{a,b} |\langle a|\mathbf{D}|b\rangle|^2,$$

$$-f(\gamma\gamma') = \frac{2m}{3\hbar e^2}\frac{\omega_{\gamma\gamma'}}{g(\gamma)} S(\gamma\gamma'),\qquad(9.57)$$

$$W(\gamma\gamma') = \frac{4\omega^3}{3\hbar c^3}\frac{1}{g} S(\gamma\gamma') = \frac{2\omega^2 e^2}{mc^3} |f(\gamma\gamma')|.\qquad(9.58)$$

We shall consider as an example the transition between the two terms γSL and $\gamma'SL'$, neglecting the fine splitting of these terms. In this case $g = (2L+1)(2S+1)$, $g' = (2L'+1)(2S+1)$, and

$$\begin{aligned}S(\gamma SL;\,\gamma'SL') &= \sum_{JJ'}\sum_{MM'} |\langle \gamma SLJM|\mathbf{D}|\gamma'SL'J'M'\rangle|^2 \\ &= \sum_{JJ'} S(\gamma SLJ;\,\gamma'SL'J').\end{aligned}\qquad(9.59)$$

Thus the total probability of the transition $\gamma SL \to \gamma'SL'$ is determined by (9.58), in which the line strength is equal to the sum of the line strengths of all the multiplet components. Similarly, if the splitting on terms γSL is neglected, the total probability of the transition between two electron configurations $\gamma \to \gamma'$ is also determined by (9.58), where

$$S(\gamma\gamma') = \sum_{\alpha\alpha'}\sum_{JJ'} S(\alpha J;\,\alpha'J') = \sum_{\alpha\alpha'} S(\alpha\alpha').\qquad(9.60)$$

Here α and α' denote the set of quantum numbers describing the terms of the configurations γ and γ'. The properties of additivity (9.59, 60) are an important feature of line strengths. The corresponding relations between the transition probabilities or oscillator strengths are more complex. For example, from (9.58, 59) it follows that

$$W(\gamma SL;\,\gamma'SL')$$
$$= \frac{1}{(2L+1)(2S+1)}\sum_{JJ'}(2J+1)\,W(\gamma SLJ;\,\gamma'SL'J'),\qquad(9.61)$$

$$f(\gamma SL;\,\gamma'SL') = \frac{1}{(2L+1)(2S+1)}\sum_{JJ'}(2J+1)f(\gamma J;\,\gamma'J').$$

Describing transitions by line strengths is also convenient for the reason that the intensities of lines are proportional to line strengths. In fact, the intensity of a line is proportional to the probability of the transition and to the number of atoms participating in the transition. The population of the level γ is in turn proportional to the statistical weight of this level g. Therefore,

$$I \propto gW \propto g|f| \propto S.$$

9.2.3 LS Coupling Approximation. Relative Intensities of Multiplet Components

In the LS coupling approximation the line strength is determined by the expression

$$S(\gamma SLJ; \gamma'S'L'J') = |(\gamma SLJ\|D\|\gamma'S'L'J')|^2. \tag{9.62}$$

The dependence of the line strength on J can be found explicitly. Since the operator D commutes with S,

$$(\gamma SLJ\|D\|\gamma'S'L'J')$$

$$= (-1)^{S+1+L'+J}(\gamma L\|D\|\gamma'L')\sqrt{(2J+1)(2J'+1)} \begin{Bmatrix} L & J & S \\ J' & L' & 1 \end{Bmatrix} \delta_{SS'}. \tag{9.63}$$

From this relation and also from the triangle condition $\Delta(LL'1)$ for the $6j$ symbol follow the selection rules

$$\begin{aligned}\Delta S &= 0, \\ \Delta L &= 0, \pm 1, L + L' \geq 1.\end{aligned} \tag{9.64}$$

Thus in the LS coupling approximation, besides the general selection rules (9.34, 36), there are additional selection rules (9.64). According to (9.63),

$$S(\gamma SLJ; \gamma'SL'J') = (2J+1)(2J'+1) \begin{Bmatrix} L & J & S \\ J' & L' & 1 \end{Bmatrix}^2 |(\gamma L\|D\|\gamma'L')|^2. \tag{9.65}$$

It is convenient to transform this expression in such a way that the line strength (9.62) is expressed in terms of the total strength of the multiplet line

$$S(\gamma SL; \gamma'SL') = \sum_{JJ'} S(\gamma SLJ; \gamma'SL'J').$$

The $6j$ symbols satisfy the following sum rule

$$\sum_{J'} (2J'+1) \begin{Bmatrix} L & J & S \\ J' & L' & 1 \end{Bmatrix}^2 = \frac{1}{2L+1}. \tag{9.66}$$

Besides,

$$\sum_J (2J+1) = (2L+1)(2S+1)$$

[see (2.6)]. Therefore,

$$S(\gamma SL; \gamma'SL') = (2S+1)|(\gamma L\|D\|\gamma'L')|^2, \qquad (9.67)$$

and

$$S(\gamma SLJ; \gamma'SL'J') = \frac{(2J+1)}{(2L+1)(2S+1)} q_1(SLJ; SL'J') S(\gamma SL; \gamma'SL'), \quad (9.68)$$

$$q_1(SLJ; SL'J') = (2L+1)(2J'+1) \begin{Bmatrix} L & J & S \\ J' & L' & 1 \end{Bmatrix}^2. \qquad (9.69)$$

The quantities q_1 in (9.69) satisfy the sum rules

$$\sum_{J'} q_1(SLJ; SL'J') = 1, \qquad (9.70)$$

$$\sum_{JJ'} (2J+1) q_1(SLJ; SL'J') = (2L+1)(2S+1). \qquad (9.71)$$

The relative intensities of the multiplet components are determined by the quantities $(2J+1)q_1$. Analysis of (9.68–71) shows that among multiplet components the most intense are those for which the change in J and L is the same. These lines are usually called the principal lines. The most intense principal lines correspond to the largest values of J of the initial level. The intensities of the principal lines decrease with decreasing J. The remaining multiplet components are called satellites. The satellites, for which J and L change in opposite directions, are the lowest intensity ones.

According to (9.68, 70), the sum $\sum_{J'} S(\gamma J; \gamma'J') \propto (2J+1)$, and the sum $\sum_{J'} W(\gamma J; \gamma'J')$ does not depend on J. Thus the total probability of all transitions (and also the sum of the oscillator strengths) within the given multiplet beginning from the level γJ does not depend on J, and the sum of the line strengths is proportional to $(2J+1)$. Therefore, when

$$N_1 : N_2 = (2J_1+1) : (2J_2+1)$$

(this occurs, for example, with a Boltzmann distribution with temperature $kT \gg \Delta E_{J_1 J_2}$), it is possible to formulate the following rule for the relative intensities of the multiplet components:

The sum of the intensities of all lines of a multiplet, having one and the same initial level, is proportional to the statistical weight of the given level. A similar rule also holds for all lines of a multiplet having one and the same final level.

We shall find, further, the total probability of all transitions within the given multiplet $W(\gamma SL; \gamma'SL')$. Let us assume that all levels J of the term γSL are equally populated. Then the probability of finding an atom in the level J is $(2J+1)[(2L+1)(2S+1)]^{-1}$ and

$$W(\gamma SL; \gamma'SL') = \frac{1}{(2L+1)(2S+1)} \sum_{JJ'} (2J+1) W(JJ')$$

$$= \frac{4}{3\hbar c^3} \frac{1}{(2L+1)(2S+1)} \sum_{JJ'} \omega_{JJ'}^3 S(JJ') .$$

If one neglects the difference in frequencies of the different multiplet components and assumes $\omega_{JJ'} = \omega_0$, then this probability is the same as in the absence of fine splitting and is determined by the line strength $S(\gamma SL; \gamma'SL')$.

9.2.4 One Electron Outside Closed Shells

In this case the quantum numbers SLJ coincide with the quantum numbers slj of a valence electron; therefore (9.65, 68) give

$$S(nlj; n'l'j') = \frac{2j+1}{2l+1} q_1\left(\frac{1}{2} lj; \frac{1}{2} l'j'\right) |(nl\|D\|n'l')|^2 .$$

Because $\boldsymbol{D} = -e\boldsymbol{r} = -er\boldsymbol{n}$, where $\boldsymbol{n}$ is the unit vector directed along $\boldsymbol{r}$,

$$(nl\|D\|n'l') = -e \int R_{nl} R_{n'l'} rr^2 dr (l\|n\|l') . \tag{9.72}$$

Introducing the notation

$$R_{n'l'}^{nl} = \int R_{nl} R_{n'l'} rr^2 dr , \tag{9.73}$$

we obtain

$$S(nlj; n'l'j') = \frac{2j+1}{2l+1} q_1\left(\frac{1}{2} lj; \frac{1}{2} l'j'\right) l_{\max} (eR_{n'l'}^{nl})^2 ,\tag{9.74}$$

$$-f(nlj; n'l'j')$$
$$= \frac{2m}{3\hbar} \frac{\omega}{2l+1} q_1\left(\frac{1}{2} lj; \frac{1}{2} l'j'\right) l_{\max} (R_{n'l'}^{nl})^2 , \tag{9.75}$$

$$q_1\left(\frac{1}{2} lj; \frac{1}{2} l'j'\right) = (2l+1)(2j'+1) \begin{Bmatrix} l & j & 1/2 \\ j' & l' & 1 \end{Bmatrix}^2 .$$

According to (9.71), the total line strength and the total oscillator strength of a multiplet are

$$S(nl;\ n'l') = 2l_{\max}\,(eR^{nl}_{n'l'})^2\,,$$

$$-f(nl;\ n'l') = \frac{2m\omega}{3\hbar}\frac{l_{\max}}{2l+1}\,(R^{nl}_{n'l'})^2\,. \tag{9.76}$$

These quantities define, obviously, the total intensity of all components of a multiplet if the small difference in the frequencies $\omega_{jj'}$ is neglected. In this approximation the spin-orbit interaction leads to the splitting of the line $nl \to n'l'$ into components $nlj \to n'l'j'$ but does not affect the total intensity of the transition. It is convenient to introduce the symmetrical quantity $F_1(nl;\ n'l')$ instead of $f(nl;\ n'l')$

$$F_1(nl;\ n'l') = -(2l+1)f(nl;n'l') = (2l'+1)f(n'l',\ nl). \tag{9.77}$$

According to (9.58, 76),

$$F_1(nl;\ n'l') = \frac{1}{3}\frac{\hbar\omega}{\text{Ry}}\,l_{\max}\,(R^{nl}_{n'l'})^2\,a_0^{-2}\,, \tag{9.78}$$

$$W(nl;\ n'l') = \frac{2\omega^2 e^2}{mc^3}\frac{F_1(nl;\ n'l')}{(2l+1)}\,. \tag{9.79}$$

9.2.5 Multielectron Configurations. Different Coupling Schemes

Assuming a coupling scheme, it is possible to express the line strength and the probability of any transition in the form of the product of the square of the radial integral $(R_{\gamma\gamma'})^2$ and the factor Q_1 depending on the angular momenta quantum numbers. The latter can be calculated using general methods given in Sects. 4.3 and 5.2. In the general case the probability of the transition $\gamma \to \gamma'$ can be expressed in a form similar to (9.79)

$$W(\gamma,\gamma') = \frac{2\omega^2 e^2}{mc^3}\,Q_1(\gamma,\gamma')\,\frac{F_1(nl;n'l')}{(2l+1)}\,, \tag{9.80}$$

$$F_1(nl;\ n'l') = \frac{1}{3}\frac{\hbar\omega}{\text{Ry}}\,l_{\max}\,(R_{\gamma\gamma'})^2\,a_0^{-2}\,. \tag{9.81}$$

In the notation of the radial integral $R_{\gamma\gamma'}$ it is taken into consideration that, in the general case, the single-electron wave functions depend not only on quantum numbers nl, $n'l'$ but also on other quantum numbers included in the set γ, γ'. The Q factors for the most interesting cases are given in Sect. 9.6.

9.2.6 Relative Intensities of Zeeman and Stark Components of Lines

When investigating the Zeeman splitting of spectral lines, observations are usually conducted along two directions – along the field (along the z axis), and perpendicular to the field (along the x axis). In the first case the vector $\mathbf{k}$ is directed along the z axis and the polarization vectors lie in the x, y plane. The directions x and y can be chosen as the two independent directions of polarization $p = 1, 2$. In this case we have from (9.30)

$$dW = dW_1 + dW_2 \propto (|\langle \gamma JM | D_x | \gamma' J' M' \rangle|^2 + |\langle \gamma JM | D_y | \gamma' J' M' \rangle|^2) \, dO$$

or

$$dW \propto \sum_{q=\pm 1} |\langle \gamma JM | D_q | \gamma' J' M' \rangle|^2 \, dO \, . \tag{9.82}$$

Thus, along the z axis radiation with right-circular polarization (transitions: $\Delta M = 1$) and left-circular (transitions: $\Delta M = -1$) polarization is observed. The intensities of the corresponding line components, which are called σ components, are according to (9.33) proportional to the squares of the $3j$ symbols

$$\Delta M = 1, \, dW \propto \begin{pmatrix} J & 1 & J' \\ -M & 1 & M-1 \end{pmatrix}^2 dO \, ,$$

$$\Delta M = -1, \, dW \propto \begin{pmatrix} J & 1 & J' \\ -M-1 & M+1 \end{pmatrix}^2 dO \, . \tag{9.83}$$

In transverse observation (along the x axis) the directions y and z can be chosen as the two independent directions of polarizations

$$dW = dW_1 + dW_2 \propto \{|\langle \gamma JM | D_z | \gamma' J' M' \rangle|^2 + |\langle \gamma JM | D_y | \gamma' J' M' \rangle|^2\} \, dO$$

or

$$dW \propto \left\{ |\langle \gamma JM | D_0 | \gamma' J' M' \rangle|^2 + \frac{1}{2} \sum_{q \pm 1} |\langle \gamma JM | D_q | \gamma' J' M' \rangle|^2 \right\} dO \, . \tag{9.84}$$

Thus, in addition to the σ components, π components (transitions $\Delta M = 0$), polarized along the z axis, are also observed in the direction perpendicular to $\mathbf{H}$. The intensity of these components is determined by the expression

$$\Delta M = 0 \quad dW \propto \begin{pmatrix} J & 1 & J' \\ -M & 0 & M \end{pmatrix}^2 . \tag{9.85}$$

As regards the σ components, their intensities are half as large as in observation along the z axis. The 3j symbols in (9.83, 85) are calculated by the formulas of Sect. 4.2. The results of these calculations are collected in Table 8.1.

The relative intensities of Stark π and σ components (we have in mind the quadratic Stark effect) are calculated in exactly the same way. The difference lies only in that all levels, with the exception of the $M = 0$ level, are twofold degenerate. Two states, M and $-M$, belong to each level. Therefore the intensities of π components are proportional to

$$2\begin{pmatrix} J & 1 & J' \\ -M & 0 & M \end{pmatrix}^2 \tag{9.86}$$

and the intensities of the σ components are proportional to

$$\begin{pmatrix} J & 1 & J' \\ -M & 1 & M-1 \end{pmatrix}^2 + \begin{pmatrix} J & 1 & J' \\ M & -1 & M+1 \end{pmatrix}^2 = 2\begin{pmatrix} J & 1 & J' \\ -M & 1 & M-1 \end{pmatrix}^2 \tag{9.87}$$

in observation along the z axis and

$$\frac{1}{2}\left[\begin{pmatrix} J & 1 & J' \\ -M & 1 & M-1 \end{pmatrix}^2 + \begin{pmatrix} J & 1 & J' \\ M & -1 & -M+1 \end{pmatrix}^2\right] = \begin{pmatrix} J & 1 & J' \\ -M & 1 & M-1 \end{pmatrix}^2 \tag{9.88}$$

in transverse observation. If the splitting of one of the levels is considerably less than that of the other and different σ components are not resolved, then instead of (9.87) it is easy to obtain

$$2\left[\begin{pmatrix} J & 1 & J' \\ -M & 1 & M-1 \end{pmatrix}^2 + \begin{pmatrix} J & 1 & J' \\ -M & -1 & M+1 \end{pmatrix}^2\right], \quad |M| \neq 0,$$

$$2\begin{pmatrix} J & 1 & J' \\ -M & 1 & M-1 \end{pmatrix}^2, \quad M = 0. \tag{9.89}$$

These equations relate to observation along the z axis. In transverse observation the intensity of the σ components, as already noted above, is half as large. Relative intensities calculated by (9.86, 89) are given in Table 7.1.

9.3 Multipole Radiation

9.3.1 Fields of Electric and Magnetic Multipole Moments

It has already been noted in Sect. 9.1 that radiation of higher multipoles can be obtained from (9.16) by continuing the expansion of the factor $\exp[i\mathbf{k}\cdot\mathbf{r}]$ in powers of $\mathbf{k}\cdot\mathbf{r}$. In this way, however, it is difficult to separate the fields of the electric and

9.3 Multipole Radiation

magnetic multipole moments and it is therefore more advisable to determine these fields directly from the wave equation.

In a space free of charges, the field strengths E and H, as well as the vector potential A, satisfy the wave equation

$$\Delta G + k^2 G = 0. \tag{9.90}$$

The solutions of this equation can be obtained by acting with the operator of the angular momentum $L = -i[R, p]$ on the function Φ satisfying the scalar wave equation

$$\Delta \Phi + k^2 \Phi = 0. \tag{9.91}$$

This follows from the fact that the operators L and Δ commute

$$\Delta L\Phi + k^2 L\Phi = L(\Delta \Phi + k^2 \Phi) = 0.$$

We shall seek the solutions of (9.90) in the form of outgoing spherical waves. Such solutions can be constructed by giving Φ the form $\Phi_{lm}(R, \theta, \varphi) = R_l(R) Y_{lm}(\theta, \varphi)$, where

$$R_l(R) = (-1)^l \left(\frac{R}{k}\right)^l \left(\frac{1}{R}\frac{d}{dR}\right)^l \frac{e^{ikR}}{R} = \begin{cases} (-i)^l \dfrac{e^{ikR}}{R}, & kR \gg l, \\ \dfrac{(2l-1)!!}{2^{l-1}k^l} \dfrac{1}{R^{l+1}}, & kR \ll 1 \end{cases} \tag{9.92}[1]$$

We shall introduce the notation: $L Y_{lm} = \mathbf{Y}_{lm}$. The vector functions $\mathbf{Y}_{lm}$, as can be readily proved, satisfy the orthogonality condition

$$\int \mathbf{Y}_{lm}^* \cdot \mathbf{Y}_{l'm'} \, dO = l(l+1) \, \delta_{ll'} \delta_{mm'}. \tag{9.93}$$

Since the operator L acts only on the angular variables, we have $G_{lm} = L\Phi_{lm} = R_l Y_{lm}$. Thus,

$$G_{lm} = R_l(R)\, \mathbf{Y}_{lm}(\theta, \varphi) = \begin{cases} (-i)^l \dfrac{e^{ikR}}{R} \mathbf{Y}_{lm}, & kR \gg l, \\ \dfrac{(2l-1)!!}{2^{l+1} k^l} \dfrac{1}{R^{l+1}} \mathbf{Y}_{lm}, & kR \ll 1. \end{cases} \tag{9.94}$$

By means of (9.94) we can determine E and H in two different ways

---

[1] Note the convention that $\alpha!! = 2 \cdot 4 \cdot 6 \ldots \alpha$ if α is even, and $\alpha!! = 1 \cdot 3 \cdot 5 \ldots \alpha$ if α is odd.

$$\boldsymbol{H}_{lm} = -a_{lm}\boldsymbol{G}_{lm}, \quad \boldsymbol{E}_{lm} = -a_{lm}\frac{\mathrm{i}}{k}\operatorname{rot}\boldsymbol{G}_{lm} \tag{9.95}$$

and

$$\boldsymbol{E}_{lm} = a_{lm}\boldsymbol{G}_{lm}, \quad \boldsymbol{H}_{lm} = -a_{lm}\frac{\mathrm{i}}{k}\operatorname{rot}\boldsymbol{G}_{lm}. \tag{9.96}$$

Here a_{lm} are arbitrary constants. The choice of signs in (9.95) and (9.96) is dictated by convenience in writing down the subsequent formulas. We shall consider both possibilities of determining the field. According to (9.94)

$$\boldsymbol{e}_R \cdot \boldsymbol{G}_{lm} = -\mathrm{i}\boldsymbol{e}_R \cdot [\boldsymbol{R}, \nabla]\boldsymbol{\Phi}_{lm} = 0. \tag{9.97}$$

Thus in the case of (9.95), $\boldsymbol{e}_R \cdot \boldsymbol{H}_{lm} = 0$, i.e., the magnetic field does not have a radial component. But the radial component of $\boldsymbol{E}$ is nonvanishing and for $kR \ll 1$, $E_R \propto R^{-l-2}$. Thus at close distances there is the same dependence on R as for the electric multipole static field. In the case of (9.96), on the other hand, $\boldsymbol{e}_R \cdot \boldsymbol{E}_{lm} = 0$, and $\boldsymbol{e}_R \cdot \boldsymbol{H}_{lm} \neq 0$. When $kR \ll 1$, $H_R \propto R^{-l-2}$. Such a dependence on R is characteristic of the static field of a magnetic multipole moment. We shall denote the fields (9.95) and (9.96) by $\boldsymbol{H}^{\mathrm{E}}_{lm}, \boldsymbol{E}^{\mathrm{E}}_{lm}$ and $\boldsymbol{H}^{\mathrm{M}}_{lm}, \boldsymbol{E}^{\mathrm{M}}_{lm}$, respectively.

In the general case, the radiation field of a certain system of charges can be represented in the form of the superposition of the fields $\boldsymbol{E}^{\mathrm{E}}_{lm}, \boldsymbol{E}^{\mathrm{M}}_{lm}$ and $\boldsymbol{H}^{\mathrm{E}}_{lm}, \boldsymbol{H}^{\mathrm{M}}_{lm}$

$$\boldsymbol{E} = \sum_l \sum_{m=-l}^{l} (\boldsymbol{E}^{\mathrm{E}}_{lm} + \boldsymbol{E}^{\mathrm{M}}_{lm}), \tag{9.98}$$

$$\boldsymbol{H} = \sum_l \sum_{m=-l}^{l} (\boldsymbol{H}^{\mathrm{E}}_{lm} + \boldsymbol{H}^{\mathrm{M}}_{lm}), \tag{9.99}$$

where the constants a^{E}_{lm} and a^{M}_{lm} are determined by the relations

$$a^{\mathrm{E}}_{lm} = \frac{k^{l+1}}{l(2l-1)!!}\sqrt{\frac{4\pi}{2l+1}}\, Q^*_{lm}(t), \tag{9.100}$$

$$Q_{lm}(t) = \mathrm{e}^{-\mathrm{i}\omega t}\sqrt{\frac{4\pi}{2l+1}}\, e\int r^l Y_{lm}(\theta, \varphi)\, \rho(\boldsymbol{r})\, d\boldsymbol{r}, \tag{9.101}$$

$$a^{\mathrm{M}}_{lm} = -\frac{k^{l+1}}{l(2l-1)!!}\sqrt{\frac{4\pi}{2l+1}}\, \mathfrak{M}^*_{lm}(t), \tag{9.102}$$

$$\mathfrak{M}_{lm}(t) = -\mathrm{e}^{-\mathrm{i}\omega t}\frac{1}{(l+1)c}\sqrt{\frac{4\pi}{2l+1}}\int [\operatorname{grad} r^l Y_{lm}(\theta, \varphi)][\boldsymbol{j},\boldsymbol{r}]\, d\boldsymbol{r}. \tag{9.103}$$

In the limit $\omega \to 0$, (9.101) coincides with the static electric multipole moment of order l, m. At the same time, as is easy to prove, (9.95, 100) give the field of this moment. In a similar way the static field of the magnetic multipole moment is determined in the limit $\omega \to 0$ by (9.96, 102).

We shall consider as an example the particular case $l = 1$, $m = 0$. From (9.100, 102) we have

$$a_{10}^{\text{E}} = k^2 \sqrt{\frac{4\pi}{3}} Q_{10}^* = k^2 \sqrt{\frac{4\pi}{3}} e \int z \rho(r) \, dr = k^2 \sqrt{\frac{4\pi}{3}} D_z,$$

$$a_{10}^{\text{M}} = -k^2 \sqrt{\frac{4\pi}{3}} \mathfrak{M}_{10}^* = k^2 \sqrt{\frac{4\pi}{3}} \frac{1}{2c} \int [r, j]_z \, dr = k^2 \sqrt{\frac{4\pi}{3}} \mathfrak{M}_z.$$

The fields (9.95) and (9.96) are usually called the fields of the electric and magnetic multipole moments of order l, m.

The total energy $\mathscr{E}$ of the field and the angular momentum K are determined by the expressions

$$\mathscr{E} = \frac{1}{4\pi} \int (E \cdot E^* + H \cdot H^*) \, dv, \tag{9.104}$$

$$K = \frac{1}{4\pi c} \int \{[R, [E, H^*]] + [R, [E^*, H]]\} \, dv. \tag{9.105}$$

If the expressions for E_{lm}^{E} and H_{lm}^{E} or E_{lm}^{M} and H_{lm}^{M} are substituted in (9.104, 105), then in both cases it is possible to obtain the important relations

$$K_z = \frac{m}{\omega} \mathscr{E} \, ; \, K^2 = \frac{l(l+1)}{\omega^2} \mathscr{E}^2. \tag{9.106}$$

These relations will be used below. Now, however, we shall define the concept of parity of the radiation field. This concept can be introduced since the operator $(\varDelta + k^2)$ is invariant under inversion. It is convenient to define the parity of a multipole radiation field in such a way that it coincides with the parity of the corresponding multipole moment Q_{lm} or $\mathfrak{M}_{lm}$.

This is achieved by a definition in which the parity of the field coincides with the parity of H. The radiation field is even if in the inversion operation ($X \to -X$, $Y \to -Y$, $Z \to -Z$) the intensity of the magnetic field H does not change sign, and is odd if H changes sign. Since in free space E and H are connected by the relation

$$-ikE = \text{rot } H, \tag{9.107}$$

odd E corresponds to even H and, conversely, even E corresponds to odd H. Thus

$$H(R) = H(-R), \quad E(R) = -E(-R) \quad - \text{ even wave},$$
$$H(R) = -H(-R), \quad E(R) = E(-R) \quad - \text{ odd wave}. \tag{9.108}$$

We shall now establish the parity of fields of electric and magnetic multipoles. The parity Y_{lm}, as shown in Sect. 2.1, is determined by the factor $(-1)^l$. Thus the partity of the radiation of an electric multipole is $(-1)^l$ and the parity of the radiation of magnetic multipole is $-(-1)^l$. It is not difficult to see that this definition of field parity (the field parity is defined by the parity of H and not of E) satisfies the condition formulated above. The field parity coincides with the parity of the corresponding multipole moment Q_{lm} or $\mathfrak{M}_{lm}$.

9.3.2 Intensity of Multipole Radiation

In the case of purely electric or purely magnetic multipole radiation of order l, m, the radiation intensity dI in the solid angle dO equals

$$dI = \bar{S} R^2 \sin\theta \, d\theta \, d\varphi, \tag{9.109}$$

where the density of the energy flow $\bar{S}$ averaged over time is defined by the expression

$$\bar{S} = \frac{c}{4\pi}(\text{Re}\{H\})^2 = \frac{c}{8\pi} H_{lm} \cdot H_{lm}^*. \tag{9.110}$$

Therefore,

$$dI = \frac{c}{8\pi} H_{lm} \cdot H_{lm}^* R^2 \sin\theta \, d\theta \, d\varphi. \tag{9.111}$$

By substituting in (9.111) the expression for H_{lm}^E, we obtain

$$dI = \frac{ck^{2l+2}}{2(2l+1)} \left[\frac{1}{l(2l-1)!!} \right]^2 |Q_{lm}|^2 Y_{lm} \cdot Y_{lm}^* \, dO. \tag{9.112}$$

Expression (9.112) can be integrated with respect to angles by using (9.93). Finally,

$$I_{lm}^E = \frac{c(2l+1)(l+1)}{2l} \left[\frac{k^{l+1}}{(2l+1)!!} \right]^2 |Q_{lm}|^2. \tag{9.113}$$

For the magnetic multipole radiation of order lm, we obtain in a similar way

$$I_{lm}^M = \frac{c(2l+1)(l+1)}{2l} \left[\frac{k^{l+1}}{(2l+1)!!} \right]^2 |\mathfrak{M}_{lm}|^2. \tag{9.114}$$

The intensity of radiation in the general case can be obtained by substituting (9.98, 99) in the general expression for the intensity

$$dI = \frac{c}{8\pi} \mathbf{H} \cdot \mathbf{H}^* R^2 \sin\theta \, d\theta \, d\varphi . \tag{9.115}$$

It has to be noted that fields of different electric and magnetic multipoles interfere; therefore (9.115) does not split into the sum of the independent terms. However, when integrating over all angles, the interference terms vanish as a result of the orthogonality condition (9.93). The total intensities are thus additive

$$I = \sum_{lm} (I_{lm}^E + I_{lm}^M) . \tag{9.116}$$

The order of magnitude of the terms of the sum (9.116) can be estimated by using (9.113, 114)

$$I_{lm}^E \propto \left(\frac{a}{\lambda}\right)^{2l} l \left[\frac{1}{(2l+1)!!}\right]^2 \frac{ce^2}{\lambda^2}, \quad I_{lm}^M \propto \left(\frac{v}{c}\right)^2 \left(\frac{a}{\lambda}\right)^{2l} l \left[\frac{1}{(2l+1)!!}\right]^2 \frac{ce^2}{\lambda^2}. \tag{9.117}$$

Here a is the order of magnitude of the linear dimensions of the radiating system, λ is the wavelength of the radiation, and v is the velocity of the charges. For the atom $a \sim a_0 = \hbar^2/me^2$ and for the optical region of the spectrum $\lambda = 2\pi c/\omega \sim c\hbar^3/me^4$. Consequently $a/\lambda \sim e^2/\hbar c \sim 1/137$; therefore I_{lm}^E and I_{lm}^M decrease very rapidly with increase of l. The ratio v/c is approximately the same as the ratio a/λ. Hence it follows that the terms I_{lm}^M and $I_{l+1,m}^E$ are of the same order of magnitude.

The total intensity of the radiation of a multipole moment of order l can be obtained by summing (9.113, 114) with respect to m

$$I_l^E = \frac{c(2l+1)(l+1)}{2l} \left[\frac{k^{l+1}}{(2l+1)!!}\right]^2 \sum_{m=-l}^{l} |\mathcal{Q}_{lm}|^2 ,$$

$$I_l^M = \frac{c(2l+1)(l+1)}{2l} \left[\frac{k^{l+1}}{(2l+1)!!}\right]^2 \sum_{m=-l}^{l} |\mathfrak{M}_{lm}|^2 . \tag{9.118}$$

As is easy to verify from (9.118), when $l = 1$ the necessary expression for dipole radiation follows:

$$I^E = \frac{ck^4}{3} \sum_{m=-1}^{1} |\mathcal{Q}_{1m}|^2 = \frac{ck^4}{3} |\mathbf{D}|^2, \quad I^M = \frac{ck^4}{3} |\mathfrak{M}|^2 . \tag{9.119}$$

Quantum formulas for the intensity of spontaneous multipole emission can be obtained by using the correspondence principle formulated above. It is necessary to make the replacement

$$|Q_{lm}|^2 \to 4|\langle a|Q_{lm}|b\rangle|^2, \ |\mathfrak{M}_{lm}|^2 \to 4|\langle a|\mathfrak{M}_{lm}|b\rangle|^2. \tag{9.120}$$

The radiative transition probability can be obtained by dividing the intensity by the energy of the radiated quantum $\hbar\omega$. For multipole radiation of the order κq

$$W_{\kappa q}^E = \frac{2(2\kappa+1)(\kappa+1)}{[(2\kappa+1)!!]^2 \kappa} \frac{k^{2\kappa+1}}{\hbar} |\langle \gamma JM|Q_{\kappa q}|\gamma' J' M'\rangle|^2, \tag{9.121}$$

$$W_{\kappa q}^M = \frac{2(2\kappa+1)(\kappa+1)}{[(2\kappa+1)!!]^2 \kappa} \frac{k^{2\kappa+1}}{\hbar} |\langle \gamma JM|\mathfrak{M}_{\kappa q}|\gamma' J' M'\rangle|^2. \tag{9.122}$$

According to (9.101, 103) the operators $Q_{\kappa q}$ and $\mathfrak{M}_{\kappa q}$ are

$$Q_{\kappa q} = -e \sum_i r_i^\kappa C_{\kappa q}(\theta_i, \varphi_i), \tag{9.123}$$

$$\mathfrak{M}_{\kappa q} = \frac{e}{(\kappa+1)c} \sum_i [\operatorname{grad} r_i^\kappa C_{\kappa q}(\theta_i, \varphi_i)][v_i, r_i]$$

$$= -\frac{1}{\kappa+1} \frac{e\hbar}{mc} \sum_i [\operatorname{grad} r_i^\kappa C_{\kappa q}(\theta_i, \varphi_i)] l_i, \tag{9.124}$$

where $l_i = [r_i, p_i]\hbar^{-1}$ is the angular momentum operator and summing with respect to i means summing with respect to all electrons of the atom. It can be shown that, taking into account the spin magnetic moments of the electrons, it is necessary to replace the term $(\kappa+1)^{-1} l_i$ in (9.124) by $[(\kappa+1)^{-1} l_i + s_i]$. This question is discussed below in the section dealing with magnetic dipole radiation.

9.3.3 Selection Rules

From the general equation (4.120) and from the properties of $3j$ symbols it follows that the matrix elements

$$\langle \gamma JM|Q_{\kappa q}|\gamma' J' M'\rangle; \ \langle \gamma JM|\mathfrak{M}_{\kappa q}|\gamma' J' M'\rangle \tag{9.125}$$

are nonzero only when the triangle condition $\Delta(JJ'\kappa)$ is fulfilled and $M - M' = q$. Thus multipole radiation of order κ, q satisfies the following selection rules:

$$|\Delta J| = |J' - J| = \kappa, \kappa-1, \ldots, 0; \ J + J' \geqslant \kappa, \tag{9.126}$$

$$\Delta M = M' - M = q = -\kappa, -\kappa+1, \ldots, \kappa. \tag{9.127}$$

These selection rules have a simple physical meaning. Since the energy of a multipole field, the square of the angular momentum, and the z component of the angular momentum satisfy (9.106), to each quantum $\hbar\omega$ of the multipole radiation field there corresponds an angular momentum κ, q [square of the angular momentum $\hbar^2\kappa(\kappa+1)$ and z component $\hbar q$]. The selection rules (9.126, 127) express the conservation of angular momentum $\boldsymbol{J} = \boldsymbol{J}' + \boldsymbol{\kappa}$. There is a further selection rule with respect to parity. The parities of the operators of the electric and magnetic multipole moments are $(-1)^\kappa$ and $(-1)^{\kappa+1}$ respectively. Thus in an electric multipole transition of order κ the parity of the atomic state changes as the quantity

$$(-1)^\kappa \tag{9.128}$$

and in a magnetic transition the parity of the atomic state changes as the quantity

$$(-1)^{\kappa+1} \tag{9.129}$$

The parity selection rules (9.128, 129) and the selection rules (9.126, 127) are absolutely strict. In addition to these rules, in various specific cases (for example, in the LS coupling and jj coupling approximations) one can formulate additional selection rules, the fulfillment of which depends on to what extent the approximation used is valid.

9.3.4 Electric Multipole Radiation

The total probability of an electric multipole transition of order κ from the level γJ to the level $\gamma' J'$ is

$$W_\kappa^E(\gamma J; \gamma' J') = \frac{1}{2J+1} \sum_{qMM'} W_{\kappa q}^E(\gamma JM; \gamma' J'M')$$
$$= \frac{2(2\kappa+1)(\kappa+1)}{[(2\kappa+1)!!]^2 \kappa} \frac{k^{2\kappa+1}}{\hbar} \frac{1}{2J+1} \sum_{MM'q} |\langle \gamma JM | Q_{\kappa q} | \gamma' J'M' \rangle|^2. \tag{9.130}$$

We shall introduce the line strength of an electric multipole transition of order κ, defining this quantity by a relation similar to (9.54)

$$S_\kappa(\gamma J; \gamma' J') = S_\kappa(\gamma' J'; \gamma J) = \sum_{MM'q} |\langle \gamma JM | Q_{\kappa q} | \gamma' J'M' \rangle|^2, \tag{9.131}$$

$$W_\kappa^E(\gamma J; \gamma' J') = \frac{2(2\kappa+1)(\kappa+1)}{[(2\kappa+1)!!]^2 \kappa} \frac{k^{2\kappa+1}}{\hbar} \frac{1}{2J+1} S_\kappa(\gamma J; \gamma' J'). \tag{9.132}$$

In the general case of a transition between the levels γ and γ',

$$S_\kappa(\gamma\gamma') = \sum_{a,b,q} |\langle a | Q_{\kappa q} | b \rangle|^2,$$
$$W_\kappa^E(\gamma\gamma') = \frac{2(2\kappa+1)(\kappa+1)}{[(2\kappa+1)!!]^2 \kappa} \frac{k^{2\kappa+1}}{\hbar} \frac{1}{g} S_\kappa(\gamma\gamma'). \tag{9.133}$$

In the case of one electron outside closed shells,

$$\langle nlm|Q_{\kappa q}|n'l'm'\rangle = -e\sqrt{\frac{4\pi}{2\kappa+1}} \int Y^*_{lm} Y_{\kappa q} Y_{l'm'} \, dO \int R_{nl} R_{n'l'} \, r^\kappa r^2 \, dr \, . \tag{9.134}$$

Denoting the radial integral by $R^{nl}_{n'l'}(\kappa)$, we have

$$(nl||Q_\kappa||n'l') = -eR^{nl}_{n'l'}(\kappa)\,(l||C^\kappa||\,l')\,. \tag{9.135}$$

The equations for the reduced matrix elements $(l||C^\kappa||l')$ are given in Sect. 4.3. One can also introduce the oscillator strength of the transition $f_\kappa(nl; n'l')$ to define this quantity by the relation

$$-f_\kappa(nl;\,n'l') = \frac{mc}{\hbar}\frac{(2\kappa+1)(\kappa+1)}{[(2\kappa+1)!!]^2\,\kappa}\,k^{2\kappa-1}\frac{(l||C^\kappa||l')^2}{(2l+1)}\,[R^{nl}_{n'l'}(\kappa)]^2\,, \tag{9.136}$$

$$W_\kappa = \frac{2\omega^2 e^2}{mc^3}\,|f_\kappa(nl;\,n'l')|\,. \tag{9.137}$$

The relation between f_κ and the transition probability W_κ has the same form as for dipole transition [see (9.58)].

The order of magnitude of f_κ is about $\kappa\,[(2\kappa+1)!!]^{-2}(e^2/\hbar c)^{2\kappa-2} f_1$. Let us introduce also the symmetrical quantity $F_\kappa(nl; n'l')$ [see (9.77)], defining it so that

$$W_\kappa(nl;\,n'l') = \frac{me^4}{2\hbar^3}\left(\frac{e^2}{\hbar c}\right)^{2\kappa+1}\frac{\kappa+1}{2^{2\kappa-1}\kappa\,[(2\kappa-1)!!]^2}\left(\frac{\hbar\omega}{\text{Ry}}\right)^2 \zeta^{2\kappa-2}$$

$$\times \frac{F_\kappa(nl;\,n'l')}{(2l+1)}\,, \tag{9.138}$$

$$F_\kappa(nl;\,n'l') = \frac{|(l||C^\kappa||\,l')|^2}{2\kappa+1}\left(\frac{\hbar\omega}{\text{Ry}}\right)^{2\kappa-1}[R^{nl}_{n'l'}(\kappa)]^2\,a_0^{-2\kappa}\,\zeta^{-2\kappa+2}\,, \tag{9.139}$$

where $\zeta = Z - N$ is the charge of atomic core, Z is the charge of the nucleus, N is the number of electrons in the atomic core. For a neutral atom $\zeta = 1$. The quantities F_κ from (9.139) have the same order of magnitude both for atoms and for ions.

In the general case of multielectron configurations

$$W_\kappa(\gamma,\gamma') = \frac{me^4}{2\hbar^3}\left(\frac{e^2}{\hbar c}\right)^{2\kappa+1}\frac{\kappa+1}{2^{2\kappa-1}\kappa\,[(2\kappa-1)!!]^2}\left(\frac{\hbar\omega}{\text{Ry}}\right)^2\frac{Q_\kappa(\gamma\gamma')}{(2l+1)}\,\zeta^{2\kappa-2}$$

$$\times F_\kappa(nl,\,n'l')\,. \tag{9.140}$$

The tables of the quantities F_κ are given in Sect. 9.7. The formulas for the Q_κ factors are collected in Sect. 9.6.

For atomic spectroscopy, after dipole transitions, quadrupole transitions are of the greatest interest. In this case $\kappa = 2$ and the selection rules are

$$\Delta J = 0, \pm 1, \pm 2; J + J' \geqslant 2. \tag{9.141}$$

When $\kappa = 2$ $(l\|C^2\|l') \neq 0$ for $l' = l$, $l \pm 2$, see (4.141, 143). Consequently, $\Delta l = 0, 2$. This selection rule also provides the selection rule with respect to parity – the quadrupole transition is possible only between the states of the same parity.

In the LS coupling approximation it is possible to formulate an additional selection rule:

$$\Delta S = 0; \Delta L = 0, \pm 1, \pm 2; L + L' \geqslant 2. \tag{9.142}$$

In the case of jj coupling, we have

$$\Delta j = 0, \pm 1, \pm 2; j + j' \geqslant 2. \tag{9.143}$$

9.3.5 Magnetic Dipole Radiation

Magnetic multipole radiation is of greatest interest for atomic spectroscopy when $\kappa = 1$ (dipole radiation). Assuming in (9.124) that $\kappa = 1$, we have

$$\mathfrak{M}_{1q} = -\frac{e\hbar}{2mc} \sum_i [\text{grad } r_i C_q^1(\theta_i, \varphi_i)] \cdot l_i, \tag{9.144}$$

or in Cartesian coordinates

$$\mathfrak{M}_z = -\frac{e\hbar}{2mc} \sum_i l_{iz}, \quad \mathfrak{M}_x = -\frac{e\hbar}{2mc} \sum_i l_{ix}, \quad \mathfrak{M}_y = -\frac{e\hbar}{2mc} \sum_i l_{iy}, \tag{9.145}$$

$$W = \frac{4\omega^3}{3\hbar c^3} |\langle \gamma JM | \mathfrak{M} | \gamma' J' M' \rangle|^2, \tag{9.146}$$

$$\mathfrak{M} = -\frac{e\hbar}{2mc} \sum_i l_i. \tag{9.147}$$

As already noted above, (9.147) takes into account only the orbital magnetic moment of electron. Therefore, in (9.147) it is necessary to add the intrinsic magnetic moment of an electron – $(e\hbar/mc)s$. Further consideration will be based on the expression for the operator of the magnetic moment

$$\mathfrak{M} = -\frac{e\hbar}{2mc} \sum_i (l_i + 2s_i). \tag{9.148}$$

We shall again define the line strength of the transition by

$$S(\gamma J; \gamma'J') = \sum_{MM'} |\langle \gamma JM | \mathfrak{M} | \gamma'J'M'\rangle|^2 = |(\gamma J\|\mathfrak{M}\|\gamma'J')|^2, \qquad (9.149)$$

$$W(\gamma J; \gamma'J') = \frac{4\omega^3}{3\hbar c^3} \frac{1}{2J+1} S(\gamma J; \gamma'J'). \qquad (9.150)$$

In the case of the single-electron configuration,

$$(nslj\|\mathfrak{M}\|n'sl'j') = -\frac{e\hbar}{2mc}(nslj\|l\|n'sl'j') - \frac{e\hbar}{mc}(nslj\|s\|n'sl'j'). \qquad (9.151)$$

According to (4.181, 182), transitions are possible only when $n = n'$, $l = l'$, $j = j' \pm 1$, i.e., between components of the fine structure $j = l + 1/2$ and $j' = l - 1/2$

$$S(nlj; nlj-1)$$
$$= \left(\frac{e\hbar}{2mc}\right)^2 \frac{(l+1/2+j+1)(l+1/2-j+1)(1/2+j-l)(j+l-1/2)}{4j}.$$
$$(9.152)$$

Calculation of the line strength for a multielectron atom in the LS coupling approximation is carried out in exactly the same way. The operator of the magnetic moment in this case can be written

$$\mathfrak{M} = -\frac{e\hbar}{2mc}(L + 2S). \qquad (9.153)$$

The reduced matrix elements of L and S are nonzero when $L' = L$, $S' = S$, $\gamma = \gamma'$; therefore magnetic dipole transitions are possible only between components of the fine structure of one term. The expression for the line strength $S(\gamma SLJ; \gamma SLJ-1)$ can be obtained by replacing n, l, $1/2$, j in (9.152) by γ, L, S, J, respectively,

$$S(\gamma SLJ; \gamma SLJ-1)$$
$$= \left(\frac{e\hbar}{2mc}\right)^2 \frac{(L+S+J+1)(L+S-J+1)(S+J-L)(J+L-S)}{4J}.$$
$$(9.154)$$

The selection rules for magnetic dipole radiation in the LS coupling approximation have the form

$$\Delta L = 0, \quad \Delta S = 0, \quad \Delta J = \pm 1. \tag{9.155}$$

In the jj coupling approximation, calculation of the reduced matrix element $(\gamma J || \mathfrak{M} || \gamma' J')$ is considerably more complicated. We shall represent $\mathfrak{M}$

$$\mathfrak{M} = -\frac{e\hbar}{2mc} \sum_i (l_i + 2s_i) = -\frac{e\hbar}{2mc} \left(\sum_i j_i + \sum_i s_i \right) = \mathfrak{M}' + \mathfrak{M}'',$$

$$\mathfrak{M}' = -\frac{e\hbar}{2mc} J, \quad \mathfrak{M}'' = -\frac{e\hbar}{2mc} \sum_i s_i. \tag{9.156}$$

The reduced matrix element $\mathfrak{M}'$ is nonzero only when $\gamma = \gamma'$, $J = J'$. Radiative transitions are therefore determined by the term $\mathfrak{M}''$. The reduced matrix element of $\mathfrak{M}''$ is calculated by means of the general methods discussed in Sect. 4.3. For example,

$$(\gamma J_1 j J || \mathfrak{M}'' || \gamma J_1 j' J') = -\frac{e\hbar}{2mc} (\gamma J_1 j_N J || s_N || \gamma J_1 j'_N J')$$

$$= -\frac{e\hbar}{2mc} (-1)^{J_1 + 1 + j' + J} \sqrt{(2J+1)(2J'+1)} \begin{Bmatrix} j & J & J_1 \\ J' & j' & 1 \end{Bmatrix} (slj||s||slj'), \tag{9.157}$$

whence it follows that

$$S(\gamma J_1 j J; \gamma J_1 j' J')$$
$$= \frac{3}{2} \left(\frac{e\hbar}{2mc} \right)^2 (2J+1)(2J'+1) \begin{Bmatrix} j & J & J_1 \\ J' & j' & 1 \end{Bmatrix}^2 \begin{Bmatrix} s & j & l \\ j' & s & 1 \end{Bmatrix}^2 (2j+1)(2j'+1). \tag{9.158}$$

The formulas for the probability of magnetic dipole transitions do not contain radial integrals. Instead of $eR_{n'l'}^{nl}$ we have the Bohr magneton

$$\frac{e\hbar}{2mc} = \frac{1}{2} \alpha e a_0, \tag{9.159}$$

where α is the fine structure constant. Thus the probability of magnetic dipole radiation is α^2 times less than the probability of electric dipole radiation of the same frequency.

9.3.6 Transitions Between Hyperfine Structure Components. Radio Emission from Hydrogen

The line strength of an electric dipole transition between the hyperfine structure components of two different levels γJ and $\gamma' J'$ is determined by the expression

$$S(\gamma JIF; \gamma'J'IF') = \sum_{MM'} |\langle \gamma JIFM|D|\gamma'J'IF'M'\rangle|^2. \tag{9.160}$$

Since the dipole moment of the atom D commutes with the spin of the nucleus, it follows from (4.175)

$$|(\gamma JIF\|D\|\gamma'J'IF')|^2 = (2F+1)(2F'+1)\begin{Bmatrix} J & F & I \\ F' & J' & 1 \end{Bmatrix}^2 |(\gamma J\|D\|\gamma'J')|^2,$$

$$S(\gamma JIF; \gamma'J'IF') = (2F+1)(2F'+1)\begin{Bmatrix} J & F & I \\ F' & J' & 1 \end{Bmatrix}^2 S(\gamma J; \gamma'J'), \tag{9.161}$$

$$\sum_{FF'} S(\gamma JIF; \gamma'J'IF') = (2I+1) S(\gamma J; \gamma'J'). \tag{9.162}$$

If it is assumed that $I = 0$, then the line strength (9.161) coincides with the line strength $S(\gamma J; \gamma'J')$. When $I \neq 0$, the additional factor $(2I+1)$ appears in (9.162). This is due to the fact that in the case $I \neq 0$ the statistical weight of the level γJ equals $(2J+1)(2I+1)$. It is easy to see that the expression for the total probability of the transition $\gamma J; \gamma'J'$ remains as before, because

$$W(\gamma J; \gamma'J') = \frac{4\omega^3}{3\hbar c^3} \frac{1}{(2J+1)(2I+1)} \sum_{FF'} S(\gamma JIF; \gamma'J'IF')$$

$$= \frac{4\omega^3}{3\hbar c^3} \frac{1}{2J+1} S(\gamma J; \gamma'J'). \tag{9.163}$$

From (9.161, 162) follow sum rules for the relative intensities of the hyperfine structure components of a line of the same type as for fine structure components:

$$\begin{aligned} \Delta F &= 0, \pm 1, \ F + F' \geq 1, \\ \Delta M_F &= 0, \pm 1. \end{aligned} \tag{9.164}$$

Electric dipole transitions between the components of hyperfine splitting of one and the same level are forbidden by the parity selection rule. Only quadrupole and magnetic dipole transitions are allowed. Quadrupole transitions are possible only when $2J \geq 2$.

For this reason, a magnetic dipole transition is the only transition allowed between the hyperfine structure components of the ground levels $S_{1/2}$ and $P_{1/2}$. Let us consider a transition between the hyperfine structure components of a single-electron atom (hydrogen atom or alkali metal atom). In this case

$$S(\gamma jIF; \gamma jIF') = |(\gamma jIF\|\mathfrak{M}\|\gamma jIF')|^2, \tag{9.165}$$

$$(\gamma jIF\|\mathfrak{M}\|\gamma jIF') = (-1)^{I+1+j+F'} \sqrt{(2F+1)(2F'+1)} \begin{Bmatrix} j & F & I \\ F' & j & 1 \end{Bmatrix}$$

$$\times (\gamma j\|\mathfrak{M}\|\gamma j), \tag{9.166}$$

$$(nslj\|\mathfrak{M}\|nslj) = -\frac{e\hbar}{2mc}\left[1 + \frac{j(j+1) + 3/4 - l(l+1)}{2j(j+1)}\right]\sqrt{j(j+1)(2j+1)}.$$
(9.167)

The expression in square brackets equals the Landé factor g; therefore,

$$S(\gamma jIF; \gamma jIF') = g^2 \left(\frac{e\hbar}{2mc}\right)^2 (2F+1)(2F'+1)\begin{Bmatrix} j & F & I \\ F' & j & 1 \end{Bmatrix}^2$$
$$\times j(j+1)(2j+1).$$
(9.168)

Similarly, in the general case of transitions between the components F, F' of the hyperfine structure of the level γSLJ,

$$S(\gamma JIF; \gamma JIF') = g^2 \left(\frac{e\hbar}{2mc}\right)^2 (2F+1)(2F'+1)\begin{Bmatrix} J & F & I \\ F' & J & 1 \end{Bmatrix}^2$$
$$\times J(J+1)(2J+1),$$
(9.169)

where g is the Landé factor for this level. Formulas (9.168, 169) do not contain radial integrals. This simplifies obtaining numerical results. We shall consider as an example a transition between the hyperfine structure components of the ground level of hydrogen $1s_{1/2}$. By substituting in (9.168) $j = 1/2$, $I = 1/2$, $F = 1$, $F' = 0$, we obtain

$$W = \frac{4\omega^3}{3\hbar c^3}\left(\frac{e\hbar}{2mc}\right)^2.$$
(9.170)

The magnitude of the splitting in this case is 1420.4 MHz. This corresponds to $\omega/2\pi = 1.4204 \times 10^9$ s^{-1} ($\lambda = 21$ cm). Therefore,

$$W = 2.85 \times 10^{-15} \text{ s}^{-1}.$$

9.4 Calculation of Radiative Transition Probabilities[2]

9.4.1 Approximate Methods

It has been mentioned above (Sect. 9.2.5) that in the approximation of the total separation of electron variables the probabilities of radiative transitions

---

[2] For calculation of transition probabilities for hydrogenlike spectra, see [2]. Numerous data on radiative transition probabilities are collected in [22].

$\gamma \to \gamma'$ can be expressed in terms of the single-electron radial integrals $R_{\gamma\gamma'} = \int P_\gamma(r) r P_{\gamma'}(r) dr$. Thus the principal problem which arises in calculating transition probabilities is finding the radial functions $P_\gamma(r)$ and $P_{\gamma'}(r)$ $[P_\gamma(r) = rR_\gamma(r)]$.

For any atom or ion, with the exception of single-electron ones (H atom and He^+, Li^{++} ions), the radial functions can be found only by means of some approximation methods. The principal methods of approximation for calculating the radial functions are different versions of variational methods (Hartree–Fock self-consistent field method, or direct variational methods based on the use of analytic functions) and different semiempirical methods. The use of experimental values of energy levels is common in all semiempirical methods.

Variational methods are the most accurate methods of calculating the energy of an atom. This does not mean that the wave functions obtained by variational methods must give the best results in calculating other quantities. Variational methods provide good approximations for the functions $P_\gamma(r)$ in that range of values of r which gives the main contribution to the energy. At greater values of r these approximations may be not so accurate.

By means of semiempirical methods, as will be evident later, it is easier to obtain functions $P_\gamma(r)$ accurate at large values of r, i.e., just in that region which is most important in the calculation of transition probabilities. The semiempirical method will be discussed in more detail in Sect. 9.4.4. But now we shall discuss certain specific problems arising in approximate calculations of transition probabilities.

9.4.2 Three Ways of Writing Formulas for Transition Probabilities

In the nonrelativistic approximation, the interaction of the atom with the radiation field is determined by the operator

$$H' = -\frac{e}{mc} A \cdot \sum_j p_j, \tag{9.171}$$

where p_j are the operators of the electron momenta. According to (9.11, 12, 171), in the dipole approximation the matrix element of the transition $a \to b$ is proportional to $(\sum_j p_j)_{ab}$. The matrix element H'_{ab} can also be represented in another form by expressing p_j in terms of r_j or $\dot{p}_j$.

For an arbitrary operator F, which does not depend explicitly on time, and its derivative $\dot{F} = dF/dt$, we have the relation

$$\dot{F} = \frac{i}{\hbar}(HF - FH), \tag{9.172}$$

where H is the Hamiltonian of the system under consideration. Consequently

9.4 Calculation of Radiative Transition Probabilities

$$(\dot{F})_{ab} = \frac{i}{\hbar}(E_a - E_b) F_{ab} . \tag{9.173}$$

Therefore,

$$\left(\sum_j p_j\right)_{ab} = \frac{im}{\hbar}(E_a - E_b)\left(\sum_j r_j\right)_{ab}, \tag{9.174}$$

$$\left(\sum_j \dot{p}_j\right)_{ab} = \frac{i}{\hbar}(E_a - E_b)\left(\sum_j p_j\right)_{ab} . \tag{9.175}$$

Thus,

$$\frac{im}{\hbar}(E_a - E_b)\left(\sum_j r_j\right)_{ab} = \left(\sum_j p_j\right)_{ab} = -i\hbar(E_a - E_b)^{-1}\left(\sum_j \dot{p}_j\right)_{ab} . \tag{9.176}$$

Since all three operators $\sum_j r_j$, $\sum_j p_j$, $\sum_j \dot{p}_j$ are tensor operators of rank one, the calculation of the angular parts of the matrix elements H'_{ab} in all three cases is carried out in the same way. Only the radial integrals are different.

In the nonrelativistic approximation for the atomic Hamiltonian

$$H = \sum_j \frac{p_j^2}{2m} + V, \quad V = -\sum_j \frac{Ze^2}{r_j} + \sum_{j>k} \frac{e^2}{r_{jk}},$$

there exists the following relation between the operator p_j and V:

$$\dot{p}_j = \frac{i}{\hbar}(H p_j - p_j H) = \frac{i}{\hbar}(V p_j - p_j V) = \nabla_j V .$$

Using then the obvious relation $\nabla_j e^2/r_{jk} = -\nabla_k e^2/r_{jk}$, we have

$$\sum_j \dot{p}_j = \sum_j \nabla_j V = Ze^2 \sum_j r_j r_j^{-3} ;$$

therefore the radial integral can be defined by one of the three following expressions:

$$\omega_{\gamma\gamma'} R_{\gamma\gamma'} = \omega_{\gamma\gamma'} \int P_\gamma r P_{\gamma'} \, dr , \tag{9.177}$$

$$\omega_{\gamma\gamma'} R_{\gamma\gamma'} = \frac{\hbar}{m} \int P_\gamma \left(\frac{dP_{\gamma'}}{dr} \mp l_{\max} \frac{P_{\gamma'}}{r}\right) dr , \tag{9.178}$$

$$\omega_{\gamma\gamma'} R_{\gamma\gamma'} = \frac{Ze^2}{m\omega_{\gamma\gamma'}} \int P_\gamma \frac{1}{r^2} P_{\gamma'} \, dr \, . \tag{9.179}$$

The sign $\mp$ in (9.178) corresponds to the transitions $l \to l-1$, $l \to l+1$, and $l_{\max}$ is the larger of the numbers l, l'.

If the exact wave functions are used in calculating the matrix element of H', i.e., the eigenfunctions of the operator H, then all three forms of writing H'_{ab} are completely equivalent. But in the case of approximate functions the results may differ. Different ranges of values of r give the principal contribution to the radial integrals (9.177–179). It is evident that the best results are obtained in the case when the functions $P_\gamma, P_{\gamma'}$, are determined with the greatest accuracy for those particular values of r which give the principal contribution to the integrals $R_{\gamma\gamma'}$.

Let us note that when the approximate functions Ψ_γ and $\Psi_{\gamma'}$ are used, it is not the eigenvalues of H that enter into (9.177) and (9.179) but the differences $(E_\gamma - E_{\gamma'})\hbar^{-1}$ where

$$E_\gamma = \int \Psi_\gamma^* H \Psi_\gamma d\tau; \; E_{\gamma'} = \int \Psi_{\gamma'}^* H \Psi_{\gamma'} \, d\tau \, . \tag{9.180}$$

Substitution of experimental values of $\omega_{\gamma\gamma'}$ into (9.177, 179) leads to additional errors. The method of determining the frequency of a transition must be consistent with the method of calculating the matrix element.

Self-consistent field methods and also direct variational methods generally provide the wave functions to the accuracy one needs in calculating the energy. The accuracy of these functions for large r is considerably worse. Therefore, when calculating transition probabilities by means of methods of this type, one should give preference to (9.178). In semiempirical methods, one should use (9.177).

The third form of writing H'_{ab} (in terms of the operators $\dot{p}_j$) cannot be recommended for approximate calculation. In this case the wave functions must be determined in the range of small values of r with very great accuracy which can hardly be provided in calculations on multielectron atoms.

9.4.3 Theorems for Sums of Oscillator Strengths

In the calculation of the probabilities of radiative transitions it is usual to proceed from the expression for the oscillator strength of the transition f (9.52, 56). As already noted in Sect. 9.2, oscillator strengths satisfy the so-called sum rule. This rule can be formulated for an arbitrary multielectron system.

The operators p_j and r_j satisfy the commutation relations

$$p_j r_j - r_j p_j = -i3\hbar \, . \tag{9.181}$$

9.4 Calculation of Radiative Transition Probabilities

Summing (9.181) over all electrons of the system, we obtain

$$\sum_j (p_j r_j - r_j p_j) = -i3\hbar Z, \qquad (9.182)$$

where Z is the total number of electrons. Since the operators p_j and r_k of different electrons commute,

$$p_j r_k - r_k p_j = 0, \qquad (9.183)$$

(9.182) can also be written

$$(\sum_j p_j) \cdot (\sum_j r_j) - (\sum_j r_j) \cdot (\sum_j p_j) = -i3\hbar Z. \qquad (9.184)$$

The diagonal matrix element of the left-hand side of (9.184) equals

$$\sum_b [(\sum_j p_j)_{ab} (\sum_j r_j)_{ba} - (\sum_j r_j)_{ab} (\sum_j p_j)_{ba}]. \qquad (9.185)$$

But in accordance with (9.174),

$$(\sum_i p_j)_{ab} = im\omega_{ab} (\sum_i r_j)_{ab}. \qquad (9.186)$$

Therefore,

$$\sum_b \omega_{ab} |\langle a| \sum_j r_j |b\rangle|^2 = -\frac{3\hbar}{2m} Z. \qquad (9.187)$$

Let the state a be a certain arbitrary state of the atom $\gamma J M$. Then

$$-\frac{2m}{3\hbar} \sum_{\gamma' J'} \omega(\gamma J; \gamma' J') \sum_{M'} |\langle \gamma J M| \sum_j r_j |\gamma' J' M'\rangle|^2 = Z. \qquad (9.188)$$

It was shown in Sect. 9.2 that the sum with respect to M' in (9.188) does not depend on M. Therefore, the left-hand side of (9.188) can be written [compare with (9.48)]

$$-\frac{2m}{3\hbar} (2J+1)^{-1} \sum_{\gamma' J'} \omega(\gamma J; \gamma' J') \sum_{MM'} |\langle \gamma J M| \sum_i r_i |\gamma' J' M'\rangle|^2$$
$$= \sum_{\gamma' J'} f(\gamma J; \gamma' J').$$

Thus,

$$\sum_{\gamma'J'} f(\gamma J; \gamma'J') = Z. \tag{9.189}$$

Equation (9.189) is the general theorem for the sum of transition oscillator strengths. This theorem is exact because only commutation relations and (9.173) were used in its derivation.

For the hydrogen atom and single-electron ions, $Z = 1$. In the case of a multielectron atom, summing over $\gamma'J'$ in (9.189) extends to the levels of the discrete and continuous spectra of the atom, transitions of all the atomic electrons being taken into account, including the electrons belonging to inner shells.

There is no great practical value in such a general formulation of the oscillator-strength sum rule because usually only the transitions of one of the valence electrons are of interest. There is no exact sum theorem for such single-electron transitions in a multielectron atom. Nevertheless it is possible to formulate approximate rules which are useful in many applications. It is possible by rules of this type to estimate, for example, the upper limit of the most intense transitions.

Let us consider transitions from the level γSLJ of an electron configuration containing, in addition to filled shells, the group of equivalent electrons $(nl)^N$, i.e., the transitions

$$\gamma_0(nl)^N \gamma SLJ \to \gamma_0(nl)^{N-1} \gamma_1 S_1 L_1 n'l' S'L'J'.$$

We shall assume that the wave functions

$$\Psi_{\gamma SLJ}(\gamma_0(nl)^N), \quad \Psi_{S'L'J'}(\gamma_0(nl)^{N-1}\gamma_1 S_1 L_1 n'l')$$

are antisymmetrized with respect to all Z electrons of the atom. In addition, we shall assume that these functions are eigenfunctions of a certain approximate Hamiltonian H. It follows from the results of Sect. 5.2 that for any symmetric single-electron operator $\sum_{i}^{Z} g_i = G$ we have

$$\langle \gamma_0(nl)^N \gamma SLJ | \sum_{i}^{Z} g_i | \gamma_0(nl)^{N-1} \gamma_1 S_1 L_1 n'l' S'L'J' \rangle$$
$$= \langle (nl)^N \gamma SLJ | \sum_{i}^{N} g_i | (nl)^{N-1} \gamma_1 S_1 L_1 n'l' S'L'J' \rangle. \tag{9.190}$$

Therefore by repeating the derivation of (9.189), we have

$$\sum_{n'l'S'L'J'} f[\gamma_0(nl)^N \gamma SLJ; \gamma_0(nl)^{N-1} \gamma_1 S_1 L_1 n'l' S'L'J'] = N. \tag{9.191}$$

For one electron outside filled shells

$$\sum_{n'l'j'} f(nlj; n'l'j') = 1. \tag{9.192}$$

The sum rules (9.190, 191) are fulfilled only in the case when the frequencies ω, equal to the difference of the eigenvalue of the approximate Hamiltonian corresponding to the wave functions used in the calculation, are substituted in the oscillator strengths f.

If in determining f, the matrix elements of $\sum_j \mathbf{r}_j$ or $\sum_j \mathbf{p}_j$ are calculated by means of some approximate method and the transition frequencies are taken from experimental data, the sum rules (9.191, 192), generally speaking, do not have to be fulfilled. Moreover, when excluding from consideration the electrons of filled shells, replacing them by some effective field, we must extend the summation with respect to b in $\sum_b (\sum_j p_j)_{ab} (\sum_j r_j)_{ba}$ to the filled states. Thus, for example, in the case of the oscillator strength from the level np of an Na atom it is necessary to take into account nonexistent transitions to the states $1s$ and $2s$. This shows that in contrast to (9.189) the sum rules (9.191, 192) are approximate.

For an electron in a centrally symmetric field, many more additional sum rules can be established, for example, for the oscillator strengths $f(nl; n'l - 1)$ and $f(nl; n'l + 1)$ (see [2], Sects. 61, 62).

9.4.4 Semiempirical Methods of Calculating Oscillator Strengths

In the self-consistent field method, the wave functions are found simultaneously with the eigenvalues of the system of differential equations—the energy parameters ε_γ. In the semiempirical method another approach is used. One can assume that the values of ε_γ are known and seek single-electron radial functions $P_\gamma(r)$ which provide the required values of ε_γ. Usually the system of equations is replaced by one equation for the optical electron in some effective field. This equation has the form

$$\left[-\frac{1}{2} \frac{d^2}{dr^2} - \frac{Z}{r} + \frac{l(l+1)}{2r^2} + V(r) - \varepsilon_\gamma \right] P_\gamma(r) = 0 . \tag{9.193}$$

The energy parameter ε_γ is equal to the difference of the energies of the atom E_a and the "frozen" atomic core E'_i. It can be shown that $|\varepsilon_\gamma| > I_\gamma$, where $I_\gamma = |E_a - E_i|$ is the ionization energy of the electron. If the electron considered is one of the equivalent electrons of the group l^n, then by ε_γ is understood the mean value with respect to the terms of the atomic core.

It is obvious that the accuracy of the functions $P_\gamma(r)$ depends, to a great extent, on how close the chosen value ε_γ is to the true value of the difference $E_a - E'_i$. Usually in the semiempirical method, the energy parameter ε_γ is assumed to be equal to the experimental value of the ionization potential I_γ. Thereby the mean polarization of the atomic core by the optical electron is neglected. Since the difference $E_a - E'_i$ cannot be measured experimentally, the magnitude of this error can only be estimated by comparing I_γ with the Hartree–Fock value

$|E_a - E_i'|_{HF}$. It is necessary to take into account that I_y includes the instantaneous interaction of the electrons (correlation), which is not considered in the self-consistent field approximation. In principle, therefore, both cases are possible; $I_y > |E_a - E_i'|_{HF}$ and $I_y < |E_a - E_i'|_{HF}$. In the first case the correlation effect exceeds the polarization effect. In the second case the situation is reversed. Since both $|E_a - E_i'|_{HF}$ and I_y are less than the accurate value of $|E_a - E_i'|$, it is to be expected that the semiempirical method will give better results in the cases when $I > |E_a - E_i'|_{HF}$.

The polarization effect becomes greater as the wave functions of the optical electron and of the electrons of the atomic core overlap more. It is therefore most significant for the ground states of atoms having many electrons in the outer shell. For example, for the ground state of the oxygen atom (6 electrons in the states $2s^2 2p^4$), the Hartree-Fock method gives $|E_a - E_i'|_{HF} = 0.630$, whereas $I = 0.500$. In this case $|E_a - E_i'|_{HF} > I$ and the Hartree-Fock function $P_{2p}(r)$ must have better asymptotic behavior than the semiempirical one.

But for alkali-earth atoms and all the more for alkali atoms, the alternative relation holds: $|E_a - E_i'|_{HF} < I$. Thus for the ground state of the Ca atom $|E_a - E_i'|_{HF} = 0.195$ and $I = 0.225$. In such cases it is more advisable to use the semiempirical method of calculation.

The selected value of ε_y is not the eigenvalue of (9.193). Therefore this equation does not have solutions simultaneously satisfying both boundary conditions $P(0) = 0$ and $P(\infty) = 0$. This difficulty can be avoided in two ways. In the numerical integration of (9.193), one can proceed from high values of r. Since the behavior of the functions P_y and $P_{y'}$ at small distances is unimportant, the integration can be interrupted at a certain finite value of r. Another method consists in choosing the potential $V(r)$ in the form of a function of a certain parameter, the value of which is adjusted so that both boundary conditions are satisfied.

In semiempirical methods, the experimental value is the frequency of the transition in the equation for f.

9.4.5 Electric Dipole Transition Probabilities in the Coulomb Approximation

The potential $-Z/r + V(r)$ in (9.193) at large distances has the asymptotic form $-\zeta/r$, where $\zeta = Z - N$ (Z is the nuclear charge and N is the number of electrons in the atomic core). For a neutral atom $\zeta = 1$, for a singly ionized atom $\zeta = 2$, and so on. Using the fact that the principal contribution to the radial integral $R_{yy'} = \int P_y r P_{y'} \, dr$ is given by the range of high values of r, *Bates* and *Damgaard* [23] proposed a maximum simplification of the problem by assuming

$$-\frac{Z}{r} + V(r) = -\zeta/r. \tag{9.194}$$

With this assumption the solution of (9.193), satisfying the boundary condition $P_\gamma(\infty) = 0$, is given by

$$P_{nl}(r) = \left(\frac{2\zeta r}{n^*}\right)^{n^*} \exp\left(-\frac{\zeta r}{n^*}\right) \sum_{k=0}^{\infty} \frac{a_k}{r^k}, \tag{9.195}$$

where the effective principal quantum number is determined by the experimental value of energy expressed in Ry

$$n_\gamma^* = \zeta \sqrt{\frac{\text{Ry}}{|E_\gamma|}}. \tag{9.196}$$

The coefficients a_k satisfy the recursion relation

$$a_k = a_{k-1} \frac{n^*[l(l+1) - (n^* - k)(n^* - k + 1)]}{2k\zeta}, \tag{9.197}$$

$$a_0 = \frac{1}{n^*} \sqrt{\frac{\zeta}{\Gamma(n^* + l + 1)\,\Gamma(n^* - l)}}, \tag{9.198}$$

To provide convergence, the series (9.195) has to be limited. It can be done, for example, assuming that when $k \geq n^* + 1$ the coefficients $a_k = 0$.

The oscillator strengths calculated by means of the functions (9.195) are given in Table 9.6. In Section 9.7. there are given also the quantities $F_2(nl; n'l')$ for quadrupole transitions.

It is obvious that the *Bates–Damgaard* method is valid in cases when the maxima of both functions P_{nl} and $P_{n'l'}$ lie outside the atomic core. This condition can be formulated explicitly. It is necessary that the inequalities $n > n_0$ and $n' > n_0$ are fulfilled, where n_0 is the largest of the principal quantum numbers of the electrons of the atomic core. In addition, the condition $n_l^* > l + 1/2$ must be fulfilled. As a rule, both conditions are fulfilled at the same time, but the first is somewhat stricter.

9.4.6 Intercombination Transitions

In the LS coupling approximation, radiative transitions with a change of the total spin of the atom S are forbidden. However, in reality the selection rule $\Delta S = 0$ is violated because of magnetic interactions. As shown above (Sect. 5.5), the magnetic interactions increase rapidly with increasing Z. The intensities of intercombination lines behave similarly. For example, as has already been noted, such lines are practically absent in the He spectrum, but in the Hg spectrum the line 2537 Å (the transition $6s^{2\,1}S_0 - 6s6p\,^3P_1$) is very intense.

When calculating the oscillator strengths of intercombination transitions, the *LS* coupling approximation cannot be used. It is necessary to take into account the electrostatic and magnetic interactions at the same time.

In the general case the matrix elements of the dipole moment of the atom **D** in the α representation can be expressed in terms of the matrix elements of **D** in the *LS* coupling scheme $D_{\gamma\gamma'}$,

$$D_{\alpha\alpha'} = \sum (\alpha|\gamma)\, D_{\gamma\gamma'}\, (\gamma'|\alpha'). \tag{9.199}$$

To find the transformation coefficients $(\alpha|\gamma)$ it is necessary to calculate the matrix of electrostatic and magnetic interactions $H_{\gamma\gamma'}$ in the *LS* representation and bring it into diagonal form, i.e., solve the secular equation

$$|H_{\gamma\gamma'} - \varepsilon\delta_{\gamma\gamma'}| = 0. \tag{9.200}$$

After this, the transformation coefficients $(\alpha|\gamma)$ are determined by the system of equations,

$$\sum_{\gamma'} (H_{\gamma\gamma'} - \varepsilon_\alpha \delta_{\gamma\gamma'})(\gamma'|\alpha) = 0, \tag{9.201}$$

where ε_α are the roots of the secular equation (9.200). In the semiempirical method the parameters ε_α can be found from experimental data. The results of such calculations for He-like and Be-like ions are given in Tables 9.1 and 9.2 [24, 25].

Table 9.1 Intercombination transition probabilities for He-like ions

Transition Ion	$W(1^1S_0 - 2^3P_1)$ [s^{-1}]	$W(2^1S_0 - 2^3P_1)$ [s^{-1}]	$W(2^3S_1 - 2^1P_1)$ [s^{-1}]
He I	1.80×10^2	0.027	1.55
Li II	1.81×10^4	0.057	4.18×10^1
Be III	4.01×10^5	0.052	3.83×10^2
B IV	4.23×10^6	0.016	2.05×10^3
C V	2.84×10^7	6×10^{-7}	7.87×10^3
N VI	1.40×10^8	0.018	2.43×10^4
O VII	5.53×10^8	0.25	6.41×10^4
F VIII	1.85×10^9	1.36	1.50×10^5
Ne IX	5.43×10^9	4.96	3.20×10^5

Table 9.2 Probability of the transition $2\,^1S_0 - 2\,^3P_1$ in Be-like ions W [s^{-1}]

Be I	B II	C III	N IV	O V	F VI
0.71	2.0×10^1	1.9×10^2	9.2×10^2	3.6×10^3	1.1×10^4
Ne VII	Na VIII	Mg IX	Al X	Si XI	P XII
2.9×10^4	7.3×10^4	1.6×10^5	3.3×10^5	6.5×10^5	1.1×10^6
S XIII	Cl XIV	Ar XV	K XVI	Ca XVII	
2.1×10^6	3.2×10^6	5.2×10^6	8.2×10^6	1.3×10^7	

9.5 Continuous Spectrum

9.5.1 Classification of Processes

The processes responsible for the continuous spectrum which will be examined below are
1) transitions between continuous-spectrum states and discrete-spectrum states – recombination radiation
2) transitions between states of the continuous spectrum – bremsstrahlung.

Reverse processes are also possible. In the first case, photoionization, or the photoeffect, may occur, and in the second case, continuum absorption may take place.

Transitions of electrons between states of the continuous and discrete spectrum are often called bound-free transitions, and transitions between continuous spectrum states, free-free transitions.

In considering the processes enumerated above, we will focus mainly on problems which are of interest for continuous radiation in the visible, the ultraviolet, and to some extent the X-ray regions of the spectrum. We shall therefore limit ourselves to the nonrelativistic approximation and neglect the retardation in the interaction of the system with the radiation field.[3] In certain special cases two-photon transitions are also of interest.[4] For example, the transitions $2s - 1s$ of the hydrogen atom, accompanied by the radiation of two photons $\hbar\omega_1 + \hbar\omega_2 = (3/4)\mathrm{Ry}$, play an important role in the formation of the continuous spectrum adjoining the L_α line in planetary nebulae.

9.5.2 Photorecombination and Photoionization: General Expressions for Effective Cross Sections

Let us consider the single-electron system. The probability of a spontaneous radiative transition of an electron from the continuous spectrum state a to the discrete spectrum state b, accompanied by the emission of a photon with wave vector $\mathbf{k}$ and polarization vector $\mathbf{e}_{\rho k}$, can be calculated by the general equation (9.14). For the wave function ψ_a we must substitute that of the continuous spectrum in (9.14). The motion of the electron in the field of the atomic ion is described by a wave function which at great distances from the ion is a superposition of the plane wave

$$Ce^{i\mathbf{p}\cdot\mathbf{r}/\hbar} = Ce^{i\mathbf{q}\cdot\mathbf{r}} \tag{9.202}$$

where $\mathbf{p}$ is the momentum of the electron incident on the atom and $\mathbf{q} = \hbar^{-1}\mathbf{p}$ the

---

[3] For an examination of photoprocesess at relativistic velocities of electrons and a discussion of the effects of retardation, see [2].
[4] For the theory of such transitions, see [26].

9. Radiative Transitions

wave vector, and a diverging spherical wave. The latter appears as a result of the interaction of the electron with the atom. Therefore

$$\psi \propto C \left[e^{i\mathbf{q} \cdot \mathbf{r}} + \frac{f(\theta_q \cdot \mathbf{r})}{r} e^{iq \cdot r} \right]. \qquad (9.203)$$

The wave function of this type has the form [3]

$$\psi = C\psi_q^+, \quad \psi_q^+ = \frac{(2\pi)^{3/2}}{q} \sum_{\lambda\mu} i^\lambda e^{i\eta_\lambda} Y^*_{\lambda\mu}(\theta_q, \varphi_q) \, Y_{\lambda\mu}(\theta, \varphi) \, R_{q\lambda}(r) \qquad (9.204)$$

where the radial function $R_{q\lambda}(r)$ is normalized by the condition

$$\int R_{q\lambda}(r) \, R_{q'\lambda}(r) \, r^2 \, dr = \delta(q - q'). \qquad (9.205)$$

At large values of r we have

$$R_{q\lambda}(r) \to \sqrt{\frac{2}{\pi}} \frac{\sin(qr - \lambda\pi/2 + \eta_\lambda)}{r}. \qquad (9.206)$$

The phases η_λ are determined by the interaction between the electron and the atom in the whole range of r. The normalization constant C is defined in such a way that the density of the incident flow S_q is equal to unity. In this case the effective cross section of the process $d\sigma$, connected with the probability dW by the relation $d\sigma = S^{-1} dW$, becomes equal to dW. Since for the plane wave $C \exp[i\mathbf{q} \cdot \mathbf{r}]$ we have $S_q = vC^2 = C^2 p/m$, we find $C^2 = m/p$ and

$$\psi_a = \sqrt{\frac{1}{v}} \psi_q^+ = \sqrt{\frac{m}{\hbar q}} \psi_q^+. \qquad (9.207)$$

Substituting (9.207) into (9.14) and limiting ourselves to the dipole approximation, we have

$$d\sigma = dW = \frac{1}{2\pi} \frac{e^2 m k^3}{\hbar^2 q} \left| e_{pk} \int (\psi_q^+)^* \mathbf{r} \psi_b \, dr \right|^2 dO_k, \qquad (9.208)$$

where $\hbar k c = \hbar\omega = \hbar^2 q^2/2m + |E_b|$. The differential effective cross section for recombination is determined by (9.208).

Let us consider now the reverse process, i.e., a transition from the discrete spectrum state to the continuous spectrum state as a result of the absorption of a photon with wave vector $\mathbf{k}$ and polarization e_{pk}. We are interested in continuous-spectrum states in which an electron at great distance from an atom moves in a definite direction. States of this type are described by the wave functions [3]

9.5 Continuous Spectrum

$$\psi = C\psi_q^-, \quad \psi_q^- = \frac{(2\pi)^{3/2}}{q} \sum_{\lambda\mu} i^\lambda e^{-i\eta_\lambda} Y^*_{\lambda\mu}(\theta_q, \varphi_q) \, Y_{\lambda\mu}(\theta, \varphi) \, R_{q\lambda}(r). \tag{9.209}$$

In contrast to (9.204), the function ψ^- at great distance has the form of a superposition of plane and converging spherical waves. The general formula from perturbation theory for the probability of a transition from a certain state f_0 to the continuous-spectrum states f, $f + df$ is [3]

$$dW = \frac{2\pi}{\hbar} |M_{f_0 f}|^2 \, \delta(E_f - E_{f_0}) \, df. \tag{9.210}$$

It is assumed in this formula that in calculating the matrix elements of the perturbation $M_{f_0 f}$ one uses the wave function of the continuous spectrum ψ_f normalized by the condition

$$\int \psi_f^* \psi_{f'} \, df = \delta(f - f'). \tag{9.211}$$

Let us consider a transition in the interval of states q, $q + dq$. In this case it is necessary in (9.210) to replace df by dq and in accordance with (9.210) to normalize the outgoing plane waves $C\exp[i\boldsymbol{q}\cdot\boldsymbol{r}]$ to the δ function $\delta(\boldsymbol{q} - \boldsymbol{q}')$. Since $\int \exp[-i(\boldsymbol{q} - \boldsymbol{q}')\cdot\boldsymbol{r}] d\boldsymbol{r} = (2\pi)^3 \, \delta(\boldsymbol{q} - \boldsymbol{q}')$, it is necessary to put $\psi_a = (2\pi)^{-3/2} \psi_q^-$. The matrix elements of the interaction (9.11, 12) have been calculated using the assumption that $n_{k\rho}$ photons with the wave vector $\boldsymbol{k}$ and polarization $\boldsymbol{e}_{k\rho}$ are contained in the volume V, ($|M|^2 \propto n/V$). This means that a flow of photons of density cn_{pk}/V falls on the atom, and consequently $d\sigma = dW V/cn$. If it is assumed that $n = 1$, $V = 1$ in (9.12) for the interaction matrix element M, then

$$d\sigma = \frac{2\pi}{\hbar c} |M|^2 \, \delta(E - E_0) \, dq. \tag{9.212}$$

The energies of the initial and final states of the system E_0, E are $E_0 = -|E_b| + \hbar\omega$, $E = \hbar^2 q^2/2m$; therefore,

$$\delta(E - E_0) = \left(\frac{\partial E}{\partial q}\right)^{-1} \delta(q - q_0) = \frac{m}{\hbar^2 q} \delta(q - q_0)$$

$$= \frac{m}{\hbar p} \delta\left(q - \frac{1}{\hbar}\sqrt{2m(\hbar\omega - |E_b|)}\right). \tag{9.213}$$

By substituting (9.213) into (9.212) and integrating with respect to dq, we have

$$d\sigma = \frac{1}{2\pi} \frac{e^2 m}{\hbar^2} kq \left| \boldsymbol{e}_{pk} \cdot \int \psi_b^* \boldsymbol{r} \psi_q^- \, d\boldsymbol{r} \right|^2 dO_q, \quad \frac{\hbar^2 q^2}{2m} = \hbar\omega - |E_b|. \tag{9.214}$$

9. Radiative Transitions.

Comparing (9.208) with (9.214), and also (9.204, 209), it is easy to see that the differential effective cross sections of the direct and reverse processes under consideration are related by

$$\frac{1}{k^2} \frac{d\sigma_{q;bk}}{dO_k} = \frac{1}{q^2} \frac{d\sigma_{bk;q}}{dO_q} . \tag{9.215}$$

It is also not difficult to find the relation between the total cross sections. The wave functions ψ_q^+ and ψ_q^- can be expanded in spherical harmonics

$$\psi_q^\pm = \sum_{\lambda\mu} (q^\pm | \lambda\mu) \psi_{q\lambda\mu}, \quad \psi_{q\lambda\mu} = R_{q\lambda}(r) Y_{\lambda\mu}(\theta, \varphi), \tag{9.216}$$

$$(q_\pm | \lambda\mu) = \frac{(2\pi)^{3/2}}{q} i^\lambda e^{\pm i\eta_\lambda} Y_{\lambda\mu}(\theta_q, \varphi_q). \tag{9.217}$$

We shall substitute (9.216) into (9.214) and integrate over all angles. Since

$$\int (q^+ | \lambda\mu)^* (q^+ | \lambda'\mu') dO_q = \int (q^- | \lambda\mu)^* (q^- | \lambda'\mu') dO_q = \frac{(2\pi)^3}{q^2} \delta_{\lambda\lambda'} \delta_{\mu\mu'},$$

we obtain

$$\sigma_p(b; q) = 4\pi^2 \frac{e^2 m}{\hbar^2} \frac{k}{q} \sum_{\lambda\mu} |\langle b | r_p | q\lambda\mu \rangle|^2 . \tag{9.218}$$

We can integrate over dO_k in (9.208) in exactly the same way if we take into account that integration over all directions of the vector k is equivalent to integration over all directions of the vector q

$$\sigma_p(q; b) = 4\pi^2 \frac{e^2 m}{\hbar^2} \frac{k^3}{q^3} \sum_{\lambda\mu} |\langle q\lambda\mu | r_p | b \rangle|^2 . \tag{9.219}$$

From (9.218, 219) it follows

$$q^2 \sigma_p(q; b) = k^2 \sigma_p(b; q) . \tag{9.220}$$

Formulas (9.218–220) correspond to radiation or absorption of a photon of specific polarization. In the case of a multielectron atom it is necessary to add to the quantum numbers $q\lambda$ additional quantum numbers (we shall denote them by a) describing the state of the atomic core. It is further necessary to add the magnetic quantum number of the spin m_s

$$\sigma_p(b; aqm_s) = 4\pi^2 \frac{m}{\hbar^2} \frac{k}{q} \sum_{\lambda\mu} |\langle b|D_p|a, q\lambda\mu m_s\rangle|^2,$$

$$\sigma_p(aqm_s;b) = 4\pi^2 \frac{m}{\hbar^2}\left(\frac{k}{q}\right)^3 \sum_{\lambda\mu} |\langle a,q\lambda\mu m_s|D_p|b\rangle|^2 .$$
(9.221)

We shall now consider recombination of an electron with an atom into a certain definite level γ. In order to obtain the total effective cross section for this process it is necessary to sum the second of the expressions (9.221) with respect to all states b relating to the level γ, and to average over all states a of the level γ' of the initial ion and also over m_s. In addition, it is necessary to sum over the two independent directions of polarization of the emitted photon. Similarly the total effective cross section of the reverse transition $\gamma \to \gamma'q$ can be obtained by summing the first of the expressions (9.221) with respect to all final states a and m_s and averaging over all initial states b and $\rho = 1,2$. Summing with respect to a and b always includes summing with respect to magnetic quantum numbers. Thus,

$$\sum_{ab\mu} |\langle a, q\lambda\mu m_s |D_p|b\rangle|^2 = \frac{1}{3}\sum_{ab\mu} |\langle a, q\lambda\mu m_s|D|b\rangle|^2$$

(Sect. 9.2), i.e., it does not depend on ρ. Therefore summing with respect to ρ reduces to multiplying by 2. Taking this into account, we have

$$\sigma(\gamma'q;\gamma) = \frac{4\pi^2}{3}\frac{m}{\hbar^2}\left(\frac{k}{q}\right)^3 \frac{1}{g_{\gamma'}} \sum_\lambda \sum_{ab\mu m_s} |\langle a, q\lambda\mu m_s |D|b\rangle|^2,$$
(9.222)

$$\sigma(\gamma;\gamma'q) = \frac{4\pi^2}{3}\frac{m}{\hbar^2}\frac{k}{q}\frac{1}{g_\gamma}\sum_\lambda \sum_{ab\mu m_s} |\langle b|D|a, q\lambda\mu m_s\rangle|^2 .$$
(9.223)

According to (9.222, 223)

$$q^2 g_{\gamma'} \sigma(\gamma'q;\gamma) = k^2 g_\gamma \sigma(\gamma;\gamma'q) .$$
(9.224)

Relations (9.215, 220, 224) are particular cases of the principle of detailed balance.[5]

We shall represent the functions $\psi_{a,q\lambda\mu m_s}$ in terms of the eigenfunctions $\psi_{b'q\lambda}$ of the operators of the total angular momenta of the system "atomic core + electron" S, L, J. The set of quantum numbers b' includes the quantum numbers S, L, J. Using the well-known properties of unitary transformations, it is easy to obtain

---

[5] For the derivation of the general formula connecting the effective cross sections of direct and reverse processes, see [3].

9. Radiative Transitions

$$\sum_{a\mu m_s} |\langle b|D|a, q\lambda\mu m_s\rangle|^2 = \sum_{b'} |\langle b|D|b' q\lambda\rangle|^2 . \tag{9.225}$$

We shall also replace the radial functions $R_{q\lambda}$ in the integral $\langle b|D_p|b'q\lambda\rangle$ by the function $R_{E\lambda} = \hbar^{-1}(m/q)^{1/2} R_{q\lambda}$, normalized with respect to the energy scale, i.e., by the δ function $\delta(E - E')$

$$\langle b|D_p|b' q\lambda\rangle = \hbar \sqrt{\frac{q}{m}} \langle b|D_p|b' E\lambda\rangle .$$

Formulas (9.222) and (9.223) can be written

$$\frac{q^2}{\pi^2} \sigma(\gamma'E;\gamma) = \sum_{\lambda} \frac{4k^3}{3} \frac{1}{g_{\gamma'}} \sum_{bb'} |\langle b'E\lambda|D|b\rangle|^2 , \tag{9.226}$$

$$\frac{k^2}{\pi^2} \sigma(\gamma; \gamma'E) = \sum_{\lambda} \frac{4k^3}{3} \frac{1}{g_{\gamma}} \sum_{bb'} |\langle b|D|b'E\lambda\rangle|^2 . \tag{9.227}$$

When calculating the effective cross sections of the radiation transitions in the states of the continuous spectrum, it is possible to neglect fine splitting.

Let us consider photoionization. It is convenient to choose as the wave function describing the final state of the system the functions $\psi_{S_1L_1E\lambda S'L'M_S'M_L'}$, where $L' = L + \lambda$, $S' = S_1 + s$ are the total orbital angular momentum and the total spin of the system, S_1 and L_1 are the spin and orbital angular momentum of an ion. In this case in (9.227)

$$\frac{1}{g_\gamma} \sum_{bb'} |\langle b|D|b'E\lambda\rangle|^2 = \frac{1}{(2S+1)(2L+1)}$$

$$\times \sum_{S'L'} \sum_{M_LM_L'} \sum_{M_SM_S'} |\langle SLM_SM_L|D|S_1L_1E\lambda S'L'M_S'M_L'\rangle|^2$$

$$= (2L+1)^{-1} \sum_{L'} |(SL||D||S_1L_1E\lambda SL')|^2 .$$

In the parentage scheme approximation, we have

$$|(S_1L_1 lSL||D||S_1L_1, El'SL')|^2 = e^2(2L+1)(2L'+1) \begin{Bmatrix} l & L & L_1 \\ L' & l' & 1 \end{Bmatrix}^2$$

$$\times l_{\max} \left(\int R_\gamma r R_{El'} r^2 \, dr \right)^2, \tag{9.228}$$

where $l_{\max}$ is the larger of the numbers l, l'. Thus the effective cross section for the ionization proces $S_1L_1nlSL \to S_1L_1E$ is determined by the expression

$$\sigma(\gamma, \gamma' E) = \frac{4\pi^2}{3} e^2 k \sum_{l'=l\pm 1} \sum_{L'} (2L'+1) \begin{Bmatrix} l & L & L_1 \\ L' & l' & 1 \end{Bmatrix}^2 l_{\max}$$
$$\times \left(\int R_\gamma r R_{El'} r^2 \, dr \right)^2. \tag{9.229}$$

The corresponding formula for the effective cross section of the photorecombination $S_1 L_1 E \to S_1 L_1 \, nlSL$ can be obtained by means of (9.224) which in this case becomes

$$q^2 (2S_1 + 1)(2L_1 + 1) \sigma(\gamma' E; \gamma) = k^2 (2S + 1)(2L + 1) \sigma(\gamma \, \gamma' E),$$
$$q^2 = \frac{2mE}{\hbar^2}. \tag{9.230}$$

For one electron outside filled shells $S_1 = 0$, $L_1 = 0$, $L = l$, $S = 1/2$, and it follows from (9.229, 230)

$$\sigma(nl; E) = \frac{4\pi^2 e^2 k}{3(2l+1)} \sum_{l'=l\pm 1} l_{\max} \left(\int R_{nl} r R_{El'} r^2 \, dr \right)^2, \tag{9.231}$$
$$q^2 \sigma(E; nl) = 2(2l+1) k^2 \sigma(nl; E). \tag{9.232}$$

If the dependence of the radial functions R_γ and R_{El} on $S_1 L_1 SL$ is neglected, then the total effective cross sections for the single-electron transitions $nl \to E$ and $E \to nl$ for a multielectron atom coincide with (9.231, 232). If one of the electrons of the group l^N takes part in the transition, we have

$$\sigma(l^N \gamma SL; l^{N-1} \gamma_1 S_1 L_1 E) = N |G^{\gamma SL}_{\gamma_1 S_1 L_1}|^2 \sigma(\gamma_1 S_1 L_1 nlSL; \gamma_1 S_1 L_1 E), \tag{9.233}$$

$$\sigma(l^N; l^{N-1} E) = N\sigma(l; E). \tag{9.234}$$

Relations (9.230, 232) remain the same.

9.5.3 Bremsstrahlung: General Expressions for Effective Cross Sections

We shall begin by considering the simplest case of an electron in a centrally symmetric field. The effective cross section for the transition of an electron from the continuous-spectrum state q to an interval $q', q' + dq'$ as a result of the absorption of a photon $\hbar\omega$ can be obtained from (9.241) by replacing in it q by q' and ψ_b by $v^{-1/2} \psi_q^+ = (m/\hbar q)^{1/2} \psi_q^+$. Consequently, for the differential effective cross section we have

$$d\sigma_{qk;q'} = \frac{m^2 e^2 k q'}{2\pi \hbar^3 q} \left| e_{k\rho} \cdot \int (\psi_q^+)^* r \psi_{q'}^- \, dr \right|^2 dO_{q'}, \quad \hbar\omega = \frac{\hbar^2 q'^2}{2m} - \frac{\hbar^2 q^2}{2m}. \tag{9.235}$$

When calculating the effective cross section of the reverse transition $q' \to q$, accompanied by the emission of a photon with wave vector k and polarization e_{kp}, it is necessary to substitute $df = dq\, V dk/(2\pi)^3$ into the general equation (9.210) and take the functions $v^{-1/2} \psi_{q'}^+$ and $(2\pi)^{-3/2} \psi_q^-$ as the wave functions of the electron in the initial and final states. Integrating over dq, we obtain for the differential effective cross section of bremsstrahlung the expression

$$d\sigma_{q';qk} = \frac{m^2 e^2 k^3}{(2\pi)^4 \hbar^3} \frac{q}{q'} \left| e_{kp} \int (\psi_{q'}^+)^* r \psi_q^- dr \right|^2 d\omega\, dO_k\, dO_q. \tag{9.236}$$

From (9.235, 236) we have

$$(2\pi)^3 \frac{d\sigma_{q';qk}}{k^2 d\omega\, dO_k q^2 dO_q} = \frac{d\sigma_{qk;q'}}{q'^2 dO_{q'}}. \tag{9.237}$$

At a fixed value of the initial energy of the electron E', photons can be absorbed with frequency ω in the interval $0 < \omega < \infty$. But photons with frequency in the interval $0 < \omega < E'/\hbar$ can be radiated. Thus a definite high-frequency limit of bremsstrahlung corresponds to each value of E'.

We shall be interested below in cross sections integrated over all directions of motion of the electrons and photon. Having substituted (9.216) into (9.235), we shall integrate over $dO_{q'}$ and average over all possible mutual orientations of the vectors q and k.

By means of (9.216), we obtain

$$\frac{1}{4\pi} \int dO_q dO_{q'} \left| e_{pk} \cdot \int (\psi_q^+)^* r \psi_{q'}^- dr \right|^2 = \frac{1}{4\pi} \frac{(2\pi)^3}{q^2} \frac{(2\pi)^3}{q'^2} \sum_{\lambda\mu\lambda'\mu'} |\langle \lambda\mu | e_{pk} r | \lambda'\mu' \rangle|^2$$

$$= \frac{2\pi^2}{3} \frac{(2\pi)^3}{q^2 q'^2} \sum_{\lambda\mu\lambda'\mu'} |\langle \lambda\mu | r | \lambda'\mu' \rangle|^2.$$

Therefore,

$$\sigma_p(qk; q') = \frac{8\pi^4}{3} \frac{m^2 k}{\hbar^3 q^3 q'} \sum_{\lambda\lambda'} \sum_{\mu\mu'} |\langle \lambda\mu | D | \lambda'\mu' \rangle|^2. \tag{9.238}$$

Similarly, by substituting (9.216) into (9.236) and integrating over $dO_q dO$, we have

$$\frac{d\sigma_p(q'; qk)}{d\omega} = \frac{4\pi^2}{3} \frac{m^2 k^3}{\hbar^3 qq'^3} \sum_{\lambda\lambda'} \sum_{\mu\mu'} |\langle \lambda\mu | D | \lambda'\mu' \rangle|^2. \tag{9.239}$$

Expression (9.238) has to be summed with respect to the final spin states m_s' and averaged over m_s and directions of polarization of the photon, $p = 1, 2$. Ex-

9.5 Continuous Spectrum

pression (9.239) must be summed with respect to m_s and p and averaged over m_s'. Since

$$\sum_{\mu\mu'}|\langle\lambda\mu|D|\lambda'\mu'\rangle|^2 = 2e^2\lambda_{\max}\int R_{q'\lambda'}rR_{q\lambda}r^2\,dr \; ,$$

we have

$$\sigma_{qk;q'} = \frac{8\pi^4}{3}\frac{m^2 k e^2}{\hbar^3 q^3 q'}\sum_{\lambda,\lambda'=\lambda\pm1}\lambda_{\max}\left(\int R_{q'\lambda'}rR_{q\lambda}r^2\,dr\right)^2 , \qquad (9.240)$$

$$\frac{\delta\sigma_{q';qk}}{d\omega} = \frac{8\pi^2}{3}\frac{m^2 k^3 e^2}{\hbar^3 q q'^3}\sum_{\lambda,\lambda'=\lambda\pm1}\lambda_{\max}\left(\int R_{q'\lambda'}rR_{q\lambda}r^2\,dr\right)^2 . \qquad (9.241)$$

It is easy to generalize (9.240, 241) to the case of a transition in the field of an arbitrary multielectron atom in the state γ_0. Similar to (9.226, 227) it easy to obtain

$$\frac{q'^2}{\pi^2}\frac{d\sigma_{E'\cdot Ek}}{d\omega} = \frac{4k^3\hbar}{3}\sum_{\lambda,\lambda'=\lambda\pm1}\frac{1}{g}\sum_{aa'}|\langle\gamma_0 E\lambda a|D|\gamma_0 E'\lambda'a'\rangle|^2 ,$$

$$\frac{q^2}{\pi^2}\frac{k^2}{\pi^2}\sigma_{Ek;E'} = \frac{4k^3\hbar}{3}\sum_{\lambda,\lambda'=\lambda\pm1}\frac{1}{g}\sum_{aa'}|\langle\gamma_0 E'\lambda'a'|D|\gamma_0 E\lambda a\rangle|^2 , \qquad (9.242)$$

$$q'^2\frac{d\sigma_{E'\cdot Ek}}{d\omega} = \frac{k^2}{\pi^2}q^2\sigma_{Ek;E'}, \quad \frac{\hbar^2 q^2}{2m} = E, \quad \frac{\hbar^2 q'^2}{2m} = E', \quad \hbar\omega + E = E' . \qquad (9.243)$$

In these formulas $a(a')$ is the set of quantum numbers giving the state of the system "atom + electron", and g is the statistical weight of the atomic level. The functions $R_{E\lambda} = \hbar^{-1}(m/q)^{1/2} R_{q\lambda}$ are used as the radial functions.

We shall consider transitions in the field of an atom with total orbital angular momentum L_1 and total spin S_1 and choose as the wave functions $\psi_{\gamma_0 E\lambda a}$ the functions $\psi_{S_1L_1 E\lambda SLM_SM_L}$. In this case $g = (2S_1 + 1)(2L_1 + 1)$, and summing with respect to a means summing with respect to SLM_SM_L. Repeating the same reasoning as in the derivation of (9.229), we have

$$q'^2\frac{d\sigma_{E';Ek}}{d\omega} = q^2\frac{k^2}{\pi^2}\sigma_{Ek;E'} = \frac{4\pi^2}{3}e^2\hbar k^3\sum_{\lambda\lambda'}\sum_{SLL'}\frac{2S+1}{2S_1+1}\frac{(2L+1)(2L'+1)}{(2L_1+1)}$$

$$\left\{\begin{matrix}\lambda & L & L_1 \\ L' & \gamma' & 1\end{matrix}\right\}^2 \lambda_{\max}\times\left(\int R_{E\lambda}rR_{E'\lambda'}r^2\,dr\right)^2 . \qquad (9.244)$$

If the dependence of the radial functions $R_{E\lambda}$ and $R_{E'\lambda'}$ on L, L' and S is neglected, then summing with respect to L, L' and $S = S_1 \pm 1/2$ gives the same result as (9.240, 241).

9.5.4 Radiation and Absorption Coefficients

Knowing the effective cross sections for photorecombination and bremsstrahlung, it is possible to calculate the energy emitted or absorbed by a unit volume of a medium. We shall denote the energy emitted by a unit volume in 1 second resulting from the recombination of electrons with velocities v, $v+dv$ on the level γ by $Q^R_{\gamma'\gamma}(\omega)d\omega$, where γ' gives the level of the initial ion. For this quantity we have

$$Q^R_{\gamma'\gamma}(\omega)d\omega = N_e v f(v) dv N_{\gamma'} \hbar\omega \sigma(\gamma'E;\gamma) \text{ [erg/cm}^3\text{ s]}$$

where N_e is the concentration of electrons, $N_{\gamma'}$ is the concentration of ions on the level γ', $f(v)$ is the velocity distribution function, normalized to unity. Since

$$\hbar\omega = \frac{mv^2}{2} + |E_\gamma|, \quad v\,dv = \frac{\hbar}{m}d\omega, \qquad (9.245)$$

$$Q^R_{\gamma'\gamma}(\omega)d\omega = N_e N_{\gamma'} \frac{\hbar^2\omega}{m} f\left(\sqrt{\frac{2}{m}(\hbar\omega - |E_\gamma|)}\right) \sigma(\gamma'E,\gamma) d\omega, \qquad (9.246)$$

where $E = \hbar\omega - |E_\gamma|$.

The total intensity of recombination radiation $Q^R(\omega)d\omega$ can be obtained from (9.246) by summing over all levels γ' and γ for which $|E_\gamma| < \hbar\omega$

$$Q^R(\omega)\,d\omega = \sum_{\gamma\gamma'} Q^R_{\gamma\gamma'}(\omega)\,d\omega, \quad |E_\gamma| \leqslant \hbar\omega. \qquad (9.247)$$

The total energy loss due to recombination radiation is determined by the expression

$$Q^R = \int Q^R(\omega)\,d\omega = N_e \sum_{\gamma\gamma'} N_{\gamma'} \int f(v)\,v\sigma(\gamma'E,\gamma)\left[\frac{mv^2}{2} + |E_\gamma|\right] dv$$

$$= N_e \sum_{\gamma'\gamma} N_{\gamma'} \left\langle v\sigma(\gamma'E,\gamma)\left(\frac{mv^2}{2} + |E_\gamma|\right)\right\rangle. \qquad (9.248)$$

As a rule, it is sufficient to consider only the ground state of the initial ion γ'.

Similarly, it is not difficult to calculate the intensity of the bremsstrahlung $Q^B_{\gamma_0}(\omega)$ in the field of the atom on the level γ_0.

$$Q^B_{\gamma_0}(\omega) = N_e N_{\gamma_0} \hbar\omega \int_{\sqrt{2\hbar\omega/m}}^{\infty} \frac{d\sigma_{E';\,kE\gamma_0}}{d\omega} v' f(v')\,dv',$$

$$Q^B(\omega) = \sum_{\gamma_0} Q^B_{\gamma_0}(\omega). \qquad (9.249)$$

The total energy loss by bremsstrahlung obviously equals

$$Q^B = \int Q^B(\omega)\, d\omega. \tag{9.250}$$

Let us introduce the coefficient of radiation per unit volume ε_ω, defined by the relation

$$Q(\omega) = \int \varepsilon_\omega\, dO. \tag{9.251}$$

If the radiation is isotropic, $Q(\omega) = 4\pi\varepsilon_\omega$.

We shall now pass on to a calculation of the coefficient of absorption k_ω [cm^{-1}] defining the attenuation of a light beam of frequency ω. The photoionization absorption coefficient is

$$k_\omega = \sum_r N_\gamma \sigma(\gamma; \gamma'E)\ |E_\gamma| < \hbar\omega, \tag{9.252}$$

where $\sigma(\gamma; \gamma'E)$ is the effective cross section of photoionization.

The effective cross section for continuum absorption has the dimensionality [cm^{4}s] because in this case the transition probability is equal to the effective cross section multiplied by the density of flow of photons and by the density of flow of electrons. The quantity

$$N_e \int v f(v)\, \sigma_{EK;\,E'}\, dv = N_e \langle v \sigma_{EK;E'} \rangle$$

plays the role of an effective cross section for the absorption of photons (dimensionality: cm^2). Thus the continuum absorption coefficient of photons is defined by the expression

$$k_\omega = N_e \sum_{\gamma_0} N_{\gamma_0} \langle v \sigma_{EK;E'}(\gamma_0) \rangle, \tag{9.253}$$

where N_{γ_0} is the concentration of atoms in the level γ_0, and $\sigma_{EK;E'}(\gamma_0)$ is the effective cross section for continuum absorption in the field of the atom in the level γ_0. In some cases, together with spontaneous emission and absorption, it is necessary to take into account also stimulated emission. It is easy to introduce corrections for stimulated emission to the above formulas in exactly the same way as in the case of transitions between discrete spectrum states (Sect. 9.1). Thus the effective cross section for emission of a photon has to be multiplied by $[1 + (4\pi^3 c^2/\hbar\omega^3) I_k]$. If the radiation is isotropic, $I_k = (c/4\pi) \cdot U_\omega$, and the correction factor can also be written in the form $[1 + (\pi^2 c^3/\hbar\omega^3) \cdot U_\omega]$.

The correction to the absorption coefficient k_ω for stimulated emission depends on the distribution of the atoms over atomic levels and of the electrons over velocities. We shall denote below the coefficient of absorption with correction for stimulated emission by k'_ω. In conditions of thermodynamic equilibrium

$$k'_\omega = k_\omega (1 - e^{-\hbar\omega/kT}) \tag{9.254}$$

and there exists a universal relation between the emission coefficient ε_ω and the absorption coefficient k'_ω

$$\varepsilon_\omega/k'_\omega = \frac{c}{4\pi} \frac{\hbar\omega^3}{\pi^2 c^3} \cdot (e^{\hbar\omega/kT} - 1)^{-1} = \frac{c}{4\pi} U_\omega = I_\omega, \tag{9.255}$$

where U_ω is energy density of a black-body radiation. Relation (9.255) is called Kirchhoff's law.

It should be noted that (9.255) can also be fulfilled for bremsstrahlung and continuum absorption in the absence of complete thermodynamic equilibrium. It is sufficient only that the velocity distribution of the electrons is Maxwellian. Let us consider radiation and absorption of an electron in the field of an atom in the level γ_0. From (9.249, 254) it follows that

$$\varepsilon_\omega = \frac{1}{4\pi} N_e N_{\gamma_0} \hbar\omega \int_{\sqrt{2\hbar\omega/m}}^{\infty} \frac{d\sigma_{E'; Ek}}{d\omega} v' f(v') \, dv',$$

$$k'_\omega = N_e N_{\gamma_0} \int_0^\infty \sigma_{Ek; E'} v f(v) \, dv \, (1 - e^{-\hbar\omega/kT}). \tag{9.256}$$

From (9.243),

$$v'^2 \, d\sigma_{E'; Ek}/d\omega = (\omega^2/\pi^2 c^2) v^2 \sigma_{Ek; E'}, \quad v'^2 = v^2 + 2\hbar\omega/m,$$

it follows that

$$k'_\omega = N_e N_{\gamma_0}(1 - e^{-\hbar\omega/kT}) \frac{\pi^2 c^2}{\omega^2} \int_0^\infty \frac{v'^2}{v^2} \frac{d\sigma_{E'; Ek}}{d\omega} v f(v) \, dv.$$

Substituting in this expression the Maxwellian distribution function

$$f(v) \, dv = 4\pi \left(\frac{m}{2\pi kT}\right)^{3/2} v^2 e^{-mv^2/2kT} \, dv \tag{9.257}$$

and replacing the integration over v by an integration over v', we have

$$k'_\omega = N_e N_{\gamma_0}(e^{\hbar\omega/kT} - 1) \frac{\pi^2 c^2}{\omega^2} \int_{\sqrt{2\hbar\omega/m}}^{\infty} \frac{d\sigma_{E'\, Ek}}{d\omega} v' f(v') \, dv'. \tag{9.258}$$

It is easy to see that for any value of N_{γ_0} the ratio of ε_ω and k'_ω equals (9.255). There is only one assumption, which has been made above, and this is the assumption that the velocity distribution of the electrons is Maxwellian.

Relation (9.255) for the coefficients of recombination radiation ε_ω and photoionization absorption k'_ω can be obtained in a similar way by using (9.224) and assuming that: 1) the velocity distribution of electrons is Maxwellian, 2) the population of discrete levels is determined by the Boltzmann formula, 3) the ion concentration is determined by the Saha formula.

9.5.5 Photorecombination and Photoionization: Hydrogenlike Atoms

We shall consider the transitions in which the ground state of a hydrogenlike atom participates. According to (9.231, 232), we have for the effective cross sections of photoionization σ^i and photorecombination σ^r

$$\sigma^i = \frac{mc^2 E}{\hbar^2 \omega^2} \sigma^r = \frac{4\pi^2 e^2 \omega}{3c} \left(\int R_{10} r R_{E1} r^2 dr \right)^2 . \tag{9.259}$$

In the nonrelativistic approximation for the radial functions R_{10} and R_{E1}, the integral in (9.259) can be calculated exactly [Ref. 2, Sect. 71]:

$$\left(\int R_{10} r R_{E1} r^2 dr \right)^2 = \frac{2^8}{Z^4} \frac{a_0^3}{e^2} \left(\frac{\kappa^2}{1+\kappa^2} \right)^5 f(\kappa), \tag{9.260}$$

where

$$f(\kappa) = \frac{e^{-4\kappa \operatorname{arccot} \kappa}}{1 - e^{-2\pi k}}, \quad \kappa = Z \left(\frac{\mathrm{Ry}}{E} \right)^{1/2} = Z \frac{e^2}{\hbar v}, \tag{9.261}$$

and v is the velocity of the electron. The transition of an electron from the state $1s$ to a continuous spectrum is possible if the frequency of the absorbed photon $\omega \geq \omega_\Gamma = |E_{1s}|/\hbar = Z^2 \mathrm{Ry}/\hbar$ where ω_Γ is the limiting frequency for absorption. From the definition of κ it follows that ω, κ, and ω_Γ are related by

$$\hbar\omega = Z^2 \mathrm{Ry} + E = \hbar\omega_\Gamma \left(1 + \frac{1}{\kappa^2} \right), \quad \kappa^2 = \frac{\omega_\Gamma}{\omega - \omega_\Gamma}. \tag{9.262}$$

Substituting (9.260) into (9.259) and taking (9.262) into account, we have

$$\sigma^r = \frac{2^8 \pi^2}{3} \alpha^3 \frac{\omega_\Gamma^3}{\omega^2 (\omega - \omega_\Gamma)} \frac{e^{-4\kappa \operatorname{arccot} \kappa}}{1 - e^{-2\pi\kappa}} a_0^2,$$

$$\sigma^i = \frac{2^9 \pi^2}{3} \frac{\alpha}{Z^2} \left(\frac{\omega_\Gamma}{\omega} \right)^4 \frac{e^{-4\kappa \operatorname{arccot} \kappa}}{1 - e^{-2\pi\kappa}} a_0^2. \tag{9.263}$$

We shall now consider the limits of high and low values of κ.

9. Radiative Transitions

Near the limiting frequency $\kappa \gg 1$ and $\omega - \omega_r \ll \omega_r$

$$f(\kappa) \sim e^{-4} \left(\frac{\omega}{\omega_r}\right)^{4/3} \approx (2.72)^{-4} \left(\frac{\omega}{\omega_r}\right)^{4/3}$$

and consequently

$$\sigma^r \approx \left(\frac{4}{2.72}\right)^4 \frac{\pi^2}{3} \alpha^3 \frac{\omega_r^2}{\omega(\omega - \omega_r)} \left(\frac{\omega}{\omega_r}\right)^{1/3} a_0^2, \qquad (9.264)$$

$$\sigma^i \approx 2 \left(\frac{4}{2.72}\right)^4 \frac{\pi^2}{3} \frac{\alpha}{Z^2} \left(\frac{\omega_r}{\omega}\right)^{8/3} a_0^2.$$

For $\kappa \sim 1$,

$$f(\kappa) \sim (2.72)^{-4} \left(1 + \frac{4}{3} \frac{\omega - \omega_r}{\omega_r}\right) \approx (2.72)^{-4} \left(\frac{4}{3} \frac{\omega}{\omega_r} - \frac{1}{3}\right).$$

Finally, far from the limiting frequency, $\kappa \ll 1$ and $\omega - \omega_r \gg \omega_r$

$$f(\kappa) \sim \frac{1}{2\pi} \sqrt{\frac{\omega - \omega_r}{\omega_r}} \approx \frac{1}{2\pi} \left(\frac{\omega}{\omega_r}\right)^{1/2}$$

$$\sigma^r \approx \frac{2^7 \pi}{3} \alpha^3 \frac{\omega_r^2}{\omega(\omega - \omega_r)} \left(\frac{\omega_r}{\omega}\right)^{1/2} a_0^2 \approx \frac{2^7 \pi}{3} \alpha^3 \left(\frac{\omega_r}{\omega}\right)^{5/2} a_0^2, \qquad (9.265)$$

$$\sigma^i \approx \frac{2^8 \pi}{3} \frac{\alpha}{Z^2} \left(\frac{\omega_r}{\omega}\right)^{7/2} a_0^2.$$

Thus the effective cross section for the photoeffect is maximum at the photoabsorption limit

$$\sigma^i = \frac{2\pi^2}{3} \left(\frac{4}{2.72}\right)^4 \frac{\alpha}{Z^2} a_0^2.$$

With increasing ω, σ^i decreases at first as $\omega^{-8/3}$ and then when $\omega \gg \omega_r$ as $\omega^{-7/2}$. The cross section σ^r when $\omega \gg \omega_r$ decreases as $\omega^{-5/2}$. In approaching the limiting frequency ω_r, $\sigma^r \to \infty$.

Formulas (9.265) coincide with the results of the Born approximation which can be obtained from (9.259) by substituting for R_{E1} the radial function of the free motion. The Born approximation consists in taking plane waves as the continuous-spectrum functions. The condition of validity of the Born approximation to the scattering of electrons in a Coulomb field $- Ze^2/r$ is $Ze^2/\hbar v \ll 1$, i.e., $\kappa \ll 1$. In the nonrelativistic approximation, exact analytical expressions can also be obtained for the effective cross sections of photoprocesses corresponding to the levels $n = 2, 3, 4 \ldots$ [2, 27].

However for $n > 2$, these formulas are cumbersome and not very suitable for calculations. The simple quasi-classical formulas, first obtained by Kramers, are usually used for various approximations. The condition of validity of these formulas for the Coulomb field is the opposite of the condition for the Born condition, $Ze^2/\hbar v \gg 1$, or $\omega \lesssim \omega_\Gamma$. For the effective cross sections σ_n^r and σ_n^i connected in accordance with (9.224) by the relation

$$\sigma_n^i = (q/k)^2 \cdot 1/2n^2 \cdot \sigma_n^r = 2mc^2 E/\hbar^2\omega^2 2n^2 \cdot \sigma_n^r$$

the Kramers formulas give

$$\sigma_n^r = \frac{32\pi}{3\sqrt{3}} \alpha^3 \frac{\omega_\Gamma^2}{\omega(\omega - \omega_\Gamma/n^2)} \frac{a_0^2}{n^3},$$

$$\sigma_n^i = \frac{64\pi}{3\sqrt{3}} \frac{\alpha}{Z^2} \left(\frac{\omega_\Gamma}{\omega}\right)^3 \frac{a_0^2}{n^5},$$

(9.266)

because

$$g_{\gamma'} = 1, \quad g_\gamma = 2n^2.$$

Here, as before, $\omega_\Gamma = Z^2 \text{Ry}/\hbar$. The limit for photoabsorption from the level n is determined by the condition $\omega \geqslant \omega_\Gamma/n^2 = Z^2 \text{Ry}/\hbar n^2$ or $\omega - \omega_\Gamma/n^2 = E/\hbar \geqslant 0$. Comparison of (9.266) with exact formulas shows that the quasi-classical approximation gives good results both for high and for low values of n. Thus for $n = 1$, the ratio of cross sections σ^r (9.264–266) equals $\approx 1.25 \, (\omega_\Gamma/\omega)^{1/3}$. Close to the limit of absorption $\omega \sim \omega_\Gamma$ the difference is unimportant. With increasing ω it can become noticeable.

It is interesting to compare the cross sections of the Kramers (σ_K) and Born (σ_B) approximations. When $n = 1$

$$\sigma_K/\sigma_B = \frac{1}{4\sqrt{3}} \left(\frac{\omega}{\omega_\Gamma}\right)^{1/2}.$$

Usually recombination and photoabsorption cross sections are described in the form of the Kramers cross sections multiplied by the correction factor g, the so-called Gaunt factor. This factor is close to unity for the visible and ultraviolet regions of the spectrum. When $n = 1$,

$$g = 8\pi \sqrt{3} \left(\frac{\omega_\Gamma}{\omega}\right) f\left(\sqrt{\frac{\omega_\Gamma}{\omega - \omega_\Gamma}}\right)$$

$$\omega - \omega_\Gamma \ll \omega_\Gamma \quad g \simeq 8\pi\sqrt{3} \, (2.72)^{-4} \simeq 0.8$$

$$\omega > \omega_\Gamma \quad g \simeq \frac{1}{4\sqrt{3}} \left(\frac{\omega}{\omega_\Gamma}\right)^{1/2}.$$

Consider now the photoabsorption coefficient k_ω of a hydrogenlike gas. For a certain fixed frequency ω

$$k_\omega = \sum_{n=n_0}^{\infty} \sigma_n^i(\omega) N_n, \qquad (9.267)$$

where n_0 is the lowest of the possible values of n which satisfy the condition $\omega > \omega_\Gamma/n^2 = \mathrm{Ry} Z^2/\hbar n^2$. In the case $\omega > \omega_\Gamma$, $n_0 = 1$.

Let us assume that the atomic levels have a Boltzmann distribution and let us define the energy of the levels E_n as the difference between excitation energy and the energy of the ground state level E_1 [$E_n = \mathrm{Ry} Z^2 (1 - 1/n^2)$]. Then

$$N_n = N_1 \frac{g_n}{g_1} e^{-E_n/kT} = N \frac{g_n e^{-E_n/kT}}{\sum_{n'} g_{n'} e^{-E_{n'}/kT}} = \frac{N}{S} g_n e^{-E_n/kT},$$

where N is the total concentration of atoms, g_n is the statistical weight (for a hydrogenlike atom $g_n = 2n^2$), and S is the statistical sum. The contribution of excited levels to the sum over n depends on temperature.

According to (9.266), $\sigma_n^i \propto n^{-5}$. Consequently, the terms of (9.267) decrease as $n^{-3} \exp(-\mathrm{Ry} Z^2/n^2 kT)$. When calculating k_ω at frequencies $\omega > \omega_\Gamma$, one can usually neglect all terms with $n > 2$. At low values of ω, for which $n_0 \neq 1$, a large number of levels gives approximately the same contribution to the sum over n. By substituting in (9.267) the values of the cross section σ_n^i (9.266) multiplied by the Gaunt factor $g(n, \omega)$, we have

$$k_\omega = \frac{128\pi}{3\sqrt{3}} \alpha a_0^2 \left(\frac{\mathrm{Ry}}{\hbar\omega}\right)^3 Z^4 \frac{N}{S} \sum_{n=n_0}^{\infty} \frac{g(n,\omega)}{n^3} e^{-E_n/kT}. \qquad (9.268)$$

In the range of high frequencies, $n_0 = 1$. Consequently k_ω increases with decrease of ω [when $g(n, \omega) = 1$, as ω^{-3}]. At the point $\omega = \mathrm{Ry} Z^2/\hbar$, k_ω drops by a step because in the range $\mathrm{Ry} Z^2/\hbar > \omega > \mathrm{Ry} Z^2/4\hbar$, $n_0 = 2$. With a further decrease of ω, k_ω increases right up to the limit of absorption from the level $n = 2$. Then it drops again because in the range $\mathrm{Ry} Z^2/4\hbar > \omega > \mathrm{Ry} Z^2/9\hbar$, $n_0 = 3$ and so on.

If n_0 is large (it is usually sufficient that $n_0 > 4,5$), summing over n in (9.268) can be replaced by integration and one can assume $n_0 = (\mathrm{Ry} Z^2/\hbar\omega)^{1/2}$. In the Kramers approximation this gives

$$k_\omega = \frac{64\pi}{3\sqrt{3}} \alpha a_0^2 \frac{N}{S} Z^2 \left(\frac{\mathrm{Ry}}{\hbar\omega}\right)^3 \left(\frac{kT}{\mathrm{Ry}}\right) e^{-\mathrm{Ry} Z^2/kT} (e^{\hbar\omega/kT} - 1). \qquad (9.269)$$

Multiplying (9.268, 269) by the correction factor $[1 - \exp(-\hbar\omega/kT)]$ which takes stimulated emission into account, one can find k'_ω and, by means of (9.251, 255), the recombination radiation coefficient ε_ω and $Q(\omega)d\omega$.

In the approximation (9.266) it is also easy to find the total intensity of the recombination radiation. Since for a Maxwellian distribution $\langle v^{-1}\rangle = (4/\pi)\langle v\rangle^{-1} = (2m/\pi kT)^{1/2}$, it follows from (9.248) that

$$Q^R = N_e N_i \sum_n \langle v\sigma_n^r (mv^2/2 + \mathrm{Ry}Z^2/n^2)\rangle$$

$$= \frac{64\pi\alpha^3 a_0^2 \mathrm{Ry}^2 Z^4}{3\sqrt{3}\,m} N_e N_i \langle v^{-1}\rangle \sum_{n=1}^{\infty} n^{-3} \approx 1.2 \frac{64\alpha^3 a_0^2 \mathrm{Ry}^2 Z^4}{3\sqrt{3}} \left(\frac{2\pi}{mkT}\right)^{1/2} N_e N_i \,.$$

(9.270)

A formula, similar to (9.270), can also be obtained for high values of T, i.e., for high electron velocities. Analysis of results of numerical calculations and of the Born approximation formulas shows that in the whole interval $\kappa \simeq 0\text{–}3$, i.e., for $\omega \gtrsim \omega_r$, one can use the approximate relation [28]

$$\sum_{n=1}^{\infty} \sigma_n^r \simeq (1.20 + 0.28\kappa)\,\sigma_1^r \,.$$

In the same approximation

$$\sum_n \sigma_n^r n^{-2} \simeq (1.04 + 0.04\kappa)\,\sigma_1^r \,.$$

Upon using it, we can obtain

$$Q^R = 5.10^{-24} Z^4 N_i N_e\, T^{-1/2} \;[\mathrm{erg\ cm^{-3}s^{-1}}]$$

(N_i and N_e are expressed in cm^{-3} and T in eV), which practically coincides with (9.270).

The formulas given above for σ^i are often used for approximate estimates of the effective cross sections for photoabsorption by electrons of inner shells of complex atoms. In this case it is necessary to replace Z by $Z_{\mathrm{ef}} = Z - \beta$. A number of empirical rules exist for finding the parameter β [3]. In addition, in accordance with (9.234), the cross section has to be multiplied by the numbers of electrons in the shell.

9.5.6 Photorecombination and Photoionization: Nonhydrogenlike Atoms

In the case of nonhydrogenlike atoms or ions, the radial integral in the expressions for effective cross sections cannot be calculated exactly. For approximate estimates of k_ω and ε_ω at low frequencies, to which only highly excited states contribute, one can use (9.269), replacing in it $\exp[-\mathrm{Ry}Z^2/kT]$ by $\exp[-I/kT]$, where I is the ionization potential of the atom, and assuming $Z = 1$ for a neutral atom, $Z = 2$ for a singly charged ion, and so on.

A very effective semiempirical method of calculating photorecombination and photoionization cross sections for nonhydrogenlike atoms was proposed by *Burgess* and *Seaton* [29]. This method is a generalization of the Bates and Damgaard method (Sect. 9.3) to transitions to the continuous spectrum. The discrete-spectrum radial functions R_{nl} are determined in exactly the same way as in the Bates and Damgaard method. When calculating the continuous-spectrum radial functions $R_{El'}$, the quantum defect method is used. The quantum defect $\Delta_{l'}(E)$ is determined by extrapolation of the quantum defect $\Delta_{l'}$ for the series of l' terms to the continuous spectrum, as shown in Fig. 9.1.

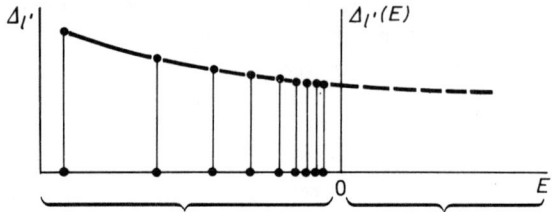

Fig. 9.1. Extrapolation of the quantum defect $\Delta_{l'}$ to the continuous spectrum states

The results of calculations of the photoionization cross section by means of the radial functions obtained in this way can be described in the form

$$\sigma = 5.45 \times 10^{-19} \frac{\nu^3}{\zeta^2 \alpha(\nu, l)(1 + \varepsilon' \nu^2)^3}$$
$$\times \sum_{l'=l\pm 1} C_{l'} |G(\nu l; \varepsilon' l') \cos\{\pi[\nu + \Delta(\varepsilon') + \chi(\nu l; \varepsilon' l')]\}|^2, \qquad (9.271)$$

where in the parentage scheme approximation for the transition

$$S_1 L_1 nlSL \rightarrow S_1 L_1 El'$$

$$C_{l'} = \sum_{L'} (2L' + 1) \begin{Bmatrix} l & L & L_1 \\ L' & l' & 1 \end{Bmatrix}^2 l_{\max}$$

and for the transition $l^N \gamma SL \rightarrow l^{N-1} \gamma_1 S_1 L_1 El'$

$$C_{l'} = N \sum_{L'} |G_{\gamma_1 S_1 L_1}^{\gamma SL}|^2 (2L' + 1) \begin{Bmatrix} l & L & L_1 \\ L' & l' & 1 \end{Bmatrix} l_{\max}$$

[see (9.229, 233)]. For one electron outside filled shells, $C_{l'} = l_{\max}/(2l + 1)$. The parameters ν (effective principal quantum number for the discrete spectrum level) and ε' are determined by the expressions

$$v^2 = \frac{\zeta^2 \mathrm{Ry}}{I_{nl}}, \quad \varepsilon' = \left(\frac{E}{\mathrm{Ry}}\right)\frac{1}{\zeta^2},$$

where I_{nl} is the ionization potential of the discrete spectrum state nl.

The functions $G(\nu l; \varepsilon' l')$ and $\chi(\nu l; \varepsilon' l')$ for a number of transitions $l - l'$ calculated in [30] are given in Table 9.3; $\Delta(\varepsilon')$ is the extrapolated value of the quantum defect $\Delta_{l'} = n_{l'} - \nu_{l'}$. For neutral atoms $\zeta = 1$, for singly charged ions $\zeta = 2$, and so on.

The function $\alpha(\nu, l)$ is determined by

$$\alpha(\nu, l) = 1 + 2\nu^{-3}\, d\Delta_l(\varepsilon)/d\varepsilon.$$

As a rule, $\alpha(\nu,l)$ is close to unity and one can assume $\alpha(\nu,l) = 1$. A comparison of (9.271) with experimental and theoretical results is given in [30]; the calculations were based on variational methods, the Hartree-Fock approximation, and other methods. This comparison and also analysis of the approximations used in calculating the radial integrals show that the method is approximately as accurate as the Bates-Damgaard method for transitions in the discrete spectrum. The conditions of validity of both methods (in particular the conditions on the magnitude of the effective quantum number for the discrete spectrum) are also the same.

9.5.7 Bremsstrahlung in a Coulomb Field

Sommerfeld calculated the effective cross section of bremsstrahlung in a Coulomb field in the nonrelativistic approximation, without taking retardation into account. The effective cross section of bremsstrahlung in the frequency interval $\omega, \omega + d\omega$ (integrated over all directions of motion of the electron and photon) is usually given in the form $d\sigma = g\, d\sigma_K$, where $d\sigma_K$ is the simplest limiting expression for $d\sigma$ – the Kramers formula

$$d\sigma_K = \frac{16\pi}{3\sqrt{3}} \frac{Z^2 e^6}{\hbar c^3 m^2 v_1^2} \frac{d\omega}{\omega} = \frac{16\pi}{3\sqrt{3}} \alpha^3 a_0^2 \left(\frac{Ze^2}{\hbar v_1}\right)^2 \frac{d\omega}{\omega} \quad (9.272)$$

(v_1 is the initial velocity of the electron) and g is again the correction factor called the Gaunt factor. According to Sommerfeld's theory

$$g(\eta_1, \eta_2) = \frac{\pi\sqrt{3}}{(e^{2\pi\eta_1} - 1)(1 - e^{-2\pi\eta_2})} x_0 \frac{d}{dx_0}|F(x_0)|^2, \quad (9.273)$$

where $F(x_0) = F(i\eta_1, i\eta_2, 1; x_0)$ is a hypergeometric function, $\eta_1 = Ze^2/\hbar v_1$, $\eta_2 = Ze^2/\hbar v_2$ ($v_2 < v_1$ is the final velocity of the electron), $x_0 = -\dfrac{4\eta_1\eta_2}{(\eta_2 - \eta_1)^2}$.

Equation (9.273) is rather complex. We shall discuss its limiting expressions. Let us introduce dimensionless variables

Table 9.3 Values of the parameters $G(vl; \varepsilon' l')$ and $\chi(vl; \varepsilon' l')$

v	ε'	$s - p$		$p - s$	
		$G(v0; \varepsilon' 1)$	$\chi(v0; \varepsilon' 1)$	$G(v1; \varepsilon' 0)$	$\chi(v1; \varepsilon' 0)$
0.6	0.00	4.978	0.160		
	0.02	4.976	0.163		
	0.04	4.971	0.167		
	0.06	4.967	0.170		
	0.08	4.963	0.173		
	0.10	4.960	0.177		
	0.15	4.959	0.185		
	0.20	4.962	0.192		
	0.30	4.976	0.206		
	0.40	5.003	0.220		
	0.50	5.036	0.233		
	0.60	5.077	0.244		
	0.70	5.123	0.255		
	0.80	5.174	0.265		
	0.90	5.227	0.275		
	1.00	5.284	0.284		
0.8	0.00	4.118	0.094	2.620	−0.347
	0.02	4.129	0.098	2.638	−0.346
	0.04	4.140	0.102	2.655	−0.346
	0.06	4.151	0.107	2.671	−0.346
	0.08	4.162	0.111	2.687	−0.346
	0.10	4.175	0.116	2.704	−0.346
	0.15	4.208	0.126	2.746	−0.346
	0.20	4.245	0.135	2.789	−0.345
	0.30	4.324	0.153	2.872	−0.345
	0.40	4.410	0.168	2.957	−0.346
	0.50	4.500	0.182	3.042	−0.346
	0.60	4.594	0.196	3.127	−0.347
	0.70	4.690	0.208	3.213	−0.348
	0.80	4.787	0.219	3.299	−0.349
	0.90	4.886	0.230	3.385	−0.350
	1.00	4.986	0.240	3.472	−0.351
1	0.00	3.436	0.045	1.975	−0.349
	0.02	3.455	0.050	1.994	−0.350
	0.04	3.477	0.056	2.012	−0.350
	0.06	3.498	0.061	2.028	−0.350
	0.08	3.519	0.066	2.044	−0.351
	0.10	3.541	0.071	2.060	−0.351
	0.15	3.599	0.083	2.101	−0.352
	0.20	3.657	0.094	2.142	−0.354
	0.30	3.777	0.114	2.222	−0.357
	0.40	3.985	0.131	2.300	−0.361
	0.50	4.014	0.146	2.377	−0.366

Table 9.3 (continued)

v	ε'	$s-p$		$p-s$	
		$G(v0;\varepsilon'1)$	$\chi(v0;\varepsilon'1)$	$G(v1;\varepsilon'0)$	$\chi(v1;\varepsilon'0)$
1	0.60	4.136	0.160	2.453	−0.370
	0.70	4.252	0.172	2.528	−0.375
	0.80	4.377	0.184	2.603	−0.380
	0.90	4.491	0.194	2.677	−0.385
	1.00	4.602	0.203	2.750	−0.391
1.2	0.00	2.897	0.012	0.704	−0.337
	0.02	2.924	0.019	0.712	−0.338
	0.04	2.951	0.025	0.719	−0.339
	0.06	2.979	0.032	0.726	−0.339
	0.08	3.007	0.038	0.733	−0.346
	0.10	3.036	0.043	0.740	−0.341
	0.15	3.107	0.057	0.756	−0.343
	0.20	3.176	0.069	0.772	−0.340
	0.30	3.312	0.091	0.803	−0.352
	0.40	3.447	0.110	0.831	−0.358
	0.50	3.576	0.126	0.857	0.365
	0.60	3.701	0.140	0.882	−0.373
	0.70	3.823	0.153	0.906	−0.381
	0.80	3.940	0.164	0.929	−0.389
	0.90	4.054	0.175	0.951	−0.397
	1.00	4.165	0.184	0.971	−0.405
1.4	0.00	2.498	−0.008	0.834	−0.325
	0.02	2.532	0.000	0.844	−0.326
	0.04	2.565	0.008	0.854	−0.327
	0.06	2.598	0.015	0.864	−0.327
	0.08	2.631	0.022	0.874	−0.328
	0.10	2.663	0.029	0.884	−0.329
	0.15	2.745	0.045	0.900	−0.332
	0.20	2.803	0.059	0.927	−0.335
	0.30	2.972	0.083	0.965	−0.342
	0.40	3.114	0.103	0.997	−0.350
	0.50	3.248	0.121	1.026	−0.359
	0.60	3.374	0.137	1.051	−0.368
	0.70	3.495	0.150	1.072	−0.377
	0.80	3.609	0.163	1.091	−0.366
	0.90	3.719	0.174	1.108	−0.396
	1.00	3.824	0.184	1.122	−0.406
1.6	0.00	2.214	−0.021	0.879	−0.314
	0.02	2.256	−0.011	0.893	−0.314
	0.04	2.296	−0.002	0.906	−0.325
	0.06	2.335	0.007	0.919	−0.316
	0.08	2.374	0.015	0.931	−0.317
	0.10	2.412	0.023	0.943	−0.317
	0.15	2.506	0.040	0.971	−0.320
	0.20	2.596	0.056	0.996	−0.323

260 9. Radiative Transitions

Table 9.3 (continued)

ν	ε'	s − p		p − s	
		$G(v0; \varepsilon' 1)$	$\chi(v0; \varepsilon' 1)$	$G(v1; \varepsilon'0)$	$\chi(v1; \varepsilon' 0)$
1.6	0.30	2.765	0.083	1.037	−0.330
	0.40	2.923	0.105	1.071	−0.337
	0.50	3.070	0.124	1.098	−0.345
	0.60	3.209	0.141	1.118	−0.353
	0.70	3.341	0.157	1.133	−0.361
	0.80	3.466	0.170	1.145	−0.369
	0.90	3.586	0.183	1.152	−0.377
	1.00	3.701	0.194	1.156	−0.386
1.8	0.00	2.011	−0.030	0.892	−0.304
	0.02	2.062	−0.019	0.911	−0.305
	0.04	2.110	−0.009	0.929	−0.305
	0.06	2.157	0.001	0.945	−0.306
	0.08	2.202	0.010	0.961	−0.306
	0.10	2.248	0.018	0.976	−0.307
	0.15	2.357	0.038	1.010	−0.309
	0.20	2.462	0.054	1.041	−0.312
	0.30	2.657	0.083	1.091	−0.318
	0.40	2.837	0.105	1.131	−0.324
	0.50	3.004	0.126	1.162	−0.329
	0.60	3.164	0.143	1.187	−0.335
	0.70	3.315	0.159	1.206	−0.341
	0.80	3.459	0.172	1.219	−0.346
	0.90	3.597	0.185	1.229	−0.351
	1.00	3.731	0.196	1.235	−0.356
2	0.00	1.855	−0.039	0.893	−0.297
	0.02	1.913	−0.027	0.915	−0.297
	0.04	1.970	−0.016	0.939	−0.298
	0.06	1.024	−0.006	0.957	−0.298
	0.08	2.077	0.004	0.977	−0.299
	0.10	2.129	0.013	0.996	−0.300
	0.15	2.253	0.033	1.040	−0.303
	0.20	2.370	0.050	1.078	−0.306
	0.30	2.586	0.079	1.140	−0.312
	0.40	2.783	0.102	1.191	−0.318
	0.50	2.964	0.121	1.235	−0.324
	0.60	3.137	0.138	1.269	−0.329
	0.70	3.300	0.153	1.301	−0.334
	0.80	3.455	0.166	1.330	−0.339
	0.90	3.605	0.178	1.352	−0.343
	1.00	3.743	0.188	1.381	−0.348
2.2	0.00	1.722	−0.049	0.882	−0.290
	0.02	1.786	−0.035	0.910	−0.291
	0.04	1.849	−0.023	0.936	−0.292
	0.06	1.909	−0.012	0.960	−0.293
	0.08	1.967	−0.002	0.983	−0.294
	0.10	2.021	0.004	1.004	−0.295

Table 9.3 (continued)

v	ε'	$s-p$		$p-s$	
		$G(v0;\varepsilon'1)$	$\chi(v0;\varepsilon'1)$	$G(v1;\varepsilon'0)$	$\chi(v1;\varepsilon'0)$
2.2	0.15	2.158	0.028	1.048	−0.299
	0.20	2.282	0.045	1.094	−0·302
	0.30	2.503	0.075	1.167	−0.311
	0.40	2.708	0.101	1.224	−0.317
	0.50	2.897	0.120	1.268	−0.323
	0.60	3.074	0.137	1.306	−0.328
	0.70	3.242	0.152	1.338	−0.333
	0.80	3.400	0.165	1.367	−0.338
	0.90	3.551	0.177	1.394	−0.342
	1.00	3.693	0.188	1.425	−0.347
2.4	0.00	1.601	−0.057	0.869	−0.286
	0.02	1.672	−0.042	0.900	−0.287
	0.04	1.739	−0.029	0.929	−0.288
	0.06	1.803	−0.017	0.956	−0.289
	0.08	1.865	−0.007	0.981	−0.291
	0.10	1.926	0.003	1.005	−0.292
	0.15	2.068	0.025	1.054	−0.296
	0.20	2.198	0.042	1.106	−0.299
	0.30	2.438	0.073	1.184	−0.310
	0.40	2.650	0.100	1.248	−0.316
	0.50	2.843	0.119	1.295	−0.322
	0.60	3.025	0.136	1.334	−0.327
	0.70	3.198	0.152	1.366	−0.332
	0.80	3.358	0.164	1.392	−0.337
	0.90	3.510	0.176	1.423	−0.342
	1.00	3.652	0.187	1.447	−0.346
2.6	0.00	1.494	−0.062	0.849	−0.283
	0.02	1.572	−0.046	0.885	−0.283
	0.04	1.645	−0.032	0.927	−0.284
	0.06	1.714	−0.019	0.946	−0.285
	0.08	1.780	−0.008	0.973	−0.287
	0.10	1.844	0.002	0.999	−0.289
	0.15	1.996	0.024	1.058	−0.293
	0.20	2.138	0.042	1.115	−0.297
	0.30	2.387	0.073	1.206	−0.308
	0.40	2.606	0.099	1.267	−0.315
	0.50	2.802	0.118	1.316	−0.321
	0.60	2.988	0.136	1.355	−0.326
	0.70	3.162	0.151	1.387	−0.331
	0.80	3.324	0.164	1.418	−0.336
	0.90	3.480	0.176	1.447	−0.341
	1.00	3.624	0.186	1.469	−0.345
2.8	0.00	1.407	−0.066	0.825	−0.278
	0.02	1.493	−0.048	0.866	−0.278
	0.04	1.572	−0.033	0.901	−0.279
	0.06	1.648	−0.020	0.934	−0.280

9. Radiative Transitions

Table 9.3 (continued)

v	ε'	$s-p$		$p-s$	
		$G(v0;\varepsilon'1)$	$\chi(v0;\varepsilon'1)$	$G(v1;\varepsilon'0)$	$\chi(v1;\varepsilon'0)$
2.8	0.08	1.720	−0.008	0.963	−0.282
	0.10	1.789	0.002	0.991	−0.285
	0.15	1.948	0.024	1.060	−0.290
	0.20	2.092	0.042	1.122	−0.295
	0.30	2.348	0.073	1.218	−0.307
	0.40	2.570	0.098	1.283	−0.315
	0.50	2.771	0.118	1.332	−0.321
	0.60	2.958	0.135	1.372	−0.326
	0.70	3.135	0.151	1.406	−0.331
	0.80	3.299	0.164	1.436	−0.336
	0.90	3.455	0.176	1.465	−0.341
	1.00	3.598	0.186	1.487	−0.345
3	0.00	1.339	−0.068	0.805	−0.272
	0.02	1.433	−0.050	0.849	−0.272
	0.04	1.519	−0.034	0.888	−0.273
	0.06	1.600	−0.021	0.925	−0.275
	0.08	1.676	−0.009	0.957	−0.277
	0.10	1.747	0.001	0.991	−0.280
	0.15	1.910	0.023	1.063	−0.287
	0.20	2.056	0.042	1.126	−0.294
	0.30	2.314	0.073	1.229	−0.366
	0.40	2.542	0.098	1.297	−0.314
	0.50	2.745	0.118	1.345	−0.320
	0.60	2.932	0.135	1.387	−0.326
	0.70	3.112	0.151	1.420	−0.331
	0.80	3.276	0.164	1.450	−0.336
	0.90	3.434	0.176	1.480	−0.341
	1.00	3.578	0.186	1.501	−0.345
4	0.00	1.077	−0.090	0.730	−0.262
	0.02	1.204	−0.065	0.799	−0.263
	0.04	1.313	−0.044	0.856	−0.265
	0.06	1.412	−0.027	0.905	−0.268
	0.08	1.502	−0.013	0.952	−0.272
	0.10	1.586	−0.001	0.990	−0.275
	0.15	1.772	0.023	1.078	−0.284
	0.20	1.935	0.042	1.147	−0.291
	0.30	2.214	0.073	1.254	−0.303
	0.40	2.455	0.098	1.330	−0.312
	0.50	2.668	0.118	1.385	−0.319
	0.60	2.862	0.135	1.428	−0.325
	0.70	3.047	0.151	1.468	−0.331
	0.80	3.215	0.164	1.499	−0.336
	0.90	3.376	0.176	1.529	−0.341
	1.00	3.524	0.186	1.550	−0.345

Table 9.3 (continued)

v	ε'	$s-p$		$p-s$	
		$\overbrace{G(v0;\varepsilon'1)}$	$\overbrace{\chi(v0;\varepsilon'1)}$	$\overbrace{G(v1;\varepsilon'0)}$	$\overbrace{\chi(v1;\varepsilon'0)}$
5	0.00	0.911	−0.101	0.654	−0.250
	0.02	1.070	−0.071	0.747	−0.252
	0.04	1.200	−0.049	0.819	−0.256
	0.06	1.312	−0.031	0.878	−0.260
	0.08	1.412	−0.016	0.927	−0.264
	0.10	1.504	−0.003	0.975	−0.269
	0.15	1.704	0.023	1.070	−0.279
	0.20	1.875	0.042	1.148	−0.288
	0.30	2.166	0.073	1.256	−0.300
	0.40	2.413	0.098	1.344	−0.311
	0.50	2.631	0.118	1.407	−0.319
	0.60	2.828	0.135	1.450	−0.325
	0.70	3.016	0.152	1.491	−0.331
	0.80	3.186	0.164	1.522	−0.336
	0.90	3.349	0.176	1.552	−0.341
	1.00	3.499	0.186	1.573	−0.345
6	0.00	0.795	−0.108	0.597	−0.242
	0.02	0.985	−0.074	0.712	−0.245
	0.04	1.131	−0.052	0.794	−0.250
	0.06	1.253	−0.033	0.860	−0.255
	0.08	1.360	−0.018	0.915	−0.260
	0.10	1.456	−0.004	0.964	−0.265
	0.15	1.665	0.023	1.064	−0.276
	0.20	1.842	0.042	1.143	−0.285
	0.30	2.139	0.073	1.261	−0.299
	0.40	2.390	0.098	1.349	−0.310
	0.50	2.611	0.118	1.412	−0.318
	0.60	2.810	0.135	1.462	−0.325
	0.70	2.999	0.151	1.503	−0.331
	0.80	3.171	0.164	1.534	−0.336
	0.90	3.335	0.176	1.565	−0.341
	1.00	3.485	0.186	1.586	−0.345

264   9. Radiative Transitions

Table 9.3 (continued)

v	ε'	$p-d$		v	ε'	$p-d$	
		$G(v1;\varepsilon'2)$	$\chi(v1;\varepsilon'2)$			$G(v1;\varepsilon'2)$	$\chi(v1;\varepsilon'2)$
0.8	0.00	8.612	0.787	1.4	0.00	3.465	0.377
	0.02	8.466	0.792		0.02	3.503	0.389
	0.04	8.339	0.798		0.04	3.544	0.400
	0.06	8.233	0.804		0.06	3.590	0.410
	0.08	8.144	0.809		0.08	3.639	0.419
	0.10	8.069	0.815		0.10	3.691	0.428
	0.15	7.926	0.828		0.15	3.829	0.448
	0.20	7.828	0.841		0.20	3.974	0.466
	0.30	7.743	0.865		0.30	4.280	0.497
	0.40	7.752	0.888		0.40	4.610	0.522
	0.50	7.821	0.909		0.50	4.923	0.546
	0.60	7.940	0.928		0.60	5.267	0.562
	0.70	8.091	0.947		0.70	5.610	0.579
	0.80	8.271	0.965		0.80	5.959	0.594
	0.90	8.476	0.981		0.90	6.313	0.607
	1.00	8.702	0.997		1.00	6.670	0.618
1	0.00	7.837	0.635	1.6	0.00	3.300	0.277
	0.02	7.779	0.643		0.02	3.360	0.289
	0.04	7.735	0.650		0.04	3.424	0.302
	0.06	7.706	0.657		0.06	3.490	0.312
	0.08	7.689	0.666		0.08	3.558	0.322
	0.10	7.684	0.673		0.10	3.628	0.331
	0.15	7.702	0.689		0.15	3.806	0.351
	0.20	7.750	0.705		0.20	3.988	0.369
	0.30	7.923	0.733		0.30	4.356	0.398
	0.40	8.163	0.759		0.40	4.728	0.421
	0.50	8.446	0.782		0.50	5.107	0.442
	0.60	8.766	0.803		0.60	5.479	0.456
	0.70	9.111	0.822		0.70	5.856	0.470
	0.80	9.479	0.840		0.80	6.235	0.482
	0.90	9.867	0.856		0.90	6.614	0.492
	1.00	10.271	0.871		1.00	6.993	0.501
1.2	0.00	3.006	0.498	1.8	0.00	2.870	0.203
	0.02	3.013	0.508		0.02	2.937	0.217
	0.04	3.021	0.528		0.04	3.006	0.230
	0.06	3.038	0.527		0.06	3.075	0.241
	0.08	3.057	0.535		0.08	3.145	0.251
	0.10	3.079	0.543		0.10	3.214	0.261
	0.15	3.144	0.562		0.15	3.386	0.281
	0.20	3.218	0.579		0.20	3.557	0.298
	0.30	3.386	0.610		0.30	3.891	0.325
	0.40	3.575	0.696		0.40	4.216	0.346
	0 50	3.759	0.663		0.50	4.543	0.363
	0.60	3.986	0.680		0.60	4.846	0.375
	0.70	4.205	0.699		0.70	5.151	0.386
	0.80	4.432	0.715		0.80	5.452	0.395
	0.90	4.663	0.730		0.90	5.746	0.403
	1.00	4.901	0.744		1.00	6.035	0.410

Table 9.3 (continued)

v	ε'	$p-d$		v	ε'	$p-d$	
		$G(v1;\varepsilon'2)$	$\chi(v1;\varepsilon'2)$			$G(v1;\varepsilon'2)$	$\chi(v1;\varepsilon'2)$
2	0.00	2.426	0.160	2.2	0.30	2.879	0.312
	0.02	2.489	0.176		0.40	3.063	0.342
	0.04	2.550	0.190		0.50	3.226	0.368
	0.06	2.612	0.203		0.60	3.380	0.390
	0.08	2.675	0.215		0.70	3.511	0.409
	0.10	2.730	0.225		0.80	3.629	0.428
	0.15	2.875	0.248		0.90	3.740	0.445
	0.20	3.004	0.266		1.00	3.845	0.461
	0.30	3.251	0.295	2.4	0.00	1.955	0.131
	0.40	3.479	0.317		0.02	2.036	0.155
	0.50	3.684	0.334		0.04	2.113	0.176
	0.60	3.855	0.347		0.06	2.188	0.194
	0.70	4.044	0.359		0.08	2.261	0.211
	0.80	4.189	0.368		0.10	2.333	0.226
	0.90	4.365	0.377		0.15	2.501	0.258
	1.00	4.504	0.384		0.20	2.654	0.286
2.2	0.00	2.117	0.142		0.30	2.938	0.331
	0.02	2.182	0.162		0.40	3.211	0.367
	0.04	2.245	0.179		0.50	3.466	0.397
	0.06	2.306	0.195		0.60	3.723	0.422
	0.08	2.365	0.209		0.70	3.963	0.445
	0.10	2.421	0.222		0.80	4.201	0.467
	0.15	2.552	0.250		0.90	4.446	0.487
	0.20	2.670	0.274		1.00	4.697	0.504

Table 9.3 (continued)

v	ε'	$p-d$		$d-p$	
		$G(v1;\varepsilon'2)$	$\chi(v1;\varepsilon'2)$	$G(v2;\varepsilon'1)$	$\chi(v2;\varepsilon'1)$
2.6	0.00	1.857	0.115	0.476	−0.194
	0.02	1.956	0.141	0.511	−0.166
	0.04	2.053	0.163	0.551	−0.138
	0.06	2.147	0.183	0.598	−0.111
	0.08	2.240	0.200	0.653	−0.086
	0.10	2.329	0.215	0.716	−0.062
	0.15	2.540	0.247	0.908	−0.013
	0.20	2.735	0.274	1.143	0.023
	0.30	3.106	0.318	1.724	0.071
	0.40	3.472	0.351	2.433	0.099
	0.50	3.822	0.377	3.258	0.116
	0.60	4.153	0.397	4.190	0.128
	0.70	4.477	0.417	5.205	0.136
	0.80	4.810	0.435	6.290	0.143
	0.90	5.153	0.450	7.444	0.149
	1.00	5.494	0.462	8.669	0.153
2.8	0.00	1.753	0.095	0.506	−0.206
	0.02	1.865	0.122	0.536	−0.288
	0.04	1.974	0.145	0.565	−0.280
	0.06	2.078	0.164	0.595	−0.271
	0.08	2.178	0.180	0.627	−0.261
	0.10	2.273	0.195	0.662	−0.251
	0.15	2.492	0.225	0.761	−0.226
	0.20	2.693	0.250	0.874	−0.202
	0.30	3.082	0.290	1.139	−0.158
	0.40	3.456	0.317	1.466	−0.123
	0.50	3.792	0.337	1.872	−0.090
	0.60	4.086	0.353	2.344	−0.084
	0.70	4.388	0.369	2.859	−0.072
	0.80	4.712	0.383	3.400	−0.061
	0.90	5.039	0.394	3.973	−0.051
	1.00	5.341	0.401	4.594	−0.045
3	0.00	1.633	0.077	0.472	−0.333
	0.02	1.752	0.106	0.487	−0.338
	0.04	1.859	0.129	0.497	−0.342
	0.06	1.958	0.148	0.504	−0.346
	0.08	2.049	0.164	0.513	−0.351
	0.10	2.138	0.179	0.521	−0.356
	0.15	2.336	0.209	0.541	−0.368
	0.20	2.530	0.235	0.556	−0.378
	0.30	2.882	0.274	0.583	−0.395
	0.40	3.154	0.299	0.598	−0.407
	0.50	3.390	0.318	0.605	−0.416
	0.60	3.598	0.333	0.616	−0.424
	0.70	3.847	0.348	0.620	−0.430
	0.80	4.071	0.360	0.612	−0.434
	0.90	4.278	0.370	0.601	−0.437
	1.00	4.419	0.377	0.583	−0.439

Table 9.3 (continued)

v	ε'	$p-d$		$d-p$	
		$G(v1;\varepsilon'2)$	$\chi(v1;\varepsilon'2)$	$G(v2;\varepsilon'1)$	$\chi(v2;\varepsilon'1)$
4	0.00	1.259	0.029	0.523	−0.319
	0.02	1.417	0.069	0.555	−0.326
	0.04	1.561	0.102	0.579	−0.333
	0.06	1.691	0.128	0.599	−0.340
	0.08	1.812	0.150	0.611	−0.346
	0.10	1.920	0.167	0.628	−0.352
	0.15	2.145	0.198	0.658	−0.366
	0.20	2.421	0.235	0.684	−0.378
	0.30	2.810	0.274	0.714	−0.395
	0.40	3.109	0.299	0.729	−0.407
	0.50	3.366	0.318	0.736	−0.416
	0.60	3.590	0.333	0.748	−0.424
	0.70	3.853	0.348	0.751	−0.430
	0.80	4.089	0.360	0.741	−0.434
	0.90	4.306	0.370	0.726	−0.437
	1.00	4.457	0.377	0.704	−0.439
5	0.00	1.040	−0.001	0.506	−0.302
	0.02	1.234	0.048	0.559	−0.313
	0.04	1.398	0.085	0.596	−0.323
	0.06	1.544	0.114	0.624	−0.332
	0.08	1.674	0.137	0.646	−0.340
	0.10	1.795	0.157	0.663	−0.347
	0.15	2.054	0.194	0.707	−0.364
	0.20	2.362	0.235	0.744	−0.378
	0.30	2.770	0.274	0.774	−0.395
	0.40	3.083	0.299	0.789	−0.407
	0.50	3.350	0.318	0.796	−0.416
	0.60	3.583	0.333	0.807	−0.424
	0.70	3.852	0.348	0.810	−0.430
	0.80	4.094	0.360	0.799	−0.434
	0.90	4.317	0.370	0.782	−0.437
	1.00	4.471	0.377	0.758	−0.439
6	0.00	0.895	−0.022	0.483	−0.290
	0.02	1.123	0.036	0.555	−0.304
	0.04	1.303	0.075	0.597	−0.315
	0.06	1.457	0.105	0.628	−0.325
	0.08	1.596	0.130	0.657	−0.335
	0.10	1.723	0.151	0.682	−0.344
	0.15	1.995	0.190	0.729	−0.362
	0.20	2.327	0.235	0.777	−0.378
	0.30	2.747	0.274	0.807	−0.395
	0.40	3.068	0.299	0.822	−0.407
	0.50	3.341	0.318	0.828	−0.416
	0.60	3.578	0.333	0.839	−0.424
	0.70	3.851	0.348	0.842	−0.430
	0.80	4.096	0.360	0.830	−0.434
	0.90	4.321	0.370	0.812	−0.437
	1.00	4.479	0.377	0.787	−0.439

Table 9.3 (continued)

v	ε'	$d-f$	
		$G(v2;\varepsilon'3)$	$\chi(v2;\varepsilon'3)$
2.6	0.00	2.989	0.272
	0.02	3.156	0.291
	0.04	3.331	0.307
	0.06	3.510	0.319
	0.08	3.692	0.330
	0.10	3.874	0.340
	0.15	4.339	0.361
	0.20	4.819	0.378
	0.30	5.833	0.403
	0.40	6.907	0.421
	0.50	8.031	0.434
	0.60	9.203	0.444
	0.70	10.428	0.451
	0.80	11.699	0.455
	0.90	13.007	0.458
	1.00	14.345	0.460
2.8	0.00	2.311	0.217
	0.02	2.431	0.238
	0.04	2.550	0.254
	0.06	2.668	0.266
	0.08	2.784	0.276
	0.10	2.899	0.284
	0.15	3.188	0.300
	0.20	3.485	0.312
	0.30	4.096	0.325
	0.40	4.714	0.329
	0.50	5.337	0.331
	0.60	5.972	0.331
	0.70	6.638	0.328
	0.80	7.321	0.324
	0.90	8.012	0.318
	1.00	8.705	0.312
3	0.00	1.851	0.222
	0.02	1.940	0.252
	0.04	2.011	0.274
	0.06	2.075	0.292
	0.08	2.144	0.308
	0.10	2.201	0.321
	0.15	2.349	0.348
	0.20	2.465	0.367
	0.30	2.601	0.391
	0.40	2.668	0.406
	0.50	2.714	0.417
	0.60	2.803	0.427
	0.70	2.843	0.434
	0.80	2.842	0.439
	0.90	2.828	0.443
	1.00	2.790	0.446

9.5 Continuous Spectrum

Table 9.3 (continued)

v	ε'	$d-f$		$f-d$	
		$G(v2; \varepsilon' 3)$	$\chi(v2; \varepsilon' 3)$	$G(v3; \varepsilon' 2)$	$\chi(v3: \varepsilon' 2)$
4	0.00	1.456	0.155	0.289	−0.362
	0.02	1.631	0.203	0.291	−0.372
	0.04	1.784	0.238	0.293	−0.382
	0.06	1.914	0.264	0.291	−0.390
	0.08	2.016	0.283	0.291	−0.398
	0.10	2.100	0.298	0.287	−0.404
	0.15	2.273	0.327	0.284	−0.418
	0.20	2.409	0.348	0.287	−0.430
	0.30	2.679	0.380	0.282	−0.445
	0.40	2.898	0.401	0.284	−0.456
	0.50	3.087	0.416	0.296	−0.465
	0.60	3.243	0.427	0.303	−0.471
	0.70	3.303	0.434	0.316	−0.476
	0.80	3.312	0.439	0.332	−0.480
	0.90	3.393	0.443	0.345	−0.483
	1.00	3.265	0.446	0.375	−0.486
5	0.00	1.184	0.107	0.342	−0.342
	0.02	1.407	0.169	0.355	−0.357
	0.04	1.593	0.212	0.355	−0.368
	0.06	1.750	0.243	0.352	−0.377
	0.08	1.886	0.267	0.351	−0.386
	0.10	1.996	0.285	0.353	−0.395
	0.15	2.215	0.318	0.354	−0.412
	0.20	2.394	0.342	0.355	−0.425
	0.30	2.705	0.376	0 365	−0.444
	0.40	3.008	0.400	0.373	−0.456
	0.50	3.248	0.416	0.388	−0.465
	0.60	3.422	0.427	0.396	−0.471
	0.70	3.492	0.434	0.413	−0.476
	0.80	3.507	0.439	0.433	−0.480
	0.90	3.502	0.443	0.451	−0.483
	1.00	3.465	0.446	0.489	−0.486
6	0.00	1.004	0.073	0.360	−0.329
	0.02	1.268	0.148	0.383	−0.347
	0.04	1.476	0.196	0.389	−0.361
	0.06	1.652	0.231	0.389	−0.372
	0.08	1.801	0.257	0.389	−0.382
	0.10	1.933	0.278	0.393	−0.392
	0.15	2.185	0.314	0.394	−0.410
	0.20	2.395	0.340	0.398	−0.424
	0.30	2.756	0.376	0.406	−0.443
	0.40	3.076	0.400	0.421	−0.456
	0.50	3.330	0.416	0.438	−0.465
	0.60	3.513	0.427	0.447	−0.471
	0.70	3.589	0.434	0.465	−0.476
	0.80	3.607	0.439	0.487	−0.480
	0.90	3.605	0.443	0.507	−0.483
	1.00	3.569	0.446	0.550	−0.486

9. Radiative Transitions

$$v = \frac{Ze^2}{mv_1^3}\omega \ ; \quad y = \frac{\hbar\omega}{E_1}\left(E_1 = \frac{mv_1^2}{2}, \ 0 \leqslant y \leqslant 1\right),$$

which obey the relation $v = \eta_1 y/2$, and also

$$\xi = \eta_2 - \eta_1 = \eta_1 \left(\frac{1}{\sqrt{1-y}} - 1\right) > 0. \tag{9.274}$$

In the problem to be considered there are two independent variables, which we choose as η_1 and y. In the quasi-classical limit $\eta_1 \gg 1$ ($\eta_2 > \eta_1 \gg 1$), practically for the whole spectrum the Gaunt factor is given by[6]

$$g(v) = \tfrac{1}{4}\pi\sqrt{3}\ iv\ H^{(1)}_{iv}(iv)\ H^{(1)'}_{iv}(iv), \tag{9.275}$$

where $H^{(1)}$ and $H^{(1)'}$ are the Hankel function and its derivative. The limiting expressions for (9.275) are

$$g \simeq \begin{cases} 1 & v \gg 1 \\ \dfrac{\sqrt{3}}{\pi}\ln\left(\dfrac{2}{\gamma v}\right) & v \ll 1, \end{cases} \tag{9.276}$$

where $\gamma = 1.781$ is the Euler constant. Let us note that for almost the whole "quasi-classical" spectrum $y \gg \eta_1^{-1}$, with the exception of the narrow frequency region near $\omega = 0$, we have $g \simeq 1$.

In the limiting case of large initial velocities of the electron $\eta_1 \ll 1$, the bremsstrahlung spectrum is described by the Born-Elwert approximation

$$g_{BE} = \frac{\sqrt{3}}{\pi}\frac{\eta_2}{\eta_1}\frac{1-e^{-2\pi\eta_1}}{1-e^{-2\pi\eta_2}}\ln\frac{\eta_2+\eta_1}{\eta_2-\eta_1}. \tag{9.277}$$

If the final velocity of the electron is also large and moreover $2\pi\eta_2 \ll 1$ ($2\pi\eta_1 \ll 1$), then (9.277) gives the usual Born approximation[7]

$$g_B(\eta_1, \eta_2) = \frac{\sqrt{3}}{\pi}\ln\frac{\eta_2+\eta_1}{\eta_2-\eta_1} = \frac{\sqrt{3}}{\pi}\ln\frac{1+\sqrt{1-y}}{1-\sqrt{1-y}}. \tag{9.278}$$

When $\omega \ll E_1/\hbar$ (i.e., $\eta_2 \simeq \eta_1$), it follows from (9.277, 278) that

---

[6] For the derivation of formula (9.275) see [31].
[7] Equation (9.278) can be obtained if in calculating (9.235, 236) the wave functions ψ_q^+ and ψ_q^- are replaced by plane waves.

$$g \simeq \frac{\sqrt{3}}{\pi} \ln \frac{4}{y}. \tag{9.279}$$

If $2\pi\eta_1 \ll 1$ but $\eta_2 \gg \eta_1$, (9.277) gives

$$g \simeq 4\sqrt{3}\, \eta_1 [1 - \exp(-2\pi\eta_2)]^{-1}. \tag{9.280}$$

If the final electron velocity is large enough, $2\pi\eta_2 \ll 1$, then

$$g \simeq \frac{2\sqrt{3}}{\pi} \frac{\eta_1}{\eta_2} = \frac{2\sqrt{3}}{\pi} \sqrt{1-y} \ll 1. \tag{9.281}$$

In the opposite case $2\pi\eta_2 \gg 1$ it follows from (9.280) that

$$g \simeq 4\sqrt{3}\, \eta_1 \ll 1. \tag{9.282}$$

The more accurate expression for g near the large frequency limit when $\eta_1 \ll 1$ is

$$g \approx \frac{8\sqrt{3}\, \pi \eta_1^2}{e^{2\pi\eta_1} - 1}\left(1 + \frac{10}{3} \eta_1^2 + \ldots\right). \tag{9.283}$$

It is possible to find from (9.273) the accurate expression for g near the large frequency limit valid for arbitrary η_1. In the case $\eta_1 \gg 1$, instead of (9.283), this expression takes the form

$$g \simeq 1 + 0.1728\, \eta_1^{-2/3}. \tag{9.284}$$

We shall also give some other approximate expressions for (9.273).
First of all, when $\xi = \eta_2 - \eta_1 \ll 1$, (9.273) gives

$$g(\eta_1, \eta_2) \approx \frac{\sqrt{3}}{\pi}\left[\ln\frac{\eta_2 + \eta_1}{\eta_2 - \eta_1} + \psi(1) - \mathrm{Re}\left\{\psi\left(1 + i\frac{\eta_1 + \eta_2}{2}\right)\right\}\right], \tag{9.285}$$

where $\psi(x) = (d/dx)\ln\Gamma(x)$ and $\Gamma(x)$ is a gamma function. When $\eta_1 \ll 1$, $\eta_2 \ll 1$, (9.285) coincides with (9.278). When $y \ll 1$, (9.285) becomes

$$g \approx \frac{\sqrt{3}}{\pi}\left[\ln\frac{4}{y} + \psi(1) - \mathrm{Re}\{\psi(1 + i\eta_1)\}\right]. \tag{9.286}$$

This equation describes the Gaunt factor g in the low frequency region for arbitrary η_1. Equation (9.286) gives the same result as (9.279) in the limiting case $\eta_1 \ll 1$, and as (9.276) in the case of $\eta_1 \gg 1$, because $\mathrm{Re}\{\psi(1 + i\eta_1)\} \simeq \ln\eta_1$, $\psi(1) = -\ln\gamma$, and $y\eta_1 = 2v$.

9. Radiative Transitions

The general approximate expression for g can be obtained by means of a semiclassical method [32]

$$g(v:,\eta_1) = \frac{1}{4}\pi\sqrt{3}\,iv\left(1+\frac{1}{\gamma\eta_1}\right)H_{iv}^{(1)}\left[iv\left(1+\frac{1}{\gamma\eta_1}\right)\right]H_{iv}^{(1)'}\left[iv\left(1+\frac{1}{\gamma\eta_1}\right)\right]. \tag{9.287}$$

This formula generalizes some of the limiting approximations given above. When $\eta_1 \gg 1$, (9.287) gives the classical Gaunt factor of (9.275). When $\eta_1 \ll 1$ and $y \ll 1$, (9.287) transforms into the Born approximation given by (9.279). When $y \ll 1$ and $v \ll 1$ (η_1 is arbitrary), it is possible to obtain from (9.287) the expression

$$g \simeq \frac{\sqrt{3}}{\pi}\ln\frac{2}{(\gamma+\eta_1^{-1})v} = \frac{\sqrt{3}}{\pi}\ln\frac{4}{2\gamma v + y}, \tag{9.288}$$

which gives an interpolation between (9.276) and (9.279).

It has to be noted that the approximation given by (9.287) is valid in a wide range of parameters y and η_1, with the exception of the high frequency region of the bremsstrahlung spectrum $y \simeq 1$ at large initial velocities of the electron $\eta_1 \ll 1$. The effective cross section of continuum absorption is also usually described in the form of the Kramers formula multiplied by a correction factor — the Gaunt factor g.

The effective cross section of the radiative transition $E, \omega \to E'$ can be found by using (9.243)

$$\sigma_{E\omega, E'} = \frac{E'}{E}\frac{\pi^2 c^2}{\omega^2}\frac{d\sigma_{E', E\omega}}{d\omega}. \tag{9.289}$$

Here E is the initial energy of the electron, E' the final energy, and ω the frequency of absorbed radiation. According to (9.272)

$$\sigma = \frac{16\pi^3 c^2}{3\sqrt{3}\,\omega^3}\alpha^3 a_0^2 \eta_1^2 g = \frac{16\pi^3 Z^2 e^6}{3\sqrt{3}\,\omega^3 c\hbar m^2 v^2}g, \tag{9.290}$$

where v is the initial velocity of the electron, g the Gaunt factor. By substituting this expression in the formula for the absorption coefficient $k_\omega = N_e N \langle v\sigma\rangle$, we obtain

$$\kappa_\omega = \frac{16\pi^3 Z^2 e^6}{3\sqrt{3}\,\omega^3 m^2 c\hbar}\left\langle\frac{g}{v}\right\rangle N_i N_e. \tag{9.291}$$

In the Kramers approximation ($g = 1$) and with a Maxwellian velocity distribution for electrons [$\langle v^{-1}\rangle = (2m/\pi kT)^{1/2}$], from this formula with correction for stimulated emission follows the expression

$$k'_\omega = \frac{16\sqrt{2}\,\pi^{5/2}e^6 Z^2 N_i N_e}{3\sqrt{3}\,c\hbar m^{3/2}(kT)^{1/2}\omega^3}(1 - e^{-\hbar\omega/kT}). \tag{9.292}$$

The intensity of bremsstrahlung $Q(\omega)d\omega = \varepsilon_\omega d\omega/4\pi$ can be found from (9.255).

Let us return to (9.268) for the coefficient of photoinoization absorption and assume that the concentrations of atoms N, ions N_i, and electrons N_e are connected by the Saha formula. The statistical sum in this case equals

$$S = 2\sum_n n^2 e^{-E_n/kT}.$$

Expressing N in terms of N_i, N_e and substituting in (9.286), we obtain

$$k'_\omega = \frac{16\sqrt{2}\,\pi^{5/2}e^6 Z^2 N_i N_e}{3\sqrt{3}\,c\hbar m^{3/2}(kT)^{1/2}\omega^3}\left[\frac{2\text{Ry}\,Z^2}{kT}\sum_{n=n_0}^{\infty}\frac{1}{n^3}e^{\text{Ry}\,Z^2/n^2 kT}\right](1 - e^{-\hbar\omega/kT}). \tag{9.293}$$

This expression differs from (9.292) only by the factor in brackets, which enables one to combine (9.292, 293) and introduce a total coefficient of absorption which takes into account both transitions from levels of the discrete spectrum to the continuous spectrum and transitions between continuous-spectrum states.

$$k'_\omega = \frac{16\sqrt{2}\,\pi^{5/2}e^6 Z^2 N_i N_e}{3\sqrt{3}\,c\hbar m^{3/2}(kT)^{1/2}\omega^3}\left[\frac{2\text{Ry}\,Z^2}{kT}\sum_{n=n_0}^{\infty}e^{\text{Ry}\,Z^2/n^2 kT} + 1\right](1 - e^{-\hbar\omega/kT}). \tag{9.294}$$

If the sum over $n > n_m$ is replaced by an integral, then

$$k'_\omega = \frac{16\sqrt{2}\,\pi^{5/2}e^6 Z^2 N_i N_e}{3\sqrt{2}\,c\hbar m^{3/2}(kT)^{1/2}\omega^3}\left[\frac{2\text{Ry}\,Z^2}{kT}\sum_{n=n_0}^{n_m-1}\frac{1}{n^3}e^{\text{Ry}\,Z^2/n^2 kT} + e^{\text{Ry}\,Z^2/n_m^2 kT}\right]$$
$$\times (1 - e^{-\hbar\omega/kT}). \tag{9.295}$$

The total intensity of radiation $Q(\omega)d\omega$ can also be found by means of (9.251, 255). For a number of applications it is important to know the total intensity of radiation integrated over the whole spectrum, Q^B.

Let us assume that the velocity distribution is Maxwellian and use the Kramers approximation. In this case $Q(\omega)d\omega$ can be found either from (9.255, 292) or directly from the general formula (9.249), which in this case takes the form

$$Q(\omega) = N_e N_i \hbar\omega \int_{\sqrt{2\hbar\omega/m}}^{\infty}\frac{d\sigma}{d\omega}vf(v)\,dv.$$

After integration over dv,

$$Q(\omega)d\omega = \frac{32\pi}{3\sqrt{3}}a^3 a_0^2 Z^2 \frac{e^4}{\hbar}\left(\frac{m}{2\pi kT}\right)^{1/2} N_e N_i e^{-\hbar\omega/kT}d\omega. \tag{9.296}$$

In calculating Q^B, one can neglect the logarithmic increase of $d\sigma$ in the narrow region near the low-frequency limit and extend the Kramers formula to the whole range of frequencies. In this approximation,

$$Q^B = \int Q(\omega)\,d\omega = \frac{32\pi}{3\sqrt{3}} \alpha^3 a_0^2 Z^2 \text{ Ry } N_e N_i \left(\frac{2kT}{\pi m}\right)^{1/2}. \tag{9.297}$$

If T is given in electron volts, then

$$Q^B = 1.54 \times 10^{-25} N_e N_i Z^2 T^{1/2} \text{ [erg/cm}^3\text{s]}. \tag{9.298}$$

It is interesting to note that the calculation of Q^B in the Born approximation gives an expression differing from (9.298) only by the factor $2\sqrt{3}/\pi \simeq 1.1$.

The formulas given above can be used for approximate estimates of the effective cross section of bremsstrahlung in a field of nonhydrogenlike ions. In this case the main contribution is given by the region of large distances for which the field is close to a Coulomb field. Nevertheless, it is necessary to take into account the dependence of the effective charge Z_{ef} on v_1 and ω; see [32].

9.6 Formulas for Q Factors

9.6.1 Symmetry and Sum Rules

The probability of the electric multipole transition of the order κ has been written in Sect. 9.3 in such a form that the dependence on angular momentum quantum numbers is expressed by the factor Q_κ [see (9.140) and for the particular case $\kappa = 1$, (9.80)]. The factors $Q_\kappa(\gamma_0, \gamma)$ are defined in such a way that the following relation is fulfilled:

$$\frac{g_0}{(2l_0 + 1)} Q_\kappa(\gamma_0, \gamma) = \frac{g}{(2l + 1)} Q_\kappa(\gamma, \gamma_0). \tag{9.299}$$

The sum of the transition probabilities W with respect to a group of close levels (for example, the sum over components of fine structure J or over terms SL of the same configuration) is expressed in terms of the sum of the Q factors. This means that the Q factor for transition between two groups of levels A and B is obtained by summing the Q factors corresponding to the transitions between the levels of each of the group $Q\,(Aa, Bb)$ with respect to all final levels and averaging over initial levels

$$Q_\kappa(A,B) = \frac{1}{g(A)} \sum_{a,b} g(a) Q_\kappa\,(Aa, Bb). \tag{9.300}$$

We shall give a number of formulas for Q factors corresponding to the different electron configurations and different coupling schemes. In some cases we shall use the notation γ for a set of quantum numbers describing the term and the notation γJ for a component of fine structure of the term.

The formulas given below contain $6j$ symbols (Sect. 4.2) and $12j$ symbols. The latter is expressed in the terms of $6j$ symbols in the following way

$$\begin{bmatrix} j_1 & j_2 & j_3 & j_4 \\ l_1 & l_2 & l_3 & l_4 \\ k_1 & k_2 & k_3 & k_4 \end{bmatrix} = (-1)^{l_1-l_2-l_3+l_4} \sum_x (2x+1) \begin{Bmatrix} k_1 & k_2 & x \\ j_3 & j_1 & l_1 \end{Bmatrix} \begin{Bmatrix} k_3 & k_4 & x \\ j_3 & j_1 & l_2 \end{Bmatrix}$$

$$\times \begin{Bmatrix} k_1 & k_2 & x \\ j_4 & j_2 & l_3 \end{Bmatrix} \begin{Bmatrix} k_3 & k_4 & x \\ j_4 & j_2 & l_4 \end{Bmatrix}. \quad (9.301)$$

9.6.2 LS Coupling. Allowed Transitions

In the case of a single electron outside closed shells

$$Q_\kappa(l_0 J_0; l_1 J_1) = (2l_0 + 1)(2J_1 + 1) \begin{Bmatrix} l_0 & J_0 & 1/2 \\ J_1 & l_1 & \kappa \end{Bmatrix}^2. \quad (9.302)$$

By summing with respect to J_1, we have

$$Q_\kappa(l_0 J_0; l_1) \equiv \sum_{J_1} Q_\kappa(l_0 J_0; l_1 J_1) = 1. \quad (9.303)$$

By averaging over J_0, we have

$$Q_\kappa(l_0; l_1 J_1) \equiv \frac{1}{2(2l_0+1)} \sum_{J_0} (2J_0 + 1) Q_\kappa(l_0 J_0; l_1 J_1) = \frac{2J_1 + 1}{2(2l_1 + 1)}. \quad (9.304)$$

For the transition between two levels[8]

$$[S_p L_p] l_0 S L_0 J_0 \rightarrow [S_p L_p] l_1 S L_1 J_1,$$

where S_p and L_p are the quantum numbers of the parent ion

---

[8] Below, the component of fine structure SLJ is usually referred to as level SLJ or level J.

$$Q_\kappa(SL_0J_0; SL_1J_1) = (2L_0+1)(2J_1+1) \begin{Bmatrix} L_0 & J_0 & S \\ J_1 & L_1 & \kappa \end{Bmatrix}^2 Q_\kappa(SL_0; SL_1).$$
(9.305)

Here $Q_\kappa(SL_0; SL_1)$ is the Q factor for the transition between terms SL_0 and SL_1.

For the transition from a given level J_0 of the term SL_0 to all levels (components of fine structure) of the term SL_1, the Q factor is

$$Q_\kappa(SL_0J_0\ SL_1) \equiv \sum_{J_1} Q_\kappa(SL_0J_0; SL_1J_1) = Q_\kappa(SL_0; SL_1) \quad (9.306)$$

and does not depend on J_0.

For the transition from all levels of the term SL_0 to a given level J_1 of the term SL_1, we have

$$Q_\kappa(SL_0; SL_1J_1) \doteq [(2S+1)(2L_0+1)]^{-1} \sum_{J_0} (2J_0+1) Q_\kappa(SL_0J_0; SL_1J_1)$$
$$= [(2S+1)(2L_1+1)]^{-1}(2J_1+1) Q_\kappa(SL_0; SL_1). \quad (9.307)$$

It is convenient to group the formulas for the Q factors corresponding to the transitions between two terms as a whole in the following way.

1) Transitions in which groups of equivalent electrons are not involved:

$$\gamma_0 \equiv [S_pL_p] l_0 SL_0; \quad \gamma_1 \equiv [S_pL_p] l_1 SL_1,$$

$$Q_\kappa(\gamma_0; \gamma_1) = (2l_0+1)(2L_1+1) \begin{Bmatrix} l_0 & L_0 & L_p \\ L_1 & l_1 & \kappa \end{Bmatrix}^2. \quad (9.308)$$

By summing (9.308) with respect to L_1, we obtain

$$Q_\kappa(l_0SL_0, l_1S) = 1. \quad (9.309)$$

By averaging (9.308) over L_0, we have

$$Q_\kappa(l_0S; l_1SL_1) = \frac{2L_1+1}{(2l_1+1)(2L_p+1)}. \quad (9.310)$$

If (9.308) is summed with respect to L_1 and averaged over L_0, then

$$Q_\kappa(l_0S; l_1S) = 1. \quad (9.311)$$

2) Transitions $l_0^m \rightarrow l_0^{m-1} l_1$:

$$\gamma_0 \equiv l_0^m SL_0, \quad \gamma_1 \equiv l_0^{m-1}[S_pL_p]l_1 SL_1$$

$$Q_\kappa(\gamma_0; \gamma_1) = m |G^{SL\rho}_{S_p L_p}|^2 (2l_0 + 1)(2L_1 + 1) \begin{Bmatrix} l_0 & L_0 & L_p \\ L_1 & l_1 & \kappa \end{Bmatrix}^2 \qquad (9.312)$$

where $G^{SL\rho}_{S_p L_p}$ is the fractional parentage coefficient.[9] Summing with respect to L_1, we obtain

$$Q_\kappa(l_0^m SL_0; l_0^{m-1}[S_p L_p] l_1 S) = m |G^{SL\rho}_{S_p L_p}|^2 \qquad (9.313)$$

and summing over $S_p L_p$, we have

$$Q_\kappa(l_0^m SL_0; l_0^{m-1} l_1 S) = Q_\kappa(l_0^m S; l_0^{m-1} l_1 S) = Q_\kappa(l_0^m; l_0^{m-1} l_1) = m \ . \qquad (9.314)$$

3) Transitions $l_0^N l_1^m \to l_0^{N-1} l_1^{m+1}$:

In the general case of transitions in which two different groups of equivalent electrons are involved, the formulas for Q factors are rather complex. Because of this we give below only formulas for the particular case of a closed shell l_0^N, $N = 2(2l_0 + 1)$.

For the transition $\gamma_0 \to \gamma_1$ where

$$\gamma_0 \equiv l_0^N l_1^m SL_0, \quad \gamma_1 = l_0^{N-1} \left[\frac{1}{2} l_0\right] l_1^{m+1} [S_p L_p] SL_1,$$

we have

$$Q_\kappa(\gamma_0; \gamma_1) = (m+1)|G^{S_p L_p}_{SL_0}|^2 \frac{(2S_p + 1)(2L_p + 1)(2l_0 + 1)(2L_1 + 1)}{(2S + 1)(2L_0 + 1)}$$

$$\begin{Bmatrix} l_0 & L_1 & L_p \\ L_0 & l_1 & \kappa \end{Bmatrix}^2 . \qquad (9.315)$$

The fractional parentage coefficients for configurations l^{R-m} and l^{m+1}, where $R = 2(2l + 1)$, are related by

$$(-1)^{-S-L} \sqrt{(R-m)(2S+1)(2L+1)} \, G^{SL}_{S_p L_p}(l^{R-m})$$
$$= (-1)^{S_p + L_p - l - 1/2} \sqrt{(m+1)(2S_p+1)(2L_p+1)} \, G^{S_p L_p}_{SL}(l^{m+1}) \ . \qquad (9.316)$$

Therefore (9.315) can be written in the form

$$Q_\kappa(\gamma_0; \gamma_1) = (4l_1 + 2 - m)|G^{SL_0}_{S_p L_p}(l^{R-m})|^2 (2l_0 + 1)(2L_1 + 1) \begin{Bmatrix} l_0 & L_1 & L_p \\ L_0 & l_1 & \kappa \end{Bmatrix}^2 . \qquad (9.317)$$

---

[9] Tables for these coefficients are given in Sect. 5.1.

By summing with respect to L_1, we obtain

$$Q_\kappa(l_0^N l_1^m SL_0; l_0^{N-1} l_1^{m+1} [S_p L_p] S)$$
$$= (m+1) |G_{SL_0}^{S_p L_p}|^2 \frac{(2S_p+1)(2L_p+1)(2l_0+1)}{(2S+1)(2L_0+1)(2l_1+1)}$$
$$= (4l_1 + 2 - m) |G_{S_p L_p}^{SL_0}(l^{R-m})|^2 \frac{2l_0+1}{2l_1+1}. \tag{9.318}$$

After averaging (9.318) over L_0, it follows that

$$Q_\kappa(l_0^N l_1^m S; l_0^{N-1} l_1^{m+1} [S_p L_p] S) = \frac{(m+1)}{g(l_1^m)} \frac{(2S_p+1)(2L_p+1)(2l_0+1)}{(2l_1+1)}, \tag{9.319}$$

where $g(l_1^m)$ is the statistical weight of the configuration l_1^m given by

$$g(l_1^m) = \frac{(4l_1+2)!}{m!(4l_1+2-m)!}. \tag{9.320}$$

Summing (9.318) with respect to S_p and L_p, we obtain

$$Q_\kappa(l_0^N l_1^m SL_0; l_0^{N-1} l_1^{m+1} S)$$
$$= Q_\kappa(l_0^N l_1^m; l_0^{N-1} l_1^{m+1}) = (4l_1 + 2 - m) \frac{2l_0+1}{2l_1+1}. \tag{9.321}$$

4) Transitions between terms of the same configuration:

$$\gamma_0 \equiv l^m SL_0; \quad \gamma_1 \equiv l^m SL_1;$$
$$Q_\kappa(\gamma_0; \gamma_1) = \frac{2l+1}{2L_0+1} |(l^m SL_0 \| U^\kappa \| l^m SL_1)|^2. \tag{9.322}$$

The reduced matrix elements $(l^m SL_0 \| U^\kappa \| l^m SL_1)$ in (9.322) have even values of κ. When $\kappa = 0$,

$$(l^m SL_0 \| U^\kappa \| l^m SL_1) = m \sqrt{\frac{2L_0+1}{2l+1}} \delta_{L_0 L_1}. \tag{9.323}$$

When $L_0 \neq L_1$, $\kappa = 2$. For configurations l^{N-m}, and l^m, where $N = 2(2l+1)$ we have

$$(l^{N-m} SL_0 \| U^2 \| l^{N-m} SL_1) = -(l^m SL_0 \| U^2 \| l^m SL_1). \tag{9.324}$$

In the general case the reduced matrix element of U^κ can be calculated by means of the formula

$$(l^m SL_0 \| U^\kappa \| l^m SL_1)$$
$$= m \sum_{S_p L_p} G^{SL_0}_{S_p L_p} G^{SL_1}_{S_p L_p} (-1)^{L_p+\kappa-l-L_0} \sqrt{(2L_0+1)(2L_1+1)} \begin{Bmatrix} l & L_0 & L_p \\ L_1 & l & \kappa \end{Bmatrix}, \tag{9.325}$$

where $S_p L_p$ are the terms of the configuration l^{m-1}.

Tables of the reduced matrix elements $(l^m SL_0 \| U^2 \| l^m SL_1)$ for $l=1$ are given in Sect. 5.4.

9.6.3 jl Coupling

We shall now consider two cases: a) both the initial and the final levels are described in the scheme of jl coupling; b) one level is described in the scheme of LS coupling and the other in the scheme of jl coupling.

a) For the transition between the components of fine structure of the two doublets of jl coupling, we have

$$Q_\kappa(K_0 J_0; K_1 J_1) = (2K_0+1)(2J_1+1) \begin{Bmatrix} K_0 & J_0 & 1/2 \\ J_1 & K_1 & \kappa \end{Bmatrix}^2 Q_\kappa(K_0; K_1). \tag{9.326}$$

Summing with respect to J_1, we obtain

$$Q_\kappa(K_0 J_0; K_1) = Q_\kappa(K_0; K_1). \tag{9.327}$$

Averaging (9.326) over J_0, we obtain

$$Q_\kappa(K_0; K_1 J_1) = \frac{2J_1+1}{2(2K_1+1)} Q_\kappa(K_0; K_1). \tag{9.328}$$

For the transition between two doublets as a whole, $\gamma_0 \equiv [S_p L_p j] l_0 K_0$; $\gamma_1 \equiv [S_p L_p j] l_1 K_1$, we have

$$Q_\kappa(\gamma_0; \gamma_1) = (2l_0+1)(2K_1+1) \begin{Bmatrix} l_0 & K_0 & j \\ K_1 & l_1 & \kappa \end{Bmatrix}^2. \tag{9.329}$$

Summing (9.329) with respect to K_1, we get

$$Q_\kappa(l_0 K_0; l_1) = Q_\kappa(l_0; l_1) = 1 \tag{9.330}$$

and averaging (9.329) over K_0, we get

$$Q_\kappa(l_0; l_1 K_1) = \frac{2K_1 + 1}{(2j + 1)(2l_1 + 1)}. \tag{9.331}$$

b) If the initial level $\gamma_0 J_0$ is described in the scheme of LS coupling and the final level $\gamma_1 J_1$ in the scheme of jl coupling, i.e.,

$$\gamma_0 \equiv [S_p L_p] l_0 S_0 L_0, \quad \gamma_1 = [S_p L_p j] l_1 K_1,$$

then

$$Q_\kappa(\gamma_0 J_0; \gamma_1 J_1) = (2l_0 + 1)(2J_1 + 1)(2S_0 + 1)(2L_0 + 1)$$

$$\times (2j + 1)(2K_1 + 1) \begin{bmatrix} L_0 & S_p & \kappa & K_1 \\ l_0 & J_0 & j & 1/2 \\ L_p & l_1 & S_0 & J_1 \end{bmatrix}^2. \tag{9.332}$$

Averaging over J_0 and then summing over J_1, we obtain for the transition $\gamma_0 - \gamma_1$ as a whole

$$Q_\kappa(\gamma_0; \gamma_1) = (2l_0 + 1) \frac{(2j + 1)(2K_1 + 1)}{2S_p + 1}$$

$$\times \sum_r (2r + 1) \begin{Bmatrix} L_p & l_1 & r \\ \kappa & L_0 & l_0 \end{Bmatrix}^2 \begin{Bmatrix} L_p & l_1 & r \\ K_1 & S_p & j \end{Bmatrix}^2. \tag{9.333}$$

Summation of (9.333) with respect to K_1 gives

$$Q_\kappa([S_p L_p] l_0 S_0 L_0; [S_p L_p j] l_1) = \frac{2j + 1}{(2S_p + 1)(2L_p + 1)}. \tag{9.334}$$

Summing (9.334) over j, we obtain

$$Q_\kappa([S_p L_p] l_0 S_0 L_0; [S_p L_p] l_1) = Q_\kappa([S_p L_p] l_0; [S_p L_p] l_1) = 1. \tag{9.335}$$

Averaging (9.333) over L_0 gives

$$Q_\kappa([S_p L_p] l_0 S_0; [S_p L_p j] l_1 K_1) = \frac{2K_1 + 1}{(2S_p + 1)(2L_p + 1)(2l_1 + 1)}. \tag{9.336}$$

For transitions from groups of equivalent electrons (9.332–335) have to be multiplied by $m |G_{S_p L_p}^{S_0 L_0}|^2$.

9.7 Tables of Oscillator Strengths and Radiative Transition Probabilities

The results of numerical calculations of oscillator strengths, radiative transition probabilities, and photorecombination cross sections are given below. It is assumed that the radial functions for all levels of the same multiplicity belonging to one configuration are the same. Because of this assumption, the transition probabilities can be expressed in terms of Q factors and radial integrals. The collection of formulas for Q factors is given in Sect. 9.6.

9.7.1 Transition Probabilities for the Hydrogen Atom

The oscillator strengths and probabilities of the transitions n_0–n_1, averaged over quantum number l_0 and l_1, are given in Table 9.4. Probabilities of the transitions $n_0 l_0$–$n_1 l_1$ for $n \leqslant 6$ are given in Table 9.5.

Table 9.4 Transition probabilities in hydrogen

Transition n_1–n_0	λ[Å]	f_{01}	$W_{10}[\text{s}^{-1}]$
1—2	1,215.6	0.416	4.70×10^8
—3	1,025.7	7.91×10^{-2}	5.57×10^7
—4	972.5	2.90×10^{-2}	1.28×10^7
—5	949.7	1.39×10^{-2}	4.12×10^6
—6	937.8	7.80×10^{-3}	1.64×10^6
—7	930.7	4.81×10^{-3}	7.57×10^5
—8	926.2	3.18×10^{-3}	3.87×10^5
—9	923.1	2.21×10^{-3}	2.14×10^5
—10	920.9	1.60×10^{-3}	1.26×10^5
2—3	6,562.8	0.640	4.41×10^7
—4	4,861.3	0.119	8.42×10^6
—5	4,340.4	4.46×10^{-2}	2.53×10^6
—6	4,101.7	2.21×10^{-2}	9.73×10^5
—7	3,970.0	1.27×10^{-2}	4.39×10^5
—8	3,889.0	8.03×10^{-3}	2.21×10^5
—9	3,835.3	5.43×10^{-3}	1.22×10^5
—10	3,797.9	3.85×10^{-3}	7.12×10^4
3—4	18,751.0	0.842	8.99×10^6
—5	12,818.1	0.151	2.20×10^6
—6	10,938.1	5.58×10^{-2}	7.78×10^5
—7	10,049.4	2.77×10^{-2}	3.36×10^5
—8	9,545.9	1.60×10^{-2}	1.65×10^5
—9	9,229.0	1.02×10^{-2}	8.90×10^4
—10	9,014.9	6.98×10^{-3}	5.15×10^4
4—5	40,512	1.038	2.70×10^6
—6	26,252	0.179	7.71×10^5
—7	21,655	6.55×10^{-2}	3.04×10^5
—8	19,445	3.23×10^{-2}	1.42×10^5
—9	18,174	1.87×10^{-2}	7.46×10^4
—10	17,362	1.20×10^{-2}	4.23×10^4
5—6	74,578	1.231	1.03×10^6

9. Radiative Transitions

Table 9.5 Probabilities of the transitions $n_0 l_0 - n_1 l_1$ in hydrogen

$n_0 l_0 - n_1 l_1$	W_{10}, s^{-1}	$n_0 l_0 - n_1 l_1$	W_{10}, s^{-1}	$n_0 l_0 - n_1 l_1$	W_{10}, s^{-1}
$1s-2p$	6.26×10^8	$3s-6p$	9.55×10^5	$4d-5p$	1.88×10^5
$1s-3p$	1.67×10^8	$3p-4s$	1.83×11^6	$4d-6p$	9.42×10^4
$1s-4p$	6.82×10^7	$3p-5s$	9.05×10^5	$4d-5f$	2.58×10^6
$1s-5p$	3.44×10^7	$3p-6s$	5.07×10^5	$4d-6f$	1.29×10^6
$1s-6p$	1.97×10^7	$3p-4d$	7.04×10^6	$4f-5d$	5.05×10^4
$2p-3s$	6.31×10^6	$3p-5d$	3.39×10^6	$4f-6d$	2.14×10^4
$2p-4s$	2.58×10^6	$3p-6d$	1.88×10^6	$4f-5g$	4.25×10^6
$2p-5s$	1.29×10^6	$3d-4p$	3.47×10^5	$4f-6g$	1.37×10^6
$2p-6s$	7.35×10^5	$3d-5p$	1.49×10^5	$5s-6p$	2.43×10^5
$2s-3p$	2.24×10^7	$3d-6p$	7.82×10^4	$5p-6s$	2.68×10^5
$2s-4p$	9.67×10^6	$3d-4f$	1.38×10^7	$5p-6d$	4.49×10^5
$2s-5p$	4.95×10^6	$3d-5f$	$4\ 54 \times 10^6$	$5d-6p$	9.59×10^4
$2s-6p$	2.86×10^6	$3d-6f$	2.15×10^6	$5d-6f$	7.23×10^5
$2p-3d$	6.46×10^7	$4s-5p$	7.37×10^5	$5f-6d$	3.91×10^4
$2p-4d$	2.06×10^7	$4s-6p$	4.46×10^5	$5f-6g$	1.11×10^6
$2p-5d$	9.42×10^6	$4p-5s$	6.45×10^5	$5g-6f$	1.14×10^4
$2p-6d$	5.14×10^6	$4p-6s$	3.58×10^5	$5g-6h$	1.64×10^6
$3s-4p$	3.06×10^6	$4p-5d$	1.49×10^6		
$3s-5p$	1.64×10^6	$4p-6d$	8.62×10^5		

In the case of transitions between highly excited states $n_1 \to n_0$, $n_1 \gg 1$, $n_0 \gg 1$, $\Delta n = n_1 - n_0 \gg 1$, the Kramers approximation can be used

$$W_K(n_1; n_0) = \frac{16}{\pi 3\sqrt{3}} A_1 \frac{2}{n_1^3 n_0 (n_1^2 - n_0^2)} \tag{9.337}$$

$$A_1 = \frac{me^4}{2\hbar^3} \alpha^3 \zeta^4 \simeq 0.80 \times 10^{10} \zeta^4 \quad s^{-1}. \tag{9.338}$$

The transition probability W is often described in the form of the Kramers formula multiplied by a correction factor — the Gaunt factor g

$$W(n_1; n_0) = W_K(n_1; n_0) g(n_1; n_0). \tag{9.339}$$

For $n_1 \gg n_0$ [33]

$$g(n_1; n_0) = 1 - \frac{0.1728 [1 + (n_0/n_1)^2]}{[1 - (n_0/n_1)^2]^{2/3} n_0^{2/3}} - \frac{0.0496 [1 - 4/3 (n_0/n_1)^2 + (n_0/n_1)^4]}{(1 - (n_0/n_1)^2)^{4/3} n_0^{4/3}}. \tag{9.340}$$

In the other limiting case $\Delta n \ll n_0, n_1$ [34],

$$g(n_1; \Delta n) = \pi \sqrt{3} \, \Delta n \left(1 - \frac{\Delta n}{n_1}\right)\left(1 - \frac{1}{2}\frac{\Delta n}{n_1}\right)\left(1 + \frac{3}{2}\frac{\Delta n}{n_1}\right) J_{\Delta n}(\Delta n) \times$$

$$\times \frac{d}{d(\Delta n)} J_{\Delta n}(\Delta n), \qquad (9.341)$$

where $J_{\Delta n}(\Delta n)$ is the Bessel function.

The total probability of the radiative decay of the level n in the Kramers approximation can be described as

$$W_K(n) = \frac{3A_1}{n^5} \ln\left(\frac{n - 1/n}{2}\right). \qquad (9.342)$$

9.7.2 Radiative Transition Probabilities in the Bates–Damgaard Approximation

The probability of the electric 2κ pole transition $1 \to 0$ in an atom or ion can be written in the form

$$W_\kappa(1; 0) = A_\kappa \left(\frac{\hbar\omega_{10}}{\zeta^2 \text{Ry}}\right)^2 \frac{g_0 Q_\kappa(0; 1)}{g_1 (2l_0 + 1)} F_\kappa(l_0; l_1), \qquad (9.343)$$

$$F_\kappa(l_0; l_1) = \frac{|(l_0\|C^\kappa\|l_1)|^2}{2\kappa + 1} \left(\frac{\hbar\omega_{10}}{\zeta^2 \text{Ry}}\right)^{2\kappa-1} \left(\frac{\zeta}{a_0}\right)^{2\kappa} \left|\int P_0(r) r^\kappa P_1(r) dr\right|^2, \qquad (9.344)$$

$$A_\kappa = \frac{me^4}{2\hbar^3} \left(\frac{e^2}{\hbar c}\right)^{2\kappa+1} \frac{\kappa + 1}{2^{2\kappa-1}\kappa \, [(2\kappa - 1)1!]^2} \zeta^{2\kappa+2}. \qquad (9.345)$$

Here $\zeta = Z - N$ (Z is the nuclear charge, N is the number of electrons in the atomic core); g_0 and g_1 are the statistical weights of the initial and final levels; a_0 is the Bohr radius. The factor $Q_\kappa(0; 1)$ depends only on quantum numbers of angular momenta. For transitions between configurations $l_0^m \to l_0^{m-1} l_1$ as a whole, $Q_\kappa = m$. Formulas for the factors Q_κ are given in Sect. 9.6.

For the more interesting case of electric dipole (optically allowed) transitions $\kappa = 1$, (9.345) coincides with (9.338)

$$A_1 = 0.80 \times 10^{10} \, \zeta^4 \quad \text{s}^{-1},$$

and the oscillator strength of the transition $0 \to 1$ equals

$$f_{01} = \frac{Q_1(0; 1)}{2l_0 + 1} F_1(l_0; l_1). \qquad (9.346)$$

In the case of an electric quadrupole transition $\kappa = 2$,

$$A_2 = 0.89 \times 10^4 \, \zeta^6 \quad \text{s}^{-1}. \qquad (9.347)$$

The quantities $F_1(l_0; l_1)$ and $F_2(l_0; l_1)$ calculated in the approximation of the Coulomb field radial functions are given in Tables 9.6 and 9.7 as functions of the effective principal quantum numbers n_0^* and $\Delta n = n_1^* - n_0^*$

$$n_i^* = \sqrt{\frac{\zeta^2 \,\mathrm{Ry}}{|E_i|}}, \quad i = 0, 1. \tag{9.348}$$

The energy of the level E_i is measured from the limit of ionization. The tables give the order and the mantissa of the numbers, for example, 81–2, 20–0, 73+0, and 15+1 mean, respectively, 0.0081, 0.20, 0.73, and 1.5.

In the case of $\Delta n < 0.1$, the following expression for $F_1(n_0^* l_0; n_1^* l_1)$ can be used:

$$F_1(n_0^* l_0; n_1^* l_1) = \frac{3}{4} \frac{\hbar \omega_{10}}{\zeta^2 \,\mathrm{Ry}} l_{\max} (n_{l_{\max}}^*)^2 [(n_{l_{\max}}^*)^2 - l_{\max}^2], \tag{9.349}$$

where $l_{\max} = \max\{l_0, l_1\}$, and $n_{l_{\max}}^*$ is the value of n^* corresponding to $l_{\max}$.

9.7.3 Oscillator Strengths and Probabilities of Some Selected Transitions

The oscillator strengths and transition probabilities for a number of atoms obtained from experimental data and from especially accurate calculations are given in Table 9.8 [22]

9.7.4 Effective Cross Sections and Rates of Photorecombination

The results of numerical calculations of photorecombination cross sections σ^r and rates $\langle v \sigma^r \rangle$ are given in Tables 9.9 and 9.10 and are shown in Fig. 9.2.

The cross section of photorecombination on the level γ_0 is

$$\sigma^r(\gamma_0) = 10^{-6} \pi a_0^2 Q^r \frac{\Phi(u)}{u}, \quad u = \frac{\mathscr{E}}{|E_{\gamma_0}|}, \tag{9.350}$$

where $\mathscr{E}$ is the initial energy of the electron, E_{γ_0} is the energy of the level γ_0, Q^r is the factor depending on quantum numbers of angular momenta.

For recombination in shell l_0^m

$$A_{\zeta+1}(n_0 l_0^{m-1} S_i L_i) + e \to A_\zeta (n_0 l_0^m S_0 L_0) + \hbar \omega,$$

$$Q^r = m |G_{S_i L_i}^{S_0 L_0}|^2 \frac{(2S_0 + 1)(2L_0 + 1)}{2(2l_0 + 1)(2S_i + 1)2L_i + 1)}. \tag{9.351}$$

By summing over all terms $S_0 L_0$, we obtain

9.7 Tables of Oscillator Strengths and Radiative Transition Probabilities 285

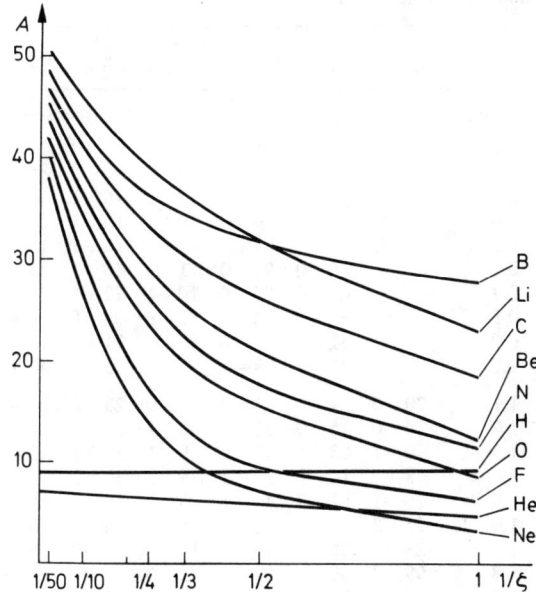

Fig. 9.2. Parameter A for determining the sum of the photorecombination rates over all levels, note (9.354)

$$Q^r = \frac{N - (m - 1)}{N}, \tag{9.352}$$

where $N = 2(2l_0 + 1)$.

The dependence of photorecombination rates $\langle v\sigma^r \rangle$ on temperature can be approximated by a simple analytical formula

$$\langle v\sigma^r \rangle = 10^{-14} Q^r \left| \frac{E_{\gamma 0}}{\text{Ry}} \right|^{1/2} G^r(\beta) \quad [\text{cm}^3\text{s}^{-1}]; \quad \beta = \frac{|E_{\gamma 0}|}{kT} \tag{9.353}$$

$$G^r(\beta) = A\beta^{3/2}(\beta + \chi)^{-1}. \tag{9.354}$$

Therefore, parameters A and χ are also given in Table 9.10.

In some cases, when the sum of the rates $G^r(\beta)$ over all atomic levels is given, one must use as $E_{\gamma 0}$ in (9.353, 354) the energy of the ground state.

Parameter A for the sum of the photorecombination rates over all levels for a number of isoelectronic sequences (from H-like to Ne-like) is shown in Fig. 9.2. Parameter χ is determined by

$$\chi = \begin{cases} 1 - a/\zeta & a/\zeta \leqslant 0.9 \\ 0.1 & a/\zeta > 0.9, \end{cases} \tag{9.355}$$

where parameter a is given in Table 9.11.

Table 9.6 Values of $F_1(l_0; l_1)$

Δn^*	n_0^*									
	0.5	1.0	1.5	2.0	2.5	3.0	3.5	4.0	4.5	5.0
					Transition s—p					
0.1			15−0	25−0	33−0	42−0	50+0	58+0	65+0	73+0
0.2			34−0	53+0	69+0	85+0	10+1	11+1	12+1	14+1
0.3			55+0	80+0	10+1	12+1	14+1	16+1	18+1	19+1
0.4			74+0	10+1	12+1	14+1	17+1	19+1	21+1	23+1
0.5			87+0	11+1	13+1	15+1	18+1	20+1	22+1	24+1
0.6		60+0	91+0	11+1	13+1	15+1	17+1	19+1	21+1	23+1
0.7		63+0	86+0	10+1	12+1	13+1	15+1	16+1	18+1	19+1
0.8		60+0	74+0	88+0	98+0	10+1	12+1	13+1	14+1	15+1
0.9		52+0	57+0	66+0	71+0	78+0	84+0	91+0	97+0	10+1
1.0		41−0	39−0	43−0	45−0	48−0	51+0	54+0	57+0	60+0
1.1	20−0	27−0	24−0	23−0	23−0	24−0	24−0	25−0	26−0	27−0
1.2	14−0	15−0	11−0	96−1	86−1	81−1	78−1	76−1	76−1	75−1
1.3	90−1	63−1	32−1	18−1	12−1	82−2	57−2	40−2	28−2	19−2
1.4	48−1	14−1	10−2	39−3	30−2	66−2	10−1	15−1	19−1	24−1
1.5	20−1	81−5	10−1	23−1	36−1	49−1	62−1	74−1	86−1	98−1
1.6	55−2	11−1	40−1	65−1	86−1	10−0	12−0	14−0	16−0	18−0
1.7	23−3	35−1	72−1	10−0	12−0	15−0	17−0	19−0	21−0	23−0
1.8	99−3	59−1	93−1	12−0	14−0	17−0	18−0	20−0	22−0	24−0
1.9	44−2	75−1	98−1	12−0	13−0	15−0	17−0	18−0	19−0	21−0
2.0	81−2	79−1	85−1	10−0	10−0	12−0	12−0	13−0	14−0	15−0
2.1	28−1	64−1	68−1	69−1	71−1	74−1	77−1	81−1	84−1	87−1
2.2	25−1	43−1	39−1	35−1	33−1	33−1	32−1	32−1	32−1	32−1
2.3	19−1	22−1	15−1	10−1	83−2	69−2	57−2	49−2	42−2	37−2
2.4	12−1	72−2	19−2	29−3	38−6	20−3	69−3	13−2	20−2	28−2
2.5	64−2	39−3	11−2	39−2	71−2	10−1	13−1	17−1	20−1	23−1
2.6	23−2	16−2	92−2	17−1	23−1	29−1	35−1	42−1	47−1	53−1
2.7	35−3	84−2	20−1	32−1	40−1	49−1	57−1	65−1	72−1	79−1
2.8	51−4	17−1	30−1	44−1	51−1	61−1	68−1	77−1	84−1	91−1
2.9	81−3	25−1	35−1	47−1	52−1	61−1	67−1	74−1	80−1	86−1
3.0	19−2	28−1	32−1	41−1	44−1	51−1	54−1	59−1	62−1	67−1
3.1	93−2	25−1	29−1	30−1	32−1	34−1	36−1	37−1	39−1	41−1
3.2	88−2	18−1	18−1	16−1	16−1	16−1	16−1	17−1	17−1	17−1
3.3	71−2	10−1	78−2	58−2	48−2	43−2	37−2	34−2	30−2	28−2
3.4	49−2	36−2	13−2	37−3	67−4	50−7	62−4	21−3	41−3	65−3
3.5	27−2	35−3	28−3	12−2	24−2	38−2	52−2	66−2	80−2	94−2
3.6	11−2	46−3	35−2	71−2	98−2	12−1	15−1	18−1	21−1	24−1
3.7	23−3	33−2	89−2	14−1	18−1	23−1	26−1	30−1	34−1	38−1
3.8	25−5	75−2	13−1	21−1	24−1	30−1	33−1	38−1	42−1	46−1
3.9	25−3	11−1	16−1	23−1	26−1	31−1	34−1	38−1	41−1	45−1
4.0	75−3	13−1	16−1	21−1	23−1	27−1	29−1	32−1	34−1	36−1
4.1	41−2	12−1	15−1	16−1	17−1	19−1	20−1	21−1	22−1	23−1
4.2	40−2	92−2	99−2	94−2	94−2	97−2	99−2	10−1	10−1	10−1
4.3	34−2	53−2	45−2	34−2	30−2	27−2	25−2	23−2	21−2	20−2
4.4	24−2	21−2	87−3	29−3	90−4	14−4	26−5	41−4	11−3	20−3

9.7 Tables of Oscillator Strengths and Radiative Transition Probabilities

Table 9.6 (continued)

Δn^*	n_0^*							
	1.2	2.0	2.5	3.0	3.5	4.0	4.5	5.0

p—s

Δn^*	1.2	2.0	2.5	3.0	3.5	4.0	4.5	5.0
0.1	11−0	20−0	29−0	37−0	45−0	53+0	61+0	68+0
0.2	19−0	36−0	52+0	67+0	82+0	96+0	11+1	12+1
0.3	24−0	45−0	66+0	86+0	10+1	12+1	14+1	16+1
0.4	25−0	48−0	71+0	93+0	11+1	13+1	15+1	17+1
0.5	23−0	45−0	67+0	88+0	11+1	13+1	15+1	17+1
0.6	18−0	38−0	56+0	75+0	94+0	11+1	13+1	15+1
0.7	13−0	28−0	42−0	57+0	72+0	87+0	10+1	11+1
0.8	81−1	18−0	28−0	38−0	49−0	59+0	70+0	81+0
0.9	38−1	10−0	15−0	22−0	28−0	34−0	41−0	47−0
1.0	11−1	40−1	64−1	96−1	12−0	15−0	18−0	22−0
1.1	26−2	86−2	16−1	24−1	34−1	44−1	55−1	66−1
1.2	27−3	18−4	41−4	35−3	90−3	17−2	26−2	37−2
1.3	51−2	71−2	88−2	96−2	10−1	10−1	11−1	11−1
1.4	12−1	21−1	28−1	35−1	41−1	47−1	52−1	58−1
1.5	18−1	33−1	48−1	61−1	74−1	87−1	99−1	11−0
1.6	19−1	39−1	58−1	77−1	95−1	11−0	13−0	14−0
1.7	17−1	38−1	57−1	77−1	96−1	11−0	13−0	15−0
1.8	12−1	30−1	46−1	64−1	81−1	99−1	11−0	13−0
1.9	60−2	19−1	29−1	43−1	55−1	69−1	81−1	95−1
2.0	15−2	91−2	13−1	22−1	28−1	37−1	44−1	52−1
2.1	65−3	22−2	42−2	68−2	95−2	12−1	15−1	18−1
2.2	64−4	70−7	45−4	21−3	47−3	83−3	12−2	17−2
2.3	13−2	18−2	21−2	22−2	23−2	23−2	23−2	23−2
2.4	36−2	60−2	82−2	10−1	11−1	13−1	14−1	15−1
2.5	55−2	10−1	14−1	19−1	23−1	26−1	30−1	34−1
2.6	63−2	13−1	19−1	25−1	31−1	37−1	43−1	48−1
2.7	57−2	13−1	20−1	27−1	34−1	41−1	47−1	54−1
2.8	40−2	11−1	16−1	23−1	30−1	37−1	43−1	50−1
2.9	19−2	74−2	11−1	17−1	21−1	27−1	32−1	37−1
3.0	42−3	36−2	53−2	91−2	11−1	15−1	18−1	21−1
3.1	25−3	95−3	18−2	29−2	41−2	54−2	68−2	82−2
3.2	27−4	27−6	23−4	11−3	23−3	42−3	62−3	87−3
3.3	60−3	77−3	93−3	96−3	99−3	97−3	95−3	92−3
3.4	15−2	26−2	36−2	44−2	52−2	58−2	64−2	70−2
3.5	24−2	47−2	68−2	88−2	10−1	12−1	14−1	15−1
3.6	28−2	61−2	91−2	12−1	15−1	17−1	20−1	23−1
3.7	26−2	63−2	96−2	13−1	16−1	20−1	23−1	27−1
3.8	18−2	54−2	82−2	11−1	15−1	18−1	21−1	25−1
3.9	88−3	37−2	55−2	86−2	10−1	14−1	16−1	19−1
4.0	15−3	18−2	26−2	47−2	59−2	80−2	94−2	11−1
4.1	13−3	49−3	94−3	15−2	22−2	29−2	37−2	45−2
4.2	14−4	21−6	13−4	63−4	13−3	24−3	36−3	50−3
4.3	31−3	40−3	50−3	51−3	52−3	51−3	50−3	47−3
4.4	85−3	14−2	20−2	24−2	28−2	32−2	35−2	38−2

Table 9.6 (continued)

Δn^*	n_0^*							
	1.5	2.0	2.5	3.0	3.5	4.0	4.5	5.0
				p—d				
0.1			36+0	58+0	79+0	98+0	11+1	13+1
0.2			87+0	13+1	17+1	20+1	23+1	27+1
0.3			14+1	20+1	25+1	30+1	35+1	39+1
0.4			20+1	27+1	33+1	38+1	43+1	48+1
0.5			25+1	32+1	38+1	43+1	48+1	52+1
0.6		22+1	28+1	35+1	39+1	44+1	48+1	52+1
0.7		24+1	29+1	34+1	37+1	41+1	44+1	47+1
0.8		25+1	27+1	30+1	32+1	35+1	36+1	39+1
0.9		24+1	23+1	25+1	25+1	26+1	27+1	28+1
1.0		20+1	18+1	18+1	18+1	18+1	18+1	18+1
1.1	14+1	15+1	13+1	11+1	10+1	10+1	10+1	10+1
1.2	12+1	10+1	76+0	61+0	52+0	46−0	42−0	39−0
1.3	95+0	57+0	34−0	22−0	16−0	12−0	92−1	72−1
1.4	65+0	24−0	91−1	32−1	94−2	11−2	26−3	33−2
1.5	38−0	50−1	96−3	82−2	29−1	56−1	84−1	11−0
1.6	15−0	51−3	40−1	10−0	15−0	20−0	25−0	30−0
1.7	32−1	56−1	14−0	24−0	30−0	37−0	42−0	47−0
1.8	21−3	16−0	26−0	36−0	41−0	48−0	52+0	57+0
1.9	31−1	28−0	35−0	43−0	46−0	50+0	53+0	56+0
2.0	93−1	36−0	37−0	41−0	42−0	44−0	45−0	47−0
2.1	19−0	36−0	34−0	33−0	32−0	32−0	32−0	32−0
2.2	23−0	30−0	25−0	22−0	20−0	18−0	17−0	16−0
2.3	22−0	21−0	15−0	11−0	88−1	73−1	61−1	52−1
2.4	19−0	11−0	60−1	31−1	18−1	98−2	49−2	21−2
2.5	14−0	41−1	83−2	42−3	70−3	43−2	96−2	15−1
2.6	75−1	36−2	26−2	15−1	29−1	45−1	59−1	74−1
2.7	26−1	59−2	31−1	62−1	83−1	10−0	12−0	14−0
2.8	23−2	39−1	77−1	11−0	13−0	16−0	18−0	20−0
2.9	30−2	87−1	12−0	15−0	17−0	19−0	20−0	22−0
3.0	20−1	13−0	14−0	16−0	17−0	18−0	19−0	20−0
3.1	60−1	14−0	14−0	14−0	14−0	14−0	14−0	15−0
3.2	82−1	13−0	12−0	10−0	10−0	95−1	90−1	87−1
3.3	91−1	10−0	77−1	60−1	50−1	43−1	37−1	32−1
3.4	85−1	61−1	35−1	20−1	13−1	88−2	55−2	33−2
3.5	69−1	26−1	71−2	13−2	35−4	45−3	18−2	39−2
3.6	40−1	43−2	22−3	45−2	95−2	16−1	22−1	29−1
3.7	16−1	95−3	11−1	25−1	34−1	46−1	55−1	65−1
3.8	28−2	14−1	33−1	53−1	63−1	77−1	87−1	98−1
3.9	40−3	38−1	57−1	77−1	84−1	97−1	10−0	11−0
4.0	72−2	64−1	73−1	87−1	90−1	99−1	10−0	10−0
4.1	26−1	74−1	80−1	80−1	81−1	83−1	84−1	85−1
4.2	38−1	71−1	67−1	61−1	57−1	56−1	54−1	52−1
4.3	45−1	57−1	45−1	35−1	30−1	27−1	24−1	21−1
4.4	45−1	37−1	22−1	13−1	95−2	67−2	46−2	31−2

Table 9.6 (continued)

Δn^*	n_0^*					
	2.5	3.0	3.5	4.0	4.5	5.0
			d—p			
0.1	24−0	44−0	63+0	81+0	99+0	11+1
0.2	39−0	74+0	10+1	14+1	17+1	20+1
0.3	46−0	90+0	13+1	17+1	21+1	25+1
0.4	45−0	92+0	13+1	18+1	22+1	27+1
0.5	38−0	83−0	12+1	16+1	21+1	25+1
0.6	29−0	66+0	10+1	13+1	17+1	21+1
0.7	19−0	47−0	73+0	10+1	13+1	15+1
0.8	10−0	29−0	46−0	65+0	84+0	10+1
0.9	44−1	15−0	23−0	35−0	45−0	57+0
1.0	10−1	55−1	88−1	13−0	18−0	24−0
1.1	11−2	84−2	15−1	27−1	39−1	54−1
1.2	19−2	10−2	82−3	27−3	51−4	27−4
1.3	10−1	17−1	23−1	27−1	31−1	34−1
1.4	19−1	40−1	59−1	76−1	92−1	10−0
1.5	24−1	59−1	88−1	11−0	14−0	17−0
1.6	23−1	65−1	10−0	13−0	17−0	21−0
1.7	18−1	59−1	91−1	13−0	16−0	20−0
1.8	11−1	44−1	68−1	10−0	13−0	16−0
1.9	43−2	26−1	40−1	64−1	83−1	10−0
2.0	59−3	11−1	16−1	29−1	38−1	51−1
2.1	57−4	18−2	34−2	65−2	96−2	13−1
2.2	77−3	25−3	22−3	47−4	22−5	42−4
2.3	31−2	45−2	64−2	73−2	82−2	87−2
2.4	56−2	11−1	17−1	22−1	26−1	30−1
2.5	68−2	18−1	27−1	36−1	45−1	54−1
2.6	66−2	21−1	32−1	45−1	57−1	69−1
2.7	51−2	20−1	30−1	45−1	57−1	71−1
2.8	28−2	15−1	23−1	37−1	47−1	60−1
2.9	91−3	96−2	14−1	24−1	31−1	40−1
3.0	23−4	42−2	59−2	11−1	14−1	20−1
3.1	56−6	72−3	13−2	26−2	38−2	55−2
3.2	44−3	11−3	11−3	25−4	30−5	15−4
3.3	14−2	20−2	28−2	32−2	36−2	38−2
3.4	25−2	51−2	77−2	99−2	12−1	13−1
3.5	29−2	81−2	12−1	17−1	21−1	25−1
3.6	28−2	97−2	15−1	21−1	27−1	33−1
3.7	20−2	94−2	14−1	21−1	27−1	34−1
3.8	10−2	75−2	11−1	18−1	23−1	29−1
3.9	26−3	47−2	69−2	12−1	15−1	20−1
4.0	35−5	21−2	28−2	57−2	73−2	10−1
4.1	24−5	36−3	64−3	13−2	19−2	28−2
4.2	28−3	66−4	73−4	17−4	36−5	59−5
4.3	85−3	10−2	15−2	17−2	20−2	21−2
4.4	13−2	28−2	42−2	55−2	67−2	77−2

Table 9.6 (continued)

Δn^*	n_0^*					
	2.5	3.0	3.5	4.0	4.5	5.0
			d—f			
0.1			58+0	96+0	13+1	16+1
0.2			14+1	22+1	28+1	34+1
0.3			25+1	36+1	45+1	53+1
0.4			37+1	50+1	60+1	69+1
0.5			48+1	62+1	70+1	80+1
0.6		49+1	56+1	69+1	76+1	84+1
0.7		56+1	61+1	70+1	75+1	81+1
0.8		59+1	60+1	66+1	68+1	72+1
0.9		57+1	54+1	56+1	56+1	58+1
1.0		50+1	45+1	44+1	42+1	42+1
1.1	50+1	40+1	34+1	30+1	28+1	26+1
1.2	43+1	28+1	22+1	18+1	15+1	13+1
1.3	33+1	16+1	12+1	83+0	63+0	48−0
1.4	23+1	81+0	47−0	22−0	12−0	60−1
1.5	14+1	24−0	81−1	48−2	24−2	26−1
1.6	62+0	13−1	73−2	84−1	15−0	25−0
1.7	14−0	52−1	16−0	34−0	45−0	58+0
1.8	43−3	26−0	42−0	63+0	73+0	87+0
1.9	86−1	54+0	67+0	85+0	91+0	10+1
2.0	29−0	78+0	81+0	93+0	93+0	98+0
2.1	65+0	84+0	85+0	83+0	81+0	79+0
2.2	80+0	77+0	72+0	63+0	57+0	53+0
2.3	82+0	59+0	50+0	38−0	32−0	26−0
2.4	74+0	37−0	27+0	16−0	11−0	78−1
2.5	57+0	17−0	92−1	31−1	11−1	14−2
2.6	32−0	41−1	67−2	23−2	13−1	34−1
2.7	12−0	18−3	16−1	64−1	96−1	14−0
2.8	20−1	46−1	96−1	17−0	21−0	27−0
2.9	38−2	15−0	20−0	29−0	31−0	37−0
3.0	58−1	26−0	29−0	36−0	37−0	40−0
3.1	19−0	33−0	35−0	36−0	36−0	36−0
3.2	28−0	33−0	33−0	30−0	28−0	26−0
3.3	33−0	28−0	25−0	20−0	17−0	15−0
3.4	33−0	20−0	15−0	10−0	79−1	57−1
3.5	28−0	10−0	64−1	28−1	15−1	54−2
3.6	18−0	35−1	10−1	75−4	12−2	78−2
3.7	86−1	13−2	24−2	20−1	33−1	54−1
3.8	22−1	12−1	35−1	73−1	91−1	12−0
3.9	15−4	61−1	90−1	13−0	15−0	18−0
4.0	17−1	12−0	14−0	18−0	18−0	21−0
4.1	81−1	16−0	18−0	19−0	19−0	20−0
4.2	13−0	18−0	18−0	17−0	16−0	15−0
4.3	17−0	16−0	15−0	12−0	11−0	97−1
4.4	18−0	12−0	97−1	66−1	53−1	40−1

9.7 Tables of Oscillator Strengths and Radiative Transition Probabilities

Table 9.6 (continued)

Δn^*	n_0^*			
	3.5	4.0	4.5	5.0
		f—d		
0.1	34−0	67+0	98+0	12+1
0.2	52+0	11+1	16+1	21+1
0.3	57+0	13+1	19+1	26+1
0.4	52+0	13+1	19+1	26+1
0.5	42−0	11+1	17+1	24+1
0.6	30−0	89+0	13+1	19+1
0.7	18−0	61+0	95+0	13+1
0.8	98−1	37−0	57+0	85+0
0.9	38−1	18−0	28−0	44−0
1.0	87−2	62−1	96−1	16−0
1.1	23−2	76−2	14−1	26−1
1.2	42−3	24−2	24−2	21−2
1.3	47−2	24−1	33−1	45−1
1.4	93−2	53−1	77−1	11−0
1.5	10−1	74−1	10−0	16−0
1.6	10−1	79−1	11−0	18−0
1.7	69−2	69−1	10−0	16−0
1.8	33−2	50−1	74−1	12−0
1.9	83−3	28−1	41−1	74−1
2.0	18−8	11−1	15−1	31−1
2.1	16−3	14−2	26−2	54−2
2.2	86−4	65−3	71−3	57−3
2.3	62−3	63−2	86−2	12−1
2.4	10−2	14−1	20−1	31−1
2.5	98−3	21−1	30−1	48−1
2.6	79−3	23−1	34−1	57−1
2.7	36−3	21−1	31−1	54−1
2.8	39−4	16−1	22−1	42−1
2.9	65−4	97−2	12−1	26−1
3.0	46−3	39−2	45−2	11−1
3.1	17−5	49−3	81−3	19−2
3.2	66−4	29−3	36−3	29−3
3.3	18−3	26−2	37−2	52−2
3.4	20−3	61−2	87−2	13−1
3.5	94−4	92−2	12−1	21−1
3.6	32−4	10−1	14−1	26−1
3.7	18−5	98−2	13−1	25−1
3.8	10−3	75−2	10−1	20−1
3.9	38−3	45−2	55−2	12−1
4.0	76−3	19−2	18−2	56−2
4.1	13−4	23−3	34−3	96−3
4.2	66−4	15−3	22−3	17−3
4.3	93−4	14−2	19−2	28−2
4.4	56−4	32−2	46−2	75−2

Table 9.7 Values $F_2(l_0; l_1)$

Δn^*	n_0^*									
	0.5	1.0	1.5	2.0	2.5	3.0	3.5	4.0	4.5	5.0
				Transition $s-d$						
0.1					15−2	18−2	19−2	18−2	18−2	16−2
0.2					14−1	16−1	16−1	15−1	14−1	13−1
0.3					55−1	58−1	56−1	53−1	49−1	45−1
0.4					13−0	14−0	13−0	12−0	11−0	10−0
0.5					27−0	26−0	24−0	21−0	19−0	17−0
0.6				44−0	46−0	42−0	37−0	33−0	29−0	26−0
0.7				70+0	68+0	59+0	51+0	45−0	39−0	35−0
0.8				99+0	89+0	74+0	63+0	54+0	47−0	41−0
0.9				12+1	10+1	85+0	71+0	59+0	51+0	44−0
1.0				14+1	11+1	89+0	72+0	59+0	50+0	43−0
1.1			19+1	15+1	11+1	84+0	66+0	54+0	45−0	38−0
1.2			21+1	14+1	10+1	72+0	55+0	43−0	35−0	29−0
1.3			21+1	13+1	81+0	55+0	40−0	30−0	24−0	20−0
1.4			19+1	10+1	57+0	36−0	25−0	18−0	13−0	11−0
1.5			15+1	68+0	33−0	19−0	12−0	80−1	57−1	41−1
1.6		25+1	10+1	37−0	14−0	67−1	33−1	18−1	99−2	56−2
1.7		20+1	59+0	14−0	31−1	65−2	58−3	11−3	11−2	23−2
1.8		14+1	23−0	19−1	43−3	75−2	15−1	20−1	23−1	24−1
1.9		86+0	41−1	73−2	39−1	54−1	61−1	62−1	61−1	59−1
2.0		42−0	39−2	82−1	11−0	12−0	11−0	10−0	10−0	92−1
2.1	18+1	84−1	74−1	19−0	20−0	18−0	16−0	14−0	12−0	11−0
2.2	10+1	11−2	21−0	29−0	25−0	21−0	17−0	15−0	13−0	11−0
2.3	45−0	93−1	36−0	36−0	27−0	20−0	16−0	13−0	11−0	96−1
2.4	13−0	26−0	47−0	36−0	24−0	17−0	13−0	10−0	81−1	67−1
2.5	13−1	42−0	51+0	30−0	17−0	11−0	82−1	60−1	45−1	35−1
2.6	57−2	52+0	42−0	21−0	10−0	60−1	37−1	24−1	16−1	11−1
2.7	45−1	53+0	29−0	11−0	40−1	17−1	80−2	35−2	14−2	50−3
2.8	88−1	46−0	16−0	37−1	51−2	32−3	24−3	13−2	25−2	35−2
2.9	11−0	34−0	55−1	14−2	32−2	83−2	12−1	15−1	16−1	17−1
3.0	11−0	20−0	39−2	12−1	29−1	34−1	37−1	37−1	36−1	35−1
3.1	39−0	66−1	70−2	54−1	68−1	66−1	63−1	59−1	54−1	50−1
3.2	24−0	39−2	53−1	10−0	10−0	91−1	81−1	71−1	63−1	56−1
3.3	12−0	13−1	12−0	15−0	12−0	10−0	84−1	71−1	61−1	53−1
3.4	47−1	69−1	18−0	17−0	12−0	92−1	73−1	59−1	48−1	41−1
3.5	80−2	13−0	23−0	15−0	99−1	69−1	51−1	39−1	31−1	24−1
3.6	30−3	19−0	21−0	12−0	64−1	40−1	27−1	19−1	13−1	10−1
3.7	10−1	21−0	15−0	73−1	29−1	15−1	84−2	47−2	25−2	14−2
3.8	24−1	20−0	96−1	29−1	63−2	15−2	17−3	16−4	29−3	65−3
3.9	35−1	16−0	40−1	35−2	25−3	18−2	38−2	52−2	64−2	70−2
4.0	39−1	11−0	63−2	28−2	11−1	13−1	16−1	17−1	17−1	17−1
4.1	14−0	42−1	71−3	22−1	31−1	32−1	31−1	30−1	29−1	27−1
4.2	10−0	51−2	19−1	52−1	54−1	48−1	44−1	40−1	36−1	33−1
4.3	55−1	29−2	55−1	80−1	69−1	57−1	49−1	42−1	37−1	33−1

9.7 Tables of Oscillator Strengths and Radiative Transition Probabilities   293

Table 9.7 (continued)

Δn^*	n_0^*							
	1.5	2.0	2.5	3.0	3.5	4.0	4.5	5.0
				p—p				
0.1	24−2	33−2	33−2	30−2	28−2	25−2	23−2	21−2
0.2	19−1	25−1	24−1	23−1	21−1	19−1	17−1	16−1
0.3	62−1	78−1	76−1	70−1	64−1	58−1	53−1	49−1
0.4	13−0	16−0	15−0	14−0	13−0	12−0	11−0	10−0
0.5	22−0	26−0	25−0	23−0	21−0	19−0	18−0	16−0
0.6	32−0	37−0	35−0	32−0	30−0	27−0	25−0	23−0
0.7	40−0	45−0	43−0	40−0	36−0	33−0	30−0	28−0
0.8	44−0	49−0	47−0	43−0	39−0	36−0	33−0	30−0
0.9	44−0	47−0	45−0	41−0	38−0	34−0	32−0	29−0
1.0	39−0	41−0	39−0	36−0	33−0	30−0	27−0	25−0
1.1	30−0	32−0	30−0	27−0	25−0	23−0	21−0	19−0
1.2	20−0	21−0	19−0	18−0	16−0	14−0	13−0	12−0
1.3	11−0	11−0	10−0	94−1	85−1	77−1	70−1	64−1
1.4	45−1	41−1	37−1	32−1	28−1	25−1	23−1	21−1
1.5	72−2	50−2	38−2	29−2	23−2	19−2	16−2	14−2
1.6	11−2	27−2	35−2	39−2	40−2	40−2	39−2	37−2
1.7	19−1	25−1	26−1	26−1	25−1	24−1	22−1	21−1
1.8	48−1	57−1	59−1	57−1	54−1	50−1	47−1	44−1
1.9	76−1	85−1	86−1	83−1	78−1	73−1	68−1	64−1
2.0	92−1	98−1	10−0	95−1	89−1	83−1	77−1	72−1
2.1	84−1	94−1	94−1	90−1	84−1	78−1	73−1	68−1
2.2	69−1	75−1	74−1	71−1	66−1	61−1	57−1	53−1
2.3	46−1	48−1	47−1	45−1	41−1	38−1	35−1	33−1
2.4	22−1	23−1	22−1	20−1	18−1	17−1	16−1	14−1
2.5	60−2	55−2	51−2	45−2	40−2	36−2	32−2	29−2
2.6	15−4	15−4	47−4	88−4	12−3	15−3	16−3	17−3
2.7	42−2	57−2	62−2	63−2	63−2	61−2	59−2	56−2
2.8	15−1	18−1	19−1	19−1	18−1	17−1	16−1	16−1
2.9	28−1	31−1	33−1	32−1	31−1	29−1	28−1	26−1
3.0	37−1	40−1	42−1	41−1	39−1	37−1	35−1	33−1
3.1	36−1	41−1	43−1	42−1	40−1	38−1	36−1	34−1
3.2	32−1	35−1	36−1	35−1	34−1	32−1	30−1	28−1
3.3	22−1	24−1	25−1	24−1	23−1	21−1	20−1	19−1
3.4	12−1	12−1	12−1	12−1	11−1	10−1	10−1	96−2
3.5	36−2	36−2	36−2	34−2	31−2	28−2	26−2	24−2
3.6	86−4	28−4	20−4	95−5	23−5	49−6	94−8	10−6
3.7	16−2	22−2	23−2	24−2	25−2	24−2	23−2	23−2
3.8	71−2	84−2	90−2	89−2	89−2	86−2	82−2	78−2
3.9	14−1	15−1	16−1	16−1	16−1	15−1	15−1	14−1
4.0	19−1	20−1	22−1	22−1	21−1	20−1	19−1	18−1
4.1	19−1	22−1	23−1	23−1	23−1	22−1	21−1	20−1
4.2	17−1	19−1	20−1	20−1	20−1	19−1	18−1	17−1
4.3	12−1	14−1	14−1	14−1	14−1	13−1	13−1	12−1

Table 9.7 (continued)

Δn^*	n_0^*							
	1.5	2.0	2.5	3.0	3.5	4.0	4.5	5.0
				p—f				
0.1					12−2	17−2	20−2	21−2
0.2					12−1	16−1	18−1	18−1
0.3					51−1	62−1	66−1	66−1
0.4					13−0	15−0	16−0	15−0
0.5					29−0	31−0	31−0	29−0
0.6				42−0	51+0	53+0	51+0	48−0
0.7				72+0	81+0	79+0	74+0	68+0
0.8				11+1	11+1	10+1	96+0	86+0
0.9				15+1	14+1	12+1	11+1	10+1
1.0				19+1	17+1	14+1	12+1	10+1
1.1			23+1	22+1	18+1	14+1	12+1	10+1
1.2			28+1	23+1	18+1	14+1	11+1	91+0
1.3			30+1	22+1	16+1	12+1	91+0	72+0
1.4			30+1	20+1	13+1	91+0	66+0	49−0
1.5			28+1	15+1	94+0	60+0	40−0	28−0
1.6		39+1	22+1	11+1	56+0	31−0	18−0	11−0
1.7		35+1	16+1	63+0	25−0	11−0	48−1	20−1
1.8		29+1	96+0	26−0	63−1	10−1	14−3	22−2
1.9		21+1	43−0	48−1	44−6	13−1	32−1	47−1
2.0		13+1	10−0	40−2	53−1	93−1	11−0	12−0
2.1	27+1	60+0	27−5	10−0	18−0	21−0	21−0	21−0
2.2	19+1	14−0	90−1	27−0	32−0	32−0	29−0	27−0
2.3	11+1	35−4	31−0	46−0	44−0	38−0	33−0	28−0
2.4	56+0	11−0	57+0	59+0	48−0	38−0	31−0	25−0
2.5	19−0	38−0	78+0	62+0	45−0	33−0	24−0	18−0
2.6	34−2	68+0	83+0	55+0	35−0	23−0	15−0	11−0
2.7	67−1	90+0	76+0	41−0	22−0	12−0	74−1	45−1
2.8	25−0	98+0	58+0	24−0	10−0	43−1	17−1	63−2
2.9	45−0	91+0	35−0	99−1	21−1	27−2	11−3	26−2
3.0	59+0	73+0	15−0	13−1	65−3	97−2	21−1	29−1
3.1	71+0	42−0	33−1	59−2	33−1	54−1	66−1	72−1
3.2	62+0	17−0	16−2	63−1	10−0	11−0	11−0	11−0
3.3	46−0	28−1	63−1	15−0	17−0	16−0	15−0	14−0
3.4	28−0	53−2	18−0	24−0	22−0	19−0	16−0	14−0
3.5	13−0	84−1	32−0	30−0	23−0	18−0	14−0	11−0
3.6	18−1	21−0	39−0	30−0	20−0	14−0	10−0	80−1
3.7	53−2	35−0	40−0	25−0	14−0	94−1	61−1	40−1
3.8	62−1	44−0	34−0	17−0	80−1	41−1	21−1	10−1
3.9	14−0	47−0	24−0	84−1	26−1	80−2	14−2	56−5
4.0	22−0	42−0	12−0	20−1	10−2	77−3	47−2	90−2
4.1	30−0	27−0	42−1	17−4	88−2	18−1	27−1	32−1
4.2	29−0	13−0	11−2	21−1	42−1	52−1	58−1	59−1
4.3	23−0	36−1	17−1	71−1	87−1	88−1	85−1	80−1

9.7 Tables of Oscillator Strengths and Radiative Transition Probabilities

Table 9.7 (continued)

Δn^*	n_0^*					
	2.5	3.0	3.5	4.0	4.5	5.0
			$d - s$			
0.1	10−2	14−2	15−2	15−2	15−2	14−2
0.2	66−2	94−2	10−1	10−1	10−1	10−1
0.3	17−1	25−1	29−1	30−1	29−1	29−1
0.4	31−1	47−1	54−1	56−1	57−1	56−1
0.5	44−1	68−1	79−1	84−1	86−1	85−1
0.6	53−1	83−1	99−1	10−0	10−0	10−0
0.7	56−1	89−1	10−0	11−0	12−0	12−0
0.8	51−1	83−1	10−0	11−0	11−0	12−0
0.9	42−1	68−1	86−1	96−1	10−0	10−0
1.0	29−1	48−1	63−1	72−1	77−1	80−1
1.1	15−1	28−1	38−1	44−1	49−1	52−1
1.2	62−2	12−1	17−1	21−1	24−1	26−1
1.3	10−2	27−2	45−2	61−2	75−2	86−2
1.4	10−3	13−4	18−4	13−3	33−3	56−3
1.5	22−2	27−2	26−2	23−2	19−2	15−2
1.6	58−2	83−2	93−2	94−2	91−2	86−2
1.7	91−2	13−1	16−1	17−1	17−1	17−1
1.8	11−1	17−1	21−1	23−1	24−1	24−1
1.9	11−1	17−1	22−1	24−1	26−1	27−1
2.0	93−2	14−1	19−1	21−1	23−1	24−1
2.1	50−2	94−2	12−1	15−1	17−1	18−1
2.2	22−2	47−2	68−2	85−2	99−2	10−1
2.3	50−3	12−2	21−2	29−2	37−2	43−2
2.4	93−5	66−5	76−4	20−3	38−3	56−3
2.5	69−3	80−3	72−3	57−3	42−3	30−3
2.6	20−2	29−2	32−2	32−2	31−2	29−2
2.7	35−2	53−2	64−2	68−2	69−2	68−2
2.8	46−2	70−2	89−2	98−2	10−1	10−1
2.9	49−2	74−2	98−2	11−1	11−1	12−1
3.0	46−2	64−2	88−2	10−1	11−1	11−1
3.1	23−2	44−2	62−2	75−2	85−2	92−2
3.2	10−2	23−2	34−2	43−2	51−2	57−2
3.3	25−3	67−3	11−2	16−2	20−2	23−2
3.4	23−5	76−5	56−4	14−3	25−3	36−3
3.5	32−3	37−3	32−3	24−3	17−3	11−3
3.6	10−2	14−2	16−2	16−2	15−2	14−2
3.7	18−2	27−2	33−2	36−2	36−2	36−2
3.8	25−2	37−2	48−2	52−2	56−2	57−2
3.9	28−2	40−2	54−2	60−2	66−2	69−2
4.0	27−2	35−2	50−2	57−2	64−2	67−2
4.1	12−2	25−2	35−2	43−2	49−2	54−2
4.2	61−3	13−2	19−2	25−2	30−2	34−2
4.3	14−3	39−3	68−3	97−3	12−2	14−2

Table 9.7 (continued)

Δn^*	n_0^*					
	2.5	3.0	3.5	4.0	4.5	5.0
			$d - d$			
0.1	11−2	18−2	21−2	21−2	21−2	20−2
0.2	92−2	14−1	16−1	16−1	16−0	15−1
0.3	30−1	44−1	49−1	51−1	50−1	48−1
0.4	67−1	94−1	10−0	10−0	10−0	10−0
0.5	11−0	15−0	17−0	17−0	17−0	16−0
0.6	17−0	22−0	24−0	24−0	24−0	23−0
0.7	21−0	28−0	30−0	30−0	29−0	28−0
0.8	25−0	31−0	33−0	33−0	32−0	31−0
0.9	25−0	31−0	33−0	33−0	32−0	30−0
1.0	23−0	28−0	29−0	29−0	28−0	26−0
1.1	18−0	22−0	23−0	23−0	22−0	20−0
1.2	13−0	15−0	16−0	15−0	14−0	14−0
1.3	80−1	92−1	92−1	87−1	81−1	75−1
1.4	35−1	39−1	37−1	34−1	31−1	28−1
1.5	77−2	81−2	68−2	53−2	42−2	34−2
1.6	31−4	14−3	55−3	10−2	15−2	18−2
1.7	80−2	10−1	13−1	15−1	16−1	17−1
1.8	25−1	30−1	36−1	39−1	40−1	40−1
1.9	43−1	51−1	59−1	61−1	62−1	61−1
2.0	55−1	63−1	72−1	74−1	74−1	72−1
2.1	54−1	65−1	73−1	74−1	73−1	70−1
2.2	47−1	56−1	61−1	61−1	60−1	57−1
2.3	33−1	39−1	42−1	41−1	40−1	38−1
2.4	18−1	21−1	22−1	21−1	20−1	18−1
2.5	56−2	71−2	71−2	63−2	56−2	50−2
2.6	26−3	38−3	25−3	11−3	37−4	66−5
2.7	16−2	17−2	24−2	29−2	34−2	37−2
2.8	83−2	90−2	11−1	12−1	13−1	13−1
2.9	17−1	18−1	22−1	23−1	24−1	24−1
3.0	24−1	25−1	30−1	31−1	32−1	32−1
3.1	25−1	29−1	33−1	34−1	35−1	34−1
3.2	23−1	26−1	30−1	31−1	31−1	30−1
3.3	17−1	20−1	22−1	23−1	22−1	21−1
3.4	99−2	11−1	13−1	13−1	12−1	11−1
3.5	34−2	45−2	50−2	46−2	42−2	39−2
3.6	27−3	46−3	46−3	35−3	23−3	15−3
3.7	66−3	55−3	71−3	90−3	11−2	12−2
3.8	41−2	40−2	48−2	54−2	60−2	62−2
3.9	90−2	90−2	10−1	11−1	12−1	12−1
4.0	13−1	13−1	16−1	17−1	17−1	17−1
4.1	14−1	15−1	18−1	19−1	20−1	19−1
4.2	13−1	15−1	17−1	18−1	18−1	18−1
4.3	10−1	11−1	13−1	14−1	14−1	13−1

9.7 Tables of Oscillator Strengths and Radiative Transition Probabilities

Table 9.8 Oscillator strengths and transition probabilities. (If quantum numbers J_0 and J_1 are omitted the values of f_{01} and A_{10} relate to the transition between terms.)

Element	Configuration	Terms	J_0—J_1	f_{01}	A_{10} [s^{-1}]
He I	$1s^2$—$1s2p$	^1S—$^1P^o$	0—1	0.276	1.80×10^9
	—$1s3p$	^1S—$^1P^o$	0—1	0.0734	5.66×10^8
	—$1s4p$	^1S—$^1P^o$	0—1	0.0302	2.46×10^8
	+$1s5p$	^1S—$^1P^o$	0—1	0.0153	1.28×10^8
	$1s2s$—$1s2p$	^1S—$^1P^o$	0—1	0.376	1.98×10^6
	—$1s3p$	^1S—$^1P^o$	0—1	0.151	1.34×10^7
	—$1s4p$	^1S—$^1P^o$	0—1	0.0507	0.72×10^7
	—$1s5p$	^1S—$^1P^o$	0—1	0.0221	3.76×10^6
	$1s2s$—$1s2p$	^3S—$^3P^o$	1—2	0.300	1.02×10^7
		^3S—$^3P^o$	1—1	0.180	1.02×10^7
		^3S—$^3P^o$	1—0	0.060	1.02×10^7
	$1s2s$—$1s3p$	^3S—$^3P^o$		0.0645	0.95×10^7
	—$1s4p$	^3S—$^3P^o$		0.0231	5.05×10^6
	—$1s5p$	^3S—$^3P^o$		0.0114	2.93×10^6
	$1s2p$—$1s3s$	$^1P^o$—^1S	1—0	0.0480	1.81×10^7
	—$1s4s$	$^1P^o$—^1S	1—0	0.834×10^{-2}	0.65×10^7
	—$1s5s$	$^1P^o$—^1S	1—0	0.308×10^{-2}	3.13×10^6
	$1s2p$—$1s3d$	$^1P^o$—^1D	1—2	0.711	0.64×10^8
	—$1s4d$	$^1P^o$—^1D	1—2	0.122	2.02×10^7
	—$1s5d$	$^1P^o$—^1D	1—2	0.0436	0.91×10^7
	$1s2p$—$1s3s$	$^3P^o$—^3S	2—1	0.0693	1.54×10^7
	—$1s3s$	$^3P^o$—^3S	1—1	0.0692	0.93×10^7
	—$1s3s$	$^3P^o$—^3S	0—1	0.0692	3.08×10^6
	—$1s4s$	$^3P^o$—^3S	—	0.0118	1.06×10^7
	—$1s5s$	$^3P^o$—^3S	—	0.365×10^{-2}	4.30×10^6
	$1s2p$—$1s3d$	$^3P^o$—^3D	—	0.609	0.70×10^8
	—$1s4d$	$^3P^o$—^3D	—	0.125	2.51×10^7
	—$1s5d$	$^3P^o$—^3D	—	0.0474	1.17×10^7
	$1s3s$—$1s3p$	^1S—$^1P^o$	0—1	0.629	2.53×10^5
	—$1s4p$	^1S—$^1P^o$	0—1	0.140	1.37×10^6
	—$1s5p$	^1S—$^1P^o$	0—1	0.0521	0.96×10^6
	$1s3s$—$1s3p$	^3S—$^3P^o$	—	0.896	1.08×10^6
	—$1s4p$	^3S—$^3P^o$	—	0.0429	0.61×10^6
	$1s3s$—$1s5p$	^3S—$^3P^o$	—	0.0245	0.61×10^6
	$1s3p$—$1s4s$	$^1P^o$—^1S	1—0	0.103	4.59×10^6
	—$1s5s$	$^1P^o$—^1S	1—0	0.182	2.02×10^6
	$1s3p$—$1s4d$	$^1P^o$—^1D	1—2	0.647	7.11×10^6
	—$1s5d$	$^1P^o$—^1D	1—2	0.139	3.31×10^6
	$1s3p$—$1s4s$	$^3P^o$—^3S	—	0.145	6.52×10^6
	—$1s5s$	$^3P^o$—^3S	—	0.0222	2.69×10^6
	$1s3p$—$1s3d$	$^3P^o$—^3D	—	0.111	1.28×10^4
	—$1s4d$	$^3P^o$—^3D	—	0.482	6.68×10^6
	—$1s5d$	$^3P^o$—^3D	—	0.123	3.43×10^6
	$1s3d$—$1s3p$	^1D—$^1P^o$	2—1	0.0139	1.68×10^2

Table 9.8 (continued)

Element	Configuration	Terms	J_0—J_1	f_{01}	A_{10} [s^{-1}]
Li I	$2s-2p$	$^2S-^2P^0$	—	0.753	3.72×10^7
	$-3p$	$^2S-^2P^0$	—	0.55×10^{-2}	1.17×10^6
	$-4p$	$^2S-^2P^0$	—	0.48×10^{-2}	1.42×10^6
	$-5p$	$^2S-^2P^0$	—	0.32×10^{-2}	1.07×10^6
	$2p-3s$	$^2P^0-^2S$	—	0.115	3.49×10^7
	$-4s$	$^2P^0-^2S$	—	0.0125	1.01×10^7
	$-5s$	$^2P^0-^2S$	—	0.42×10^{-2}	4.60×10^6
	$2p-3d$	$^2P^0-^2D$	—	0.67	7.2×10^7
	$-4d$	$^2P^0-^2D$	—	0.12	2.3×10^7
	$-5d$	$^2P^0-^2D$	—	0.045	1.06×10^7
	$3s-3p$	$^2S-^2P^0$	—	1.23	3.77×10^6
	$3p-4s$	$^2P^0-^2S$	—	0.223	7.46×10^6
	$-5s$	$^2P^0-^2S$	—	0.0254	1.44×10^6
	$3p-3d$	$^2P^0-^2D$	—	0.0743	3.81×10^3
	$-4d$	$^2P^0-^2D$	—	0.527	6.85×10^6
	$-5d$	$^2P^0-^2D$	—	0.128	3.41×10^6
Li II	$1s^2-1s2p$	$^1S-^1P^0$	0—1	0.457	2.56×10^{10}
	$-1s3p$	$^1S-^1P^0$	0—1	0.111	0.78×10^{10}
	$1s2s-1s2p$	$^1S-^1P^0$	0—1	0.213	5.18×10^6
	$-1s3p$	$^1S-^1P^0$	0—1	0.256	2.82×10^8
	$1s2s-1s2p$	$^3S-^3P^0$	—	0.308	2.26×10^7
	$-1s3p$	$^3S-^3P^0$	—	0.186	2.88×10^8
	$1s2p-1s3s$	$^1P^0-^1S$	1—0	0.031	2.04×10^8
	$-1s3d$	$^1P^0-^1D$	1—2	0.714	1.01×10^9
	$1s2p-1s3s$	$^3P^0-^3S$	—	0.039	2.85×10^8
	$-1s3d$	$^3P^0-^3D$	—	0.625	1.12×10^9
Be III	$1s^2-1s2p$	$^1S-^1P^0$	0—1	0.552	1.22×10^{11}
	$-1s3p$	$^1S-^1P^0$	0—1	0.127	3.62×10^{10}
	$1s2s-1s2p$	$^1S-^1P^0$	0—1	0.149	8.77×10^6
	$-1s3p$	$^1S-^1P^0$	0—1	0.305	4.28×10^9
	$1s2s-1s2p$	$^3S-^3P^0$	—	0.213	3.42×10^7
	$-1s3p$	$^3S-^3P^0$	—	0.252	1.65×10^9
	$1s2p-1s3d$	$^1P^0-^1D$	1—2	0.711	5.10×10^9
	$1s2p-1s3d$	$^3P^0-^3D$	—	0.640	5.61×10^9
B IV	$1s^2-1s2p$	$^1S-^1P^0$	0—1	0.609	3.72×10^{11}
	$-1s3p$	$^1S-^1P^0$	0—1	0.135	1.08×10^{11}
	$1s2s-1s2p$	$^1S-^1P^0$	0—1	0.114	1.25×10^7
	$-1s3p$	$^1S-^1P^0$	0—1	0.333	0.51×10^{10}
	$1s2s-1s2p$	$^3S-^3P^0$	—	0.163	0.45×10^8
	$-1s3p$	$^3S-^3P^0$	—	0.291	0.55×10^{10}
	$1s2p-1s3d$	$^1P^0-^1D$	1—2	0.709	1.62×10^{10}
	$1s2p-1s3d$	$^3P^0-^3D$	—	0.650	1.75×10^{10}
C V	$1s^2-1s2p$	$^1S-^1P^0$	0—1	0.647	8.87×10^{11}
	$-1s3p$	$^1S-^1P^0$	0—1	0.141	2.55×10^{11}
	$1s2s-1s2p$	$^1S-^1P^0$	0—1	0.093	1.65×10^7
	$-1s3p$	$^1S-^1P^0$	0—1	0.351	1.28×10^{10}
	$1s2s-1s2p$	$^3S-^3P^0$	—	0.132	0.56×10^8
	$-1s3p$	$^3S-^3P^0$	—	0.316	1.36×10^{10}
	$1s2p-1s3d$	$^1P^0-^1D$	1—2	0.707	3.96×10^{10}
	$1s2p-1s3d$	$^3P^0-^3D$	—	0.657	4.25×10^{10}
N VI	$1s^2-1s2p$	$^1S-^1P^0$	0—1	0.674	1.81×10^{12}
	$-1s3p$	$^1S-^1P^0$	0—1	0.144	5.16×10^{11}
	$1s2s-1s2p$	$^1S-^1P^0$	0—1	0.078	2.06×10^7
	$-1s3p$	$^1S-^1P^0$	0—1	0.364	2.69×10^{10}
	$1s2s-1s2p$	$^3S-^3P^0$	—	0.110	0.68×10^8
	$-1s3p$	$^3S-^3P^0$	—	0.334	2.85×10^{10}

9.7 Tables of Oscillator Strengths and Radiative Transition Probabilities

Table 9.8 (continued)

Element	Configuration	Terms	J_0-J_1	f_{01}	A_{10} [s^{-1}]
O VII	$1s^2-1s2p$	$^1S-^1P^0$	0–1	0.694	3.30×10^{12}
	$-1s3p$	$^1S-^1P^0$	0–1	0.146	0.94×10^{12}
Ne I	$2p^6-2p^5(^2P^0_{3/2})3s$	$^1S-[3/2]°$	0–1	0.0118	4.76×10^7
	$-2p^5(^2P^0_{1/2})3s'$	$^1S-[1/2]°$	0–1	0.162	6.64×10^8
	$2p^53s-2p^5(^2P^0_{3/2})3p$	$[3/2]°-[1/2]$	2–1	0.085	1.92×10^7
		$[3/2]°-[5/2]$	2–3	0.373	4.33×10^7
		$[3/2]°-[5/2]$	2–2	0.082	1.36×10^7
		$[3/2]°-[3/2]$	2–1	0.027	0.78×10^7
		$[3/2]°-[3/2]$	2–2	0.122	2.16×10^7
	$2p^5(^2P^0_{3/2})3s-$	$[3/2]°-[3/2]$	2–1	0.014	4.33×10^6
		$[3/2]°-[3/2]$	2–2	0.056	1.05×10^7
	$-2p^5(^2P^0_{1/2})3p'$	$[3/2]°-[1/2]$	2–1	0.040	1.28×10^7
	$2p^53s-2p^5(^2P^0_{3/2})3p$	$[3/2]°-[1/2]$	1–1	0.077	0.98×10^7
		$[3/2]°-[5/2]$	1–2	0.245	2.32×10^7
	$2p^53s-2p^5(^2P^0_{3/2})3p$	$[3/2]°-[3/2]$	1–1	0.170	2.79×10^7
		$[3/2]°-[3/2]$	1–2	0.050	0.51×10^7
		$[3/2]°-[1/2]$	1–0	0.114	0.62×10^8
	$2p^5(^2P^0_{3/2})3s-$	$[3/2]°-[3/2]$	1–1	0.018	0.33×10^7
	$-2p^5(^2P^0_{1/2})3p'$	$[3/2]°-[3/2]$	1–2	0.157	1.69×10^7
	$2p^5(^2P^0_{1/2})3s'-$	$[3/2]°-[3/2]$	1–1	0.034	0.63×10^7
	$-2p^5(^2P^0_{3/2})3p$	$[1/2]°-[1/2]$	0–1	0.073	2.92×10^6
		$[1/2]°-[3/2]$	0–1	0.246	1.28×10^7
	$2p^53s'-$	$[1/2]°-[3/2]$	0–1	0.394	2.23×10^7
	$-2p^5(^2P^0_{1/2})3p'$	$[1/2]°-[1/2]$	0–1	0.273	1.60×10^7
	$2p^53s'-2p^5(^2P^0_{3/2}$ (3p$[1/2]°-[5/2]$		1–2	0.047	3.65×10^6
		$[1/2]°-[3/2]$	1–2	0.228	1.90×10^7
	$2p^53s-2p^5(^2P^0_{1/2})3p'$	$[1/2]°-[3/2]$	1–1	0.158	2.34×10^7
		$[1/2]°-[3/2]$	1–2	0.265	2.38×10^7
		$[1/2]°-[1/2]$	1–1	0.164	2.51×10^7
		$[1/2]°-[1/2]$	1–0	0.123	7.19×10^7
Na I	$3s-3p$	$^2S-^2P^0$	1/2–3/2	0.655	0.63×10^8
			1/2–1/2	0.327	0.63×10^8
	$3s-4p$	$^2S-^2P^0$	1/2–3/2	0.94×10^{-2}	0.29×10^7
			1/2–1/2	0.48×10^{-2}	0.29×10^7
	$3s-5p$	$^2S-^2P^0$	1/2–3/2	1.47×10^{-3}	0.60×10^6
			1/2–1/2	0.74×10^{-3}	0.60×10^6
	$3p-4s$	$^2P^0-^2S$	3/2–1/2	0.163	1.67×10^7
			1/2–1/2	0.163	0.84×10^7
	$3p-5s$	$^2P^0-^2S$	3/2–1/2	0.014	0.48×10^7
			1/2–1/2	0.014	0.24×10^7
	$3p-3d$	$^2P^0-^2D$	3/2–5/2	0.75	4.95×10^7
			1/2–3/2	0.83	4.13×10^7
			3/2–3/2	0.083	0.82×10^7
	$3p-4d$	$^2P^0-^2D$	3/2–5/2	0.095	1.31×10^7
			1/2–3/2	0.106	1.09×10^7
			3/2–3/2	0.011	2.19×10^6
	$3p-5d$	$^2P^0-^2D$	3/2–5/2	0.028	0.50×10^7
			1/2–3/2	0.031	0.42×10^7
			3/2–3/2	0.31×10^{-2}	0.84×10^6

Table 9.8 (continued)

Element	Configuration	Terms	J_0—J_1	f_{01}	A_{10} [s^{-1}]
Mg I	$3s^2 - 3s3p$	$^1S - {}^1P^0$	0–1	1.81	4.95×10^8
	$- 3s4p$	$^1S - {}^1P^0$	0–1	0.22	1.22×10^8
	$3s3p - 3p^2$	$^3P^0 - {}^3P$	—	0.61	
	$- 3s3d$	$^3P^0 - {}^3D$	—	0.62	
	$- 3s4s$	$^3P^0 - {}^3S$	—	0.14	
	$- 3s4d$	$^3P^0 - {}^3D$	—	0.13	
	$3s3p - 3s3d$	$^1P^0 - {}^1D$	1–2	0.28	1.4×10^7
	$- 3s4s$	$^1P^0 - {}^1S$	1–0	0.18	2.6×10^7
	$- 3s4d$	$^1P^0 - {}^1D$	1–2	0.11	1.4×10^7
K I	$4s - 4p$	$^2S - {}^2P^0$	1/2–3/2	0.68	0.38×10^8
			1/2–1/2	0.34	0.38×10^8
	$4s - 5p$	$^2S - P^0$	1/2–3/2	0.61×10^{-2}	1.24×10^6
			1/2–1/2	0.31×10^{-2}	1.24×10^6
	$4p - 5s$	$^2P^0 - {}^2S$	3/2–1/2	0.183	1.56×10^7
			1/2–1/2	0.183	0.79×10^7
	$4p - 3d$	$^2P^0 - {}^2D$	3/2–5/2	0.81	2.59×10^7
			1/2–3/2	0.90	2.20×10^7
			3/2–3/2	0.090	4.34×10^6
	$4p - 4d$	$^2P^0 - {}^2D$	3/2–5/2	3.4×10^{-4}	3.1×10^4
			1/2–3/2	3.7×10^{-4}	2.6×10^4
			3/2–3/2	3.7×10^{-5}	0.51×10^4
	$4p - 5d$	$^2P^0 - {}^2D$	—	0.28×10^{-2}	
	$3d - 5p$	$^2D - {}^2P^0$	—	0.14	
Ca I	$4s^2 - 4s4p$	$^1S - {}^3P^0$	0–1	—	2.6×10^3
	$- 4s4p$	$^1S - {}^1P^0$	0–1	1.75	2.18×10^8
	$- 4s5p$	$^1S - {}^1P^0$	0–1	0.9×10^{-3}	2.7×10^5
	$4s4p - 4p^2$	$^3P^0 - {}^3P$	—	0.50	—
		$^1P^0 - {}^1D$	1–2	0.57	0.66×10^8
		$^1P^0 - {}^1S$	1–0	0.17	1.1×10^8
	$4s4p - 4s4d$	$^3P^0 - {}^3D$	—	0.42	—
	$- 4s5d$	$^3P^0 - {}^3D$	—	0.12	—
	$4s4p - 4s5d$	$^1P^0 - {}^1D$	1–2	0.27	4.0×10^8
	$4s4p - 4s5s$	$^3P^0 - {}^3S$	2–1	0.121	3.54×10^7
			1–1	0.130	2.31×10^7
			0–1	0.129	0.77×10^7
	$4s3d - 4s5p$	$^3D - {}^3P^0$	—	0.076	—
	$4s3d - 4s4f$	$^3D - {}^3F^0$	—	0.099	—
		$^1D - {}^1F^0$	2–3	0.094	1.88×10^7
	$4s3d - 4s5f$	$^3D - {}^3F^0$	—	0.046	—
		$^1D - {}^1F^0$	2–3	0.074	1.9×10^7

9.7 Tables of Oscillator Strengths and Radiative Transition Probabilities

Table 9.9 Photorecombination cross section; function $\Phi(u)$

Element	Level	u 0.025	0.05	0.1	0.2	0.4	0.8	1.13	1.6	2.26	3.2
H I	$1s$	1.90	1.90	1.89	1.86	1.72	1.33	1.01	0.66	0.38	0.19
	$2s$	1.11	1.11	1.11	1.11	1.10	1.03	0.93	0.77	0.56	0.36
	$2p$	2.94	2.94	2.91	2.84	2.52	1.68	1.09	0.58	0.24	0.084
He I	$1s$	0.95	0.95	0.95	0.93	0.86	0.66	0.51	0.33	0.19	0.10
Li I	$2s$	0.16	0.16	0.16	0.18	0.24	0.45	0.63	0.82	0.87	0.81
	$2p$	3.38	3.38	3.35	3.29	2.99	2.12	1.50	0.92	0.49	0.23
Be I	$2s^2$ ^1S	6.22	6.15	5.87	4.92	2.91	1.22	0.94	1.07	1.11	1.14
	$2p$ ^3P	6.50	6.43	6.12	5.19	3.06	1.05	0.52	0.20	0.013	0.006
	$2p$ ^1P	1.44	1.41	1.31	1.00	0.37	0.004	0.27	1.17	0.99	0.51
Be II	$2s$	0.16	0.16	0.16	0.17	0.18	0.22	0.24	0.25	0.22	0.17
	$2p$	0.86	0.86	0.85	0.84	0.76	0.54	0.38	0.23	0.12	0.052
Na I	$3s$	0.20×10^{-2}	0.19×10^{-2}	0.16×10^{-2}	0.05×10^{-2}	0.14×10^{-2}	3.9×10^{-2}	0.10	0.19	0.29	0.40
	$3p$	1.56	1.56	1.54	1.46	1.17	0.21	0.17	0.20×10^{-2}	0.11	0.52
	$3d$	3.13	3.11	3.09	2.99	2.58	1.56	0.93	0.45	0.19	0.089
K I	$4s$	0.14×10^{-3}	0.12×10^{-3}	0.56×10^{-4}	0.34×10^{-4}	0.28×10^{-2}	3.1×10^{-2}	0.072	0.13	0.20	0.30

9. Radiative Transitions

Table 9.10 Photorecombination rate $\langle v\sigma^r \rangle$; function $G(\beta)$.

Element	Level	β	$\frac{1}{8}$	$\frac{1}{4}$	$\frac{1}{2}$	1	2	4	8	A	χ
	$\langle v\sum_a \sigma^r(a)\rangle$ [a]		0.64	1.29	3.52	4.71	8.45	14.6	24.6	8.53	0.59
H I	$1s$		0.40	0.81	1.55	2.78	4.61	7.25	10.9	3.92	0.35
	$2s$		0.22	0.40	0.68	1.07	1.62	2.36	3.39	2.42	0.12
	$2p$		0.20	0.45	0.95	1.85	3.28	5.39	8.36	6.22	0.61
He I	$1s^2$		0.20	0.40	0.78	1.38	2.30	3.62	5.45	1.96	0.35
	$\langle v\sum_a \sigma^r(a)\rangle$		0.92	1.62	2.89	5.21	9.45	17.03	30.13	15.5	0.48
Li I	$2s$		0.36	0.48	0.59	0.66	0.71	0.76	0.86	—	—
	$2p$		0.33	0.69	1.37	2.50	4.26	6.83	10.4	7.38	0.43
	$2s^2$		0.65	1.12	1.83	3.17	5.87	10.9	19.7	7.38	0.42
Be I	$2p^3P^o$		0.21	0.52	0.21	2.64	5.40	10.3	18.6	10.7	1.63
	$2p^1P^o$		0.26	0.42	0.55	0.57	0.77	1.53	3.3	1.54	0.022
	$\langle v\sum_a \sigma^r(a)\rangle$		0.60	0.87	1.47	2.86	5.81	11.6	22.2	10.8	0.58
Na I	$3s$		0.136	0.137	0.121	0.093	0.059	0.029	0.011	—	—
	$3p$		0.21	0.24	0.40	0.78	1.50	2.66	4.34	2.83	0.26
	$3d$		0.12	0.27	0.59	1.18	2.17	3.67	5.80	6.53	0.75

[a] Summing with respect to a means summing over all atomic levels.

Table 9.11 Values of parameter a in (9.349)

Ion	H	He	Li	Be	B	C	N	O	F	Ne
a	0	1.8	0.4	0.4	1.0	1.4	1.7	1.7	1.8	1.7

Chapter 10 Relativistic Corrections in the Spectroscopy of Multicharged Ions

In the theory of atomic spectra for neutral atoms and ions with small charges the need to take relativistic effects into account very seldom arises, the effects can be included as small corrections. The situation is different for multicharged ions because relativistic effects very strongly depend on the ion charge. We give below the basic information on the Dirac equation for an electron in a Coulomb field necessary for relativistic theory of atomic spectra.

10.1 Dirac Equation. Pauli Equation

10.1.1 Dirac Equation

In the relativistic theory the stationary states of an electron in an arbitrary electromagnetic field, described by the potentials φ, A, are determined by the Dirac equation

$$[E + e\varphi - \beta E_0 - \alpha(cp + eA)]U = 0 \ . \tag{10.1}$$

In this equation $E_0 = mc^2$ is the rest-mass energy, $p = -i\hbar\nabla$ the momentum operator, and $\alpha_x, \alpha_y, \alpha_z$ and β the matrices

$$\alpha_x = \begin{pmatrix} 0 & 0 & 0 & 1 \\ 0 & 0 & 1 & 0 \\ 0 & 1 & 0 & 0 \\ 1 & 0 & 0 & 0 \end{pmatrix} \ ; \quad \alpha_y = \begin{pmatrix} 0 & 0 & 0 & -i \\ 0 & 0 & i & 0 \\ 0 & -i & 0 & 0 \\ i & 0 & 0 & 0 \end{pmatrix} \ ;$$

$$\alpha_z = \begin{pmatrix} 0 & 0 & 1 & 0 \\ 0 & 0 & 0 & -1 \\ 1 & 0 & 0 & 0 \\ 0 & -1 & 0 & 0 \end{pmatrix} \ ; \quad \beta = \begin{pmatrix} 1 & 0 & 0 & 0 \\ 0 & 1 & 0 & 0 \\ 0 & 0 & -1 & 0 \\ 0 & 0 & 0 & -1 \end{pmatrix} \ . \tag{10.2}$$

The terms E and $e\varphi$ in curly brackets in (10.1), not containing α and β, are assumed to be multiplied by the unit matrix I. The wave function U satisfying (10.1) is also the four-row matrix

$$U = \begin{pmatrix} U_1 \\ U_2 \\ U_3 \\ U_4 \end{pmatrix} \ . \tag{10.3}$$

The usual law of matrix multiplication is assumed in (10.1), for example,

$$(\beta U)_i = \sum_{k=1}^{4} \beta_{ik} U_k \; ; \quad U^*U = \sum_{k=1}^{4} U_k^* U_k \; . \tag{10.4}$$

Thus the state of an electron in the relativistic theory is described by the four functions U_1, U_2, U_3 and U_4 – the components of the wave function U. Equation (10.1) is a system of four equations involving these functions. The probability that an electron is in the element of volume dr is

$$dr \sum_{k=1}^{4} U_k^* U_k \; . \tag{10.5}$$

All other relations of the non-relativistic theory, in particular the perturbation theory formula, are generalized in a similar way. In addition to integrating over coordinates, as in the Schrödinger theory, one must sum over the components of U. Thus the matrix element of some operator H' is defined by

$$\langle U^* | H' | V \rangle = \sum_{i,k=1}^{4} \int U_i^* H'_{ik} V_k \, dr \; . \tag{10.6}$$

The operator H' is constructed with the aid of the Dirac matrices $\boldsymbol{\alpha}$, β and the unit matrix I. Thus, for example

$$I e \varphi = \begin{pmatrix} 1 & 0 & 0 & 0 \\ 0 & 1 & 0 & 0 \\ 0 & 0 & 1 & 0 \\ 0 & 0 & 0 & 1 \end{pmatrix} e\varphi \; ; \quad \alpha_z e A_z = \begin{pmatrix} 0 & 0 & 1 & 0 \\ 0 & 0 & 0 & -1 \\ 1 & 0 & 0 & 0 \\ 0 & -1 & 0 & 0 \end{pmatrix} e A_z \tag{10.7}$$

are such operators.

Equation (10.1) can also be written in a somewhat different form. We shall express the matrices α_x, α_y, α_z in terms of the Pauli two-row matrices

$$\sigma_x = \begin{pmatrix} 0 & 1 \\ 1 & 0 \end{pmatrix} \; ; \quad \sigma_y = \begin{pmatrix} 0 & -i \\ i & 0 \end{pmatrix} \; ; \quad \sigma_z = \begin{pmatrix} 1 & 0 \\ 0 & -1 \end{pmatrix} \tag{10.8}$$

and the matrix β in terms of the two-row unit matrix, which we shall denote, just like the four-row unit matrix, by I

$$\boldsymbol{\alpha} = \begin{pmatrix} 0 & \sigma \\ \sigma & 0 \end{pmatrix} \; ; \quad \beta = \begin{pmatrix} I & 0 \\ 0 & -I \end{pmatrix} \; . \tag{10.9}$$

We shall also introduce the two-component wave functions

$$\psi = \begin{pmatrix} U_1 \\ U_2 \end{pmatrix} ; \quad \chi = \begin{pmatrix} U_3 \\ U_4 \end{pmatrix} ; \quad U = \begin{pmatrix} \psi \\ \chi \end{pmatrix} . \tag{10.10}$$

Substituting (10.9, 10) into (10.1), we obtain a system of equations involving the two-component functions ψ, χ

$$(E + e\varphi - E_0)\psi + \sigma(cp + eA)\chi = 0$$
$$(E + e\varphi + E_0)\chi + \sigma(cp + eA)\psi = 0 . \tag{10.11}$$

Note that α and σ are not vectors in the ordinary sense, since $\alpha_x, \alpha_y, \alpha_z; \sigma_x, \sigma_y, \sigma_z$ do not depend on the choice of coordinate system. The designation of the operator $\alpha_x p_x + \alpha_y p_y + \alpha_z p_z$ by αp (and the similar designation of other operators of the same type) is only a convenient form of notation.

From the definition of the matrices there follows the identity

$$(\sigma G)(\sigma F) = GF + i\sigma[G, F] , \tag{10.12}$$

where G and F are arbitrary vector operators which commute with σ. In particular, when $G = F$

$$(\sigma F)(\sigma F) = F^2 . \tag{10.13}$$

10.1.2 Electron Spin

For convenience of interpretation, we shall transform (10.1) into a differential equation of second order. Operating on (10.1) with

$$[E + e\varphi + \beta E_0 + \alpha(cp + eA)]$$

and using (10.13, 12) and also the commutation relations for the matrices $\alpha_x = \alpha_1$, $\alpha_y = \alpha_2$, $\alpha_z = \alpha_3$, and $\beta = \alpha_4$,

$$\alpha_i \alpha_k + \alpha_k \alpha_i = 2\delta_{ik} , \tag{10.14}$$

it is not difficult to obtain

$$\left[E + e\varphi - E_0 - \frac{1}{2m}\left(p + \frac{e}{c}A\right)^2 + \frac{1}{2mc^2}(E + e\varphi - E_0)^2 - \frac{e\hbar}{2mc}\Sigma H \right.$$
$$\left. + i\frac{\hbar e}{2mc}\alpha E \right] U = 0 , \tag{10.15}$$

where E and H denote the strengths of the electric and magnetic fields

$$E = -\nabla\varphi \; ; \quad H = \text{rot } A$$

and

$$\Sigma = \begin{pmatrix} \sigma & 0 \\ 0 & \sigma \end{pmatrix} . \tag{10.16}$$

Compare (10.15) with the Schrödinger equation corresponding to the relativistic Hamiltonian

$$H = -e\varphi + \sqrt{c^2\left(p+\frac{e}{c}A\right)^2 + m^2c^4} . \tag{10.17}$$

Expanding the root in (10.17) in a power series in $\frac{v}{c}$, we have

$$\left[\left(E+e\varphi - \frac{p^2}{2m}\right) - \frac{1}{2m}\left(p+\frac{e}{c}A\right)^2 + \frac{1}{8m^3c^2}\left(p+\frac{e}{c}A\right)^4\right]\psi = 0 . \tag{10.18}$$

When $\frac{v}{c} \to 0$, (10.18) turns into the ordinary nonrelativistic Schrödinger equation

$$\left(W + e\varphi - \frac{p^2}{2m}\right)\psi = 0 , \quad E - mc^2 = W . \tag{10.19}$$

In the approximation (10.18)

$$c^2\left(p+\frac{e}{c}A\right)^2 = (E+e\varphi)^2 - m^2c^4 = (E+e\varphi-mc^2)(E+e\varphi+mc^2)$$

$$\approx (E+e\varphi-mc^2)2mc^2$$

$$\times \frac{1}{8m^3c^2}\left(p+\frac{e}{c}A\right)^4 \approx \frac{1}{2mc^2}(E+e\varphi-mc^2)^2 ,$$

thus the first three terms of (10.15) are contained in the relativistic Schrödinger equation (10.18). The last two terms

$$-\frac{e\hbar}{2mc}\Sigma H , \quad i\frac{e\hbar}{2mc}\alpha E \tag{10.20}$$

are characteristic of the Dirac theory. Only these terms contain the matrices Σ and α. The first of these can be interpreted as the interaction of the magnetic moment μ

$$\mu = -\frac{e\hbar}{2mc}\Sigma = -2\mu_0 \frac{1}{2}\Sigma \tag{10.21}$$

with the magnetic field, and the second as the interaction of the electric moment $i\frac{e\hbar}{2mc}\alpha$ with the electric field.

Consider in somewhat more detail the first of the terms in (10.20) for whcih we introduce the matrices s_x, s_y, s_z, defined by

$$s = \frac{1}{2}\Sigma = \frac{1}{2}\begin{pmatrix} \sigma & 0 \\ 0 & \sigma \end{pmatrix} . \tag{10.22}$$

These matrices satisfy the commutation relations

$$s_x s_y - s_y s_x = is_z$$
$$s_y s_z - s_z s_y = is_x \tag{10.23}$$
$$s_z s_x - s_x s_z = is_y$$

which are the same as the commutation relations for the components of the angular momentum. It can be shown, in addition, that upon rotating the coordinate system through an angle $\delta\theta$ around an axis directed along the unit vector n, the wave function $U(0)$ (the particle is at the origin) is transformed according to the law

$$U(0) = (1 + i\delta\theta n s) U'(0) . \tag{10.24}$$

For a system with angular momentum J the operator of an infinitesimal rotation is

$$1 + i\delta\theta n J .$$

Since the orbital angular momentum in the case under consideration is zero, it follows from (10.24) that the matrices $s = \frac{1}{2}\Sigma$ are the operators of the intrinsic electron-spin angular momentum. Substituting in (10.24) the two-component functions ψ and χ, we obtain

$$\psi = \left(1 + \frac{i}{2}\delta\theta n \sigma\right)\psi'$$

$$\chi = \left(1 + \frac{i}{2}\delta\theta n \sigma\right)\chi' . \tag{10.25}$$

308   10. Relativistic Corrections in the Spectroscopy of Multicharged Ions

Thus the components U_1 and U_2 of the function ψ transform upon rotation of the coordinate system into each other without affecting the components U_3 and U_4 of the function χ. The latter in turn transform into each other independently of the components U_1 and U_2. A two-component function which transforms upon rotation of the coordinate system in accordance with (10.25) is called a spinor. The wave function U which is the set of the two spinors ψ and χ is called a bispinor.

Comparison of (10.21) and (10.22) shows that the ratio of the magnetic moment of an electron to its spin angular momentum is $-2\mu_0$, i.e., it is twice as large as the ordinary value for orbital momentum.

10.1.3 Non-Relativistic Approximation. Pauli Equation

In a weak field $|e\varphi| \ll mc^2$ there exist stationary states in which $v \ll c$. Here the total energy E is close to the rest energy E_0, therefore

$$(E + e\varphi - E_0) \sim mv^2 \ll mc^2$$

$$(E + e\varphi + E_0) \sim 2mc^2$$

$$\sigma(cp + eA) \sim mvc \ll mc^2$$

and from the second equation of (10.11) it follows that

$$\chi \approx \frac{1}{2mc^2} \sigma \left(p + \frac{e}{c} A \right) \psi \sim \frac{v}{c} \psi \; . \tag{10.26}$$

Thus when $v \ll c$ the components U_3 and U_4 are small in comparison with U_1 and U_2. This enables us to obtain an approximate equation involving large components U_1 and U_2 alone. It is simplest to do this by proceeding from (10.15).

Substituting (10.10) into (10.15), and denoting the energy of the electron less its rest mass, $E - E_0$, by W, we obtain

$$\left[W + e\varphi - \frac{1}{2m} \left(p + \frac{e}{c} A \right)^2 + \frac{1}{2mc^2} (W + e\varphi)^2 - \frac{e\hbar}{2mc} \sigma H \right] \psi$$

$$+ i \frac{e\hbar}{2mc} \sigma E \chi = 0 \tag{10.27}$$

(we do not write out the second equation relating the functions ψ and χ). The term $i \frac{e\hbar}{2mc} \sigma E \chi$ is of order of magnitude

$$\frac{e}{mc} i\hbar \nabla \varphi \cdot \chi \approx \frac{P}{mc} e\varphi \frac{v}{c} \psi \approx e\varphi \left(\frac{v}{c} \right)^2 \psi \; .$$

Thus to first order in $\frac{v}{c}$ we have

$$\left[W + e\varphi - \frac{1}{2m}\left(p + \frac{e}{c}A\right)^2 - \mu_0 \sigma H\right]\psi = 0 . \tag{10.28}$$

This equation is called the Pauli equation. It is the fundamental equation of the non-relativistic theory. It differs from the Schrödinger equation in that (10.28) contains the term $-\mu_0 \sigma H$ due to the electron spin. Thus in the non-relativistic approximation an electron behaves as a particle having intrinsic angular momentum

$$s = \tfrac{1}{2}\sigma \tag{10.29}$$

and an intrinsic magnetic moment $-2\mu_0 s$. The states of motion of the electron are described by the two-component spinor ψ ($\chi \to 0$). The components U_1 and U_2 of the spinor function ψ have a simple physical meaning. Assuming $U_2 = 0$, we have

$$s_z \psi = \frac{1}{2}\begin{pmatrix} 1 & 0 \\ 0 & -1 \end{pmatrix}\begin{pmatrix} U_1 \\ 0 \end{pmatrix} = \frac{1}{2}\begin{pmatrix} U_1 \\ 0 \end{pmatrix} = \frac{1}{2}\psi . \tag{10.30}$$

But if $U_1 = 0$, then

$$s_z \psi = \frac{1}{2}\begin{pmatrix} 1 & 0 \\ 0 & -1 \end{pmatrix}\begin{pmatrix} 0 \\ U_2 \end{pmatrix} = -\frac{1}{2}\begin{pmatrix} 0 \\ U_2 \end{pmatrix} = -\frac{1}{2}\psi . \tag{10.31}$$

In the first case the function ψ describes the state in which the eigenvalue of the operator s_z is $\tfrac{1}{2}$. The quantity

$$\psi^* \psi \, dr = U_1^*(r) U_1(r) dr$$

defines the probability that the electron is in the element of volume dr and the z component of its spin equals $\tfrac{1}{2}$. In the second case the function ψ describes the state in which the z component of the spin is $-\tfrac{1}{2}$.

The functions U_1 and U_2 satisfy the equations

$$\left[W + e\varphi - \frac{1}{2m}\left(p + \frac{e}{c}A\right)^2 - \mu_0 H\right] U_1 = 0$$

$$\left[W + e\varphi - \frac{1}{2m}\left(p + \frac{e}{c}A\right)^2 + \mu_0 H\right] U_2 = 0 . \tag{10.32}$$

In the general case $U_1 \neq 0$, $U_2 \neq 0$, the probability that the electron spin is directed parallel to the z axis is

$$\int U_1^*(r) U_1(r) dr$$

and the probability that the electron spin is directed anti-parallel to the z axis is

$$\int U_2^*(r) U_2(r) dr \; .$$

Thus the index of the spinor components U_1 and U_2 is a fourth variable, determining the spin direction. In contrast to the coordinates of the electron r this variable is discrete and only takes on two values.

With this interpretation, instead of the two-component function ψ, it is possible to describe the states of an electron by the ordinary wave function $\psi(r,\mu)$, depending on r and the additional spin variable μ. It is convenient to choose the magnitude of the z component of the spin as this variable. Thus μ takes on two values $\frac{1}{2}$ and $-\frac{1}{2}$. If the electron is in a state with a specific value $\mu = \mu_0$, then

$$\psi(r,\mu) = \psi(r)\delta(\mu,\mu_0) \; . \tag{10.33}$$

The values of the function $\psi(r,\mu)$ with $\mu = \frac{1}{2}, -\frac{1}{2}$ are obviously related to the components U_1, U_2 by

$$\psi(r,\tfrac{1}{2}) = U_1(r) \; ; \quad \psi(r,-\tfrac{1}{2}) = U_2(r) \; .$$

The spin functions $\delta(\frac{1}{2},\mu)$ and $\delta(-\frac{1}{2},\mu)$ are mutually orthogonal

$$\sum_\mu \delta(\tfrac{1}{2},\mu)\delta(-\tfrac{1}{2},\mu) = \delta(\tfrac{1}{2},\tfrac{1}{2})\delta(-\tfrac{1}{2},\tfrac{1}{2}) + \delta(\tfrac{1}{2},-\tfrac{1}{2})\delta(-\tfrac{1}{2},-\tfrac{1}{2}) = 0 \; ;$$

thus an arbitrary wave function $\psi(r,\mu)$ can be represented in the form of a linear combination of the functions

$$\psi(r)\delta(\tfrac{1}{2},\mu) \; ; \quad \psi(r)\delta(-\tfrac{1}{2},\mu) \; .$$

10.2 Central Field

10.2.1 Non-Relativistic Approximation

Assuming in the Pauli equation $-e\varphi = V(r)$, $A = 0$, $H = 0$, we obtain

$$\left[W - V(r) - \frac{p^2}{2m} \right] \psi = 0 \; . \tag{10.34}$$

This is equivalent to two independent equations for the two components of ψ

$$\left[W - V(r) - \frac{p^2}{2m}\right] U_1 = 0 ,$$

$$\left[W - V(r) - \frac{p^2}{2m}\right] U_2 = 0 . \tag{10.35}$$

In the absence of an external magnetic field, U_1 and U_2 satisfy the same Schrödinger equation. This is due to the fact that the Hamiltonian

$$H = \frac{p^2}{2m} + V(r)$$

does not contain the spin operators σ. Therefore U_1 and U_2 can be obtained by multiplying the solution of (10.35)

$$R_{nl}(r) Y_{lm}(\theta\varphi)$$

by $\delta(\frac{1}{2}, \mu)$ and $\delta(-\frac{1}{2}, \mu)$ respectively:

$$U_1 = R_{nl}(r) Y_{lm}(\theta\varphi) \delta(\tfrac{1}{2}, \mu) ,$$

$$U_2 = R_{nl}(r) Y_{lm}(\theta\varphi) \delta(-\tfrac{1}{2}, \mu) .$$

The general solution of (10.35) has the form

$$\psi = R_{nl}(r) Y_{lm}(\theta\varphi) \begin{pmatrix} C_1 \delta(\tfrac{1}{2}, \mu) \\ C_2 \delta(-\tfrac{1}{2}, \mu) \end{pmatrix} . \tag{10.36}$$

The coefficients C_1 and C_2 are subject to the condition

$$|C_1|^2 + |C_2|^2 = 1 .$$

When $C_1 = 1$ and $C_2 = 0$, (10.36) defines the wave function of the state in which the z components of the orbital angular momentum m_l and spin μ are fixed, with $\mu = \tfrac{1}{2}$

$$\psi_{m_l, \mu = 1/2} = R_{nl}(r) Y_{lm_l}(\theta\varphi) \begin{pmatrix} 1 \\ 0 \end{pmatrix} . \tag{10.37}$$

If $C_1 = 0$ and $C_2 = 1$, then

$$\psi_{m_l, \mu = -1/2} = R_{nl}(r) Y_{lm_l}(\theta\varphi) \begin{pmatrix} 0 \\ 1 \end{pmatrix} . \tag{10.38}$$

10. Relativistic Corrections in the Spectroscopy of Multicharged Ions

In a general form one can write

$$\psi_{m_l,\mu} = R_{nl}(r) Y_{lm_l}(\theta\varphi) q_\mu ,\qquad(10.39)$$

where the q_μ are spin functions which are eigenfunctions of the operator s_z. These functions have the form

$$q_{1/2} = \begin{pmatrix} 1 \\ 0 \end{pmatrix} ; \quad q_{-1/2} = \begin{pmatrix} 0 \\ 1 \end{pmatrix} .\qquad(10.40)$$

The wave functions ψ_{ljm}, eigenfunctions of the operators l^2, s^2, j^2 and j_z (the total angular momentum of the electron is denoted by j) are also particular cases (10.36). Using the general rule for constructing wave functions, upon adding the angular momenta we obtain

$$\psi_{ljm} = R_{nl}(r) \left[C^j_{m-1/2, 1/2} Y_{lm-1/2} \cdot \begin{pmatrix} 1 \\ 0 \end{pmatrix} + C^j_{m+1/2, -1/2} Y_{lm+1/2} \begin{pmatrix} 0 \\ 1 \end{pmatrix} \right]$$

$$\psi_{ljm} = R_{nl}(r) \begin{pmatrix} C^j_{m-1/2, 1/2} Y_{lm-1/2}(\theta,\varphi) \\ C^j_{m+1/2, -1/2} Y_{lm+1/2}(\theta,\varphi) \end{pmatrix} .\qquad(10.41)$$

The Clebsch-Gordan coefficients entering into (10.41) are given in the following formulas:

$$C^j_{m-1/2, 1/2} = \left(l\tfrac{1}{2} m - \tfrac{1}{2}, \tfrac{1}{2} \Big| l\tfrac{1}{2} jm \right) = \begin{cases} \sqrt{\dfrac{l+m+1/2}{2l+1}} & j = l+\tfrac{1}{2} \\[2mm] \sqrt{\dfrac{l-m+1/2}{2l+1}} & j = l-1/2 \end{cases}$$

$$C^j_{m+1/2, -1/2} = \left(l\tfrac{1}{2} m + \tfrac{1}{2}, -\tfrac{1}{2} \Big| l\tfrac{1}{2} jm \right) = \begin{cases} -\sqrt{\dfrac{l-m+1/2}{2l+1}} & j = l+\tfrac{1}{2} \\[2mm] \sqrt{\dfrac{l+m+1/2}{2l+1}} & j = l-1/2 \end{cases}$$

$$(10.42)$$

10.2.2 Second Approximation with Respect to v/c. Fine Splitting

By substituting the function χ from (10.26) into (10.27) and keeping terms of order $\left(\dfrac{v}{c}\right)^2 W$, one can obtain[1]

$$\left\{W+e\varphi-\frac{1}{2m}\left(p+\frac{e}{c}A\right)^2-\frac{e\hbar}{2mc}\sigma H+\frac{1}{8m^3c^2}\left(p+\frac{e}{c}A\right)^4\right.$$

$$\left.-\frac{\mu_0}{2mc}\sigma[E,p]+\frac{\mu_0\hbar}{4mc}\nabla^2\varphi\right\}\psi=0 \ . \tag{10.43}$$

In the case of a central field this equation becomes ($-e\varphi = V$)

$$\left\{W-V(r)-\frac{p^2}{2m}+\frac{p^4}{8m^3c^2}-\frac{\mu_0}{2mc}\sigma[E,p]-\frac{\hbar^2}{8m^2c^2}\nabla^2 V(r)\right\}\psi=0 \ . \tag{10.44}$$

The last three terms in (10.44),

$$\frac{p^4}{8m^3c^2}-\frac{\mu_0}{2mc}\sigma[E,p]-\frac{\hbar^2}{8m^2c^2}\nabla^2 V \ , \tag{10.45}$$

define corrections to the non-relativistic theory of order $\left(\dfrac{v}{c}\right)^2$. The first of these terms takes into account the dependence of the electron mass on velocity. The second term

$$-\frac{\mu_0}{2mc}\sigma[E,p] = -2\mu_0 s\frac{1}{2c}[E,v] \tag{10.46}$$

gives the spin-orbit interaction. Substituting in (10.46)

$$E=-\nabla\varphi=-\frac{\partial\varphi}{\partial r}\cdot\frac{r}{r}$$

and using the definition of the orbital momentum $\hbar l = [r,p]$, we obtain

$$-\frac{\mu_0}{mc}s[E,p]=\frac{\mu_0\hbar}{mc}\frac{\partial\varphi}{\partial r}\frac{1}{r}ls \ . \tag{10.47}$$

---

[1] In the derivation of this equation the normalization condition of the exact theory, $\int(\psi^*\psi+\chi^*\chi)d\tau = 1$ should be used. The term $\chi^*\chi$ is of order $(v/c)^2$ and thus cannot be omitted in the approximation under consideration.

10. Relativistic Corrections in the Spectroscopy of Multicharged Ions

The last term in (10.45) does not have a classical analogue. This term, proportional to ΔV, cannot be interpreted by means of any pictorial representation.

The operator (10.45) commutes with the operators l^2, s^2, j^2, j_z but does not commute with l; thus (10.44) does not have solutions of the type of (10.36) with arbitrary coefficients C_1 and C_2. In particular, stationary states $\psi_{m_l,\mu}$, in which the z components m_l, μ of the orbital angular momentum and electron spin are uniquely defined, are impossible. Only for a well defined choice of these coefficients, such as that for which the function ψ is an eigenfunction of the operators j^2, j_z

$$\psi_{ljm} = R(r) \begin{pmatrix} C^j_{m-1/2,\,1/2}\, Y_{lm-1/2} \\ C^j_{m+1/2,\,-1/2}\, Y_{lm+1/2} \end{pmatrix} \tag{10.48}$$

it is possible to satisfy (10.44). The function (10.48) describes stationary states in which the absolute magnitudes of the angular momenta s, l, j and the z component of the total angular momentum m are given.

Substituting (10.48) into (10.44), it is not difficult to obtain the radial equation for determining $R(r)$. This equation differs from the radial equation in the first approximation (Pauli equation) for the functions $R_{nl}(r)$ by terms of order $\left(\dfrac{v}{c}\right)^2 W$. Thus it is possible to use perturbation theory to determine $R(r)$ and also the corresponding energy levels.

It is easy to see that different energy levels correspond to the states $j = l + \tfrac{1}{2}$ and $j = l - \tfrac{1}{2}$. This follows just from the fact that the radial equations for these states are different. In fact,

$$ls\,\psi_{ljm} = \frac{1}{2}[j^2 - l^2 - s^2]\,\psi_{ljm} = \frac{1}{2}[j(j+1) - l(l+1) - s(s+1)]\,\psi_{ljm}$$

$$= \frac{1}{2} \begin{cases} l\,\psi_{ljm}, & j = l + 1/2 \\ -(l+1)\,\psi_{ljm}, & j = l - 1/2 \end{cases} \tag{10.49}$$

Thus, the level nl splits into the sublevels $j = l \pm \tfrac{1}{2}$. This is called fine splitting. The magnitude of the fine splitting is determined, obviously, by the difference of the corrections $\Delta E_{nlj=l+1/2}$ and $\Delta E_{nlj=l-1/2}$, where

$$\Delta E_{nlj} = \Delta E^{(1)}_{nlj} + \Delta E^{(2)}_{nlj} + \Delta E^{(3)}_{nlj}, \tag{10.50}$$

$$\Delta E^{(1)}_{nlj} = -\frac{1}{8m^3c^2} \int \psi^*_{nljm}\, p^4\, \psi_{nljm}\, d\mathbf{r}, \tag{10.51}$$

$$\Delta E^{(2)}_{nlj} = -\frac{\mu_0 \hbar}{mc} \int \psi^*_{nljm}\, \frac{\partial \varphi}{\partial r}\frac{1}{r}\, ls\, \psi_{nljm}\, d\mathbf{r}, \tag{10.52}$$

$$\Delta E_{nlj}^{(3)} = \frac{\hbar^2}{8m^2c^2} \int \psi_{nljm}^* \nabla^2 V \psi_{nljm} d\mathbf{r} \ . \tag{10.53}$$

Before calculating these corrections, we shall show that $\Delta E_{nlj}^{(3)}$ is non-vanishing only for s-states ($l=0$). In fact, $\Delta E_{nlj}^{(3)}$ is proportional to the matrix element of $\nabla^2 \varphi = -4\pi\varrho$, where ϱ is the density of the charges producing the field. If the field is produced by a nucleus with charge Ze, then $\varrho = Ze\delta(\mathbf{r})$. Therefore

$$\Delta E_{nlj}^{(3)} = \frac{\pi e^2 Z \hbar^2}{2m^2c^2} \int \psi_{nljm}^* \delta(\mathbf{r}) \psi_{nljm} d\mathbf{r} = \frac{\pi e^2 Z \hbar^2}{2m^2c^2} |\psi_{nljm}(0)|^2 \tag{10.54}$$

and $|\psi_{nljm}(0)|^2 \neq 0$ only when $l=0$. Thus in the case $l \neq 0$,

$$\Delta E_{nlj} = \Delta E_{nlj}^{(1)} + \Delta E_{nlj}^{(2)} \ . \tag{10.55}$$

We shall calculate the corrections (10.50) in the case of a Coulomb field $V(r) = -\frac{Ze^2}{r}$ (hydrogen atom and hydrogenlike ions). Calculation of (10.55) has already been carried out [see (1.24)],

$$\Delta E_{nlj} = \Delta E_{nlj}^{(1)} + \Delta E_{nlj}^{(2)} = \alpha^2 \left(\frac{3}{4n} - \frac{1}{j+1/2} \right) \frac{Z^4}{n^3} \text{Ry} \ . \tag{10.56}$$

When $l=0$, (10.54) is added to (10.55) which in this case equals [see (6.52)],

$$\Delta E_{nlj}^{(3)} = \frac{\pi e^2 Z \hbar^2}{2m^2c^2} |\psi_{n0}(0)|^2 = \frac{\pi e^2 Z \hbar^2}{2m^2c^2} \cdot \frac{Z^3}{\pi n^3} \left(\frac{me^2}{\hbar^2} \right)^3 = \alpha^2 \frac{Z^4}{n^3} \text{Ry} \ . \tag{10.57}$$

In addition, (1.23) for $\Delta E_{nlj}^{(2)}$ loses its meaning in this case since both the numerator and the denominator of (1.23) vanish. It is not difficult to remove this indeterminacy. In deriving (10.44, 52) we used the approximate expression

$$\chi \approx \frac{\sigma(c\mathbf{p}+e\mathbf{A})}{2mc^2} \psi \ ,$$

whereas the exact expression has the form

$$\chi = \frac{\sigma(c\mathbf{p}+e\mathbf{A})}{E+E_0-V} \psi \ .$$

If the principal contribution to the integral is given by the range of small values of r, for which the condition $mc^2 \geq \dfrac{Ze^2}{r}$ is not fulfilled, it is not possible to neglect the term $V(r)$ in the denominator. Retaining this term, we obtain

$$\Delta E_{nlj}^{(2)} = -\mu_0 \hbar c \int \psi_{nljm}^* \frac{\partial \varphi}{\partial r} \frac{1}{r} \frac{2ls}{[2mc^2 - V(r)]} \psi_{nljm} \, dr$$

$$= -\frac{\mu_0 \hbar}{mc} \int \psi_{nljm} \frac{\partial \varphi}{\partial r} \frac{1}{r} ls \left(1 + \frac{\alpha^2 Z a_0}{2 r}\right)^{-1} \psi_{nljm} \, dr \, . \qquad (10.58)$$

The radial integral in (10.58) is finite and therefore when $l = 0$, (10.58) vanishes. Consequently, when $l = 0$ we have

$$\Delta E_{nlj}^{(1)} + \Delta E_{nlj}^{(3)} = -\alpha^2 \left(2 - \frac{3}{4n}\right) \frac{Z^4}{n^3} \text{Ry} + \alpha^2 \frac{Z^4}{n^3} \text{Ry} = -\alpha^2 \left(1 - \frac{3}{4n}\right) \frac{Z^4}{n^3} \text{Ry} \, . \qquad (10.59)$$

This expression can be obtained by substituting in (10.56) $l = 0$, $j = \tfrac{1}{2}$. Thus, for all values of l, including $l = 0$, we get

$$\Delta E_{nlj} = \alpha^2 \left(\frac{3}{4n} - \frac{1}{j + 1/2}\right) \frac{Z^4}{n^3} \text{Ry} \, . \qquad (10.60)$$

The essential feature of this expression is the independence of l. Relativistic corrections of the order of $\left(\dfrac{v}{c}\right)^2$ lead to a splitting with respect to j, but do not remove the degeneracy with respect to l which is peculiar to the Coulomb field.

10.2.3 Dirac Equation for a Central Field

In the case of a central field, (10.1) becomes

$$[E - V(r) - \beta E_0 - \boldsymbol{\alpha} c \boldsymbol{p}] U = 0 \, . \qquad (10.61)$$

The Hamiltonian

$$H = \beta E_0 + V(r) + \boldsymbol{\alpha} c \boldsymbol{p} \qquad (10.62)$$

does not commute either with the components of the orbital angular momentum $\boldsymbol{l}$ or with $\boldsymbol{l}^2$. Thus the Dirac equation (10.61) does not have solutions which are eigenfunctions of the operator $\boldsymbol{l}^2$. At the same time the Hamiltonian (10.62) commutes with the operators j^2, j_z and the inversion operator. This indicates the existence of solutions U_{jm} describing stationary states with given values of the square of the total angular momentum j and its z component m. Each such state is also characterized by a specific parity. It is convenient to describe the parity of the state by the index l, as in the non-relativistic theory, which for a given value j takes two values $j+\frac{1}{2}$ and $j-\frac{1}{2}$, one odd and one even. The wave functions U_{ljm} have the form

$$U_{ljm} = \begin{pmatrix} \psi_{ljm} \\ \chi_{ljm} \end{pmatrix} \tag{10.63}$$

$$\psi_{ljm} = g(r) \begin{pmatrix} C^j_{m-1/2,\,1/2}\, Y_{lm-1/2}(\theta,\varphi) \\ C^j_{m+1/2,\,-1/2}\, Y_{lm+1/2}(\theta,\varphi) \end{pmatrix} \tag{10.64}$$

$$\chi_{ljm} = if(r) \begin{pmatrix} C^j_{m-1/2,\,1/2}\, Y_{\tilde{l}m-1/2}(\theta,\varphi) \\ C^j_{m+1/2,\,-1/2}\, Y_{\tilde{l}m+1/2}(\theta,\varphi) \end{pmatrix},$$

where $\tilde{l} = 2j - l$. When $j = l+\frac{1}{2}$, $\tilde{l} = l+1$, and when $j = l-\frac{1}{2}$, $\tilde{l} = l-1$. It is easy to see that the wave function (10.63) is not an eigenfunction of the operator l^2. In fact,

$$l^2 \psi_{ljm} = l(l+1)\psi_{ljm}\,; \quad l^2 \chi_{ljm} = \tilde{l}(\tilde{l}+1)\chi_{ljm}\,.$$

Representing the wave function U_{ljm} by

$$U_{ljm} = \begin{pmatrix} \psi_{ljm} \\ 0 \end{pmatrix} + \begin{pmatrix} 0 \\ \chi_{ljm} \end{pmatrix},$$

we have

$$l^2 U_{ljm} = l(l+1)\begin{pmatrix} \psi_{ljm} \\ 0 \end{pmatrix} + \tilde{l}(\tilde{l}+1)\begin{pmatrix} 0 \\ \chi_{ljm} \end{pmatrix}$$

$$= l(l+1)U_{ljm} + [l(l+1) - \tilde{l}(\tilde{l}+1)]\begin{pmatrix} 0 \\ \chi_{ljm} \end{pmatrix}. \tag{10.65}$$

Therefore at small velocities of the electron, $\frac{v}{c} \ll 1$, the second term in (10.65) is small (approximately $\frac{v}{c}$ times less than the first), and to an accuracy which corresponds to neglecting the small components χ in comparison with ψ we have conservation of the absolute magnitude of the orbital angular momentum. Thus, in the non-relativistic approximation the index l acquires the meaning of orbital angular momentum. But in the general case of relativistic velocities the concept of orbital angular momentum does not have physical meaning. On the other hand, the concept of spin is not connected with any approximation, because the operator $s^2 = \frac{3}{4}$, like any constant, commutes with any operator, including the Hamiltonian (10.62). As regards the z component of the spin, it is conserved only in the non-relativistic approximation.

The radial functions $g(r)$ and $f(r)$ in (10.64) satisfy the system of differential equations

$$\left(\frac{d}{dr} + \frac{\varkappa}{r}\right) rg(r) = \frac{1}{\hbar c}(E + E_0 - V)rf(r)$$

$$\left(\frac{d}{dr} - \frac{\varkappa}{r}\right) rf(r) = -\frac{1}{\hbar c}(E - E_0 - V)rg(r)$$

(10.66)

$$\varkappa = \begin{cases} -(j+1/2) = -l-1, & j = l+1/2 \\ (j+1/2) = l, & j = l-1/2, \end{cases}$$

(10.67)

which can be obtained by substituting (10.64) into (10.61).

$$V(r) \to 0 \quad \text{when} \quad r \to \infty \qquad V(r) \to -\frac{Ze^2}{r} \quad \text{when} \quad r \to 0 \ ;$$

the functions $g(r)$ and $f(r)$ must satisfy the boundary conditions

$$\left. \begin{array}{c} rg(r) \\ rf(r) \end{array} \right\} \to 0 \quad \text{when} \quad r \to 0$$

and when $r \to \infty$ $rg(r)$, $rf(r)$ are finite.

From the normalization condition

$$\int (\psi^* \psi + \chi^* \chi) dr = 1$$

(10.68)

and (10.64) it follows that

$$\int (g^2 + f^2) r^2 dr = 1 \ .$$

(10.69)

The boundary conditions for g, f ensure the existence of this integral.

10.2.4 Coulomb Field. Energy Levels, Fine Splitting

Substituting $V = -\dfrac{Ze^2}{r}$ in (10.66), we have

$$\left(\frac{d}{dr} + \frac{\varkappa}{r}\right) rg(r) = \left[\frac{1}{\hbar c}(E_0 + E) + \alpha \frac{Z}{r}\right] rf(r)$$

$$\left(\frac{d}{dr} - \frac{\varkappa}{r}\right) rf(r) = \left[\frac{1}{\hbar c}(E_0 - E) - \alpha \frac{Z}{r}\right] rg(r) \;, \quad (10.70)$$

where $\alpha = \dfrac{e^2}{\hbar c} \approx \dfrac{1}{137}$ is the fine structure constant. Solutions of the system (10.70) which satisfy the necessary conditions of finiteness exist at the following values of E:

$$E = E_0 \left[1 + \left(\frac{\alpha Z}{n - |\varkappa| + \sqrt{\varkappa^2 - \alpha^2 Z^2}}\right)^2 \right]^{-1/2}, \quad (10.71)$$

where $n = 1, 2, 3, \ldots$.

It will be shown below that in the non-relativistic limit the quantum number n coincides with the principal quantum number. It is convenient to introduce the quantum number

$$K = |\varkappa| = j + \tfrac{1}{2} \quad (10.72)$$

because only $|\varkappa|$ is contained in (10.71). According to (10.67), to the given value of j correspond two values of $l = j - \tfrac{1}{2}$, $l = j + \tfrac{1}{2}$ and the following values of $\varkappa$ and K:

$$l = j - \tfrac{1}{2} \quad \varkappa = -(j + \tfrac{1}{2}) = -l - 1 \;, \quad K = -\varkappa = l + 1 \;,$$

$$l = j + \tfrac{1}{2} \quad \varkappa = j + \tfrac{1}{2} = l \;, \qquad\qquad K = \varkappa = l \;. \quad (10.73)$$

At a given value of n

$$l = 0, 1, \ldots n - 1 \;, \quad K = 1, 2, \ldots n \;. \quad (10.74)$$

Since in the non-relativistic approximation the quantum number l defines the orbital angular momentum of the electron, the state $n, \varkappa, l$ and the energy level nK can be described in the usual spectroscopic notation

	$s_{1/2}$	$p_{1/2}$	$p_{3/2}$	$d_{3/2}$	$d_{5/2}$	$f_{5/2}$	$f_{7/2}$	$g_{7/2}$
$\varkappa =$	-1	1	-2	2	-3	3	-4	4
$K =$	1	1	2	2	3	3	4	4

Thus

$n = 1$, $K = 1$ $\quad 1s_{1/2}$

$n = 2\quad K = 1\quad 2s_{1/2}\ 2p_{1/2}$

$\qquad\quad\ K = 2\qquad\qquad\quad 2p_{3/2}$

$n = 3\quad K = 1\quad 3s_{1/2}\ 3p_{1/2}$

$\qquad\quad\ K = 2\qquad\qquad\quad 3p_{3/2}\ 3d_{3/2}$

$\qquad\quad\ K = 3\qquad\qquad\qquad\qquad\ \ 3d_{5/2}$

$n = 4\quad K = 1\quad 4s_{1/2}\ 4p_{1/2}$

$\qquad\quad\ K = 2\qquad\qquad\quad 4p_{3/2}\ 4d_{3/2}$

$\qquad\quad\ K = 3\qquad\qquad\qquad\qquad\ \ 4d_{5/2}\ 4f_{5/2}$

$\qquad\quad\ K = 4\qquad\qquad\qquad\qquad\qquad\qquad\ \ 4f_{7/2}$.

It is convenient to define the atomic energy level E_{nK} as the difference between the total energy E from (10.71) and the rest energy $E_0 = mc^2$. This difference is equal to ($\alpha^2 mc^2 = 2\,\text{Ry}$)

$$E_{nK} = -\frac{2Z^2}{N(N+n-K+\gamma)}\,\text{Ry}\,, \tag{10.75}$$

where

$$\gamma = \sqrt{K^2 - (\alpha Z)^2}\,, \tag{10.76}$$

$$N = \sqrt{(n+K+\gamma)^2 + (\alpha Z)^2} = \sqrt{n^2 - 2(n-K)(K-\gamma)}\,. \tag{10.77}$$

For light nuclei, $Z \ll 137$, an approximate expression for the energy E_{nK} is obtained by expanding (10.75) in a series in powers of αZ. At $\alpha Z \ll 1$

$$\gamma \simeq K\left[1 - \frac{(\alpha Z)^2}{2K^2}\right] = K - \frac{(\alpha Z)^2}{2K},$$

$$N \simeq n - \frac{n-K}{2nK}(\alpha Z)^2,$$

and

$$E_{nK} = -\left[1 + \frac{(\alpha Z)^2}{n}\left(\frac{1}{K} - \frac{3}{4n}\right)\right]\frac{Z^2}{n^2}\text{Ry}$$

$$= -\left[1 + \frac{(\alpha Z)^2}{n}\left(\frac{1}{j+1/2} - \frac{3}{4n}\right)\right]\frac{Z^2}{n^2}\text{Ry}. \quad (10.78)$$

The terms of order higher than $\alpha^2 Z^2$ are neglected. The first term in (10.78) is the non-relativistic expression for the energy (Balmer formula). The fine splitting of levels is given by the second term. Fine splitting, as has been noted already, does not depend on l. It is significant that degeneracy with respect to l is not connected with the approximate character of (10.78) since (10.71) and (10.75) also depend only on j (on K) and not on l. All levels n, $K(K \neq n)$ are twofold degenerate with respect to l.

The difference between (10.78) and (10.75) becomes substantial for heavy nuclei. In the approximation (10.78) the splitting of the levels $j' = l + \frac{1}{2}$ and $j'' = l - \frac{1}{2}$ is

$$\delta E_{j'j''} = \frac{\alpha^2 Z^4}{n^3 l(l+1)} \text{Ry}.$$

At the same time from the exact formula (10.71) it follows that $\varkappa' = l+1$, $\varkappa'' = -l$

$$E_{nK'} - E_{nK''} = \left[\frac{1}{\sqrt{1 + \frac{\alpha^2 Z^2}{(n-K'+\gamma')^2}}} - \frac{1}{\sqrt{1 + \frac{\alpha^2 Z^2}{(n-K''+\gamma'')^2}}}\right] \cdot E_0.$$

The difference $\gamma - K$ is small in comparison with n, therefore

$$\frac{1}{[n+(\gamma-K)]^2} \simeq \frac{1}{n^2 + 2n(\gamma-K)} \simeq \frac{n^2 - 2n(\gamma-K)}{n^4}.$$

$$E_{nK'} - E_{nK''} \simeq \frac{Z^2}{n^3}(\gamma' - \gamma'' - K' + K'')\alpha^2 E_0 = \frac{Z^2}{n^3}(\gamma' - \gamma'' - 1)\alpha^2 E_0$$

and

$$\frac{E_{nK'} - E_{nK''}}{\delta E_{j'j''}} = \frac{2l(l+1)(\gamma' - \gamma'' - 1)}{\alpha^2 Z^2} = H_r(lZ) \ . \tag{10.79}$$

Thus, the resulting expressions for the splitting of the components $j = l \pm 1/2$ and corresponding splitting constant ζ_{nl} defined in the Sect. 5.5.3 are

$$\Delta E_{j,j-1} = E_{nK'} - E_{nK''} = \delta E_{j'j''} H_r(lZ) = \frac{\alpha^2 Z^4 H_r(lZ)}{n^3 l(l+1)} \text{Ry} \ ,$$

$$\zeta_{nl} = \frac{\alpha^2 Z^4 H_r(lZ)}{n^3 l(l+1)(l+1/2)} \text{Ry} \ . \tag{10.88}$$

The quantity $H_r(lZ)$ is called the relativistic correction to the doublet splitting formula.

The value of $H_r(lZ)$ when $l = 1$ is given in Table 6.3 (see also Fig. 6.1). For small values of Z, $H_r(lZ)$ is nearly unity. Thus for $Z = 1, 2, 10$ the values of $H_r(lZ)$ are respectively 1.0000; 1.0001; 1.0023. With a further increase in Z, H_r increases, reaching values of the order of 1.25 for heavy nuclei.

10.2.5 Coulomb Field. Radial Functions

For the discrete spectrum the radial functions $g_{n\varkappa}$ and $f_{n\varkappa}$ satisfying the system of Eqs. (10.70) and the boundary conditions (10.68) have the form [2]

$$g_{n\varkappa}(r) = -C_0 \sqrt{4n^2 - \alpha^2 Z^2} \cdot e^{-1/2\varrho} \varrho^{\gamma-1} [-(n-K)F(-n+K+1, 2\gamma+1, \varrho)$$
$$+ (N-\varkappa)F(-n+K, 2\gamma+1, \varrho)] \ ,$$

$$f_{n\varkappa}(r) = -C_0 \alpha Z e^{-1/2\varrho} \varrho^{\gamma-1} [(n-K)F(-n+K+1, 2\gamma+1, \varrho)$$
$$+ (N-\varkappa)F(-n+K, 2\gamma+1, \varrho)] \ , \tag{10.81}$$

where

$$\varrho = \frac{2Zr}{Na_0} \ , \quad K = |\varkappa|$$

$$C_0 = \frac{\varkappa}{|\varkappa|} \cdot \frac{1}{n\Gamma(2\gamma+1)} \sqrt{\frac{\Gamma(2\gamma+n-K+1)}{8N(N-\varkappa)(n-K)!}} \cdot \left(\frac{2Z}{Na_0}\right)^{3/2} . \tag{10.82}$$

γ and N are determined by (10.76, 77):

$$\gamma = \sqrt{K^2 - (\alpha Z)^2} \ , \quad N = \sqrt{n^2 - 2(n-K)(K-\gamma)} \ .$$

The functions $F(\alpha,\beta,x)$ are the confluent hypergeometric functions determined by the series

$$F(\alpha,\beta,x) = 1 + \frac{\alpha}{\beta 1!}x + \frac{\alpha(\alpha+1)}{\beta(\beta+1)2!}x^2 + \ldots . \tag{10.83}$$

In the non-relativistic limit $\alpha Z = 0$, $\gamma = K$ and $N = n$ in both cases $\varkappa = K = l$, $\frac{\varkappa}{|\varkappa|} = 1$ and $\varkappa = K = -l-1$, $\frac{\varkappa}{|\varkappa|} = -1$ using the following recurrence relations for the functions $F(\alpha,\beta,x)$:

$$\frac{x}{\beta}F(\alpha+1,\beta+1,x) = F(\alpha+1,\beta,x) - F(\alpha,\beta,x) ,$$

$$\alpha F(\alpha+1,\beta,x) = (\alpha-\beta+1)F(\alpha,\beta,x) + (\beta-1)F(\alpha,\beta-1,x) .$$

It is not difficult to obtain

$$g_{n\varkappa} \to R_{nl}(r) = \frac{1}{(2l+1)!}\sqrt{\frac{(n+l)!}{(n-l-1)!2n}} \cdot \left(\frac{2Z}{na_0}\right)^{3/2}$$

$$\cdot e^{-(Zr)/(na_0)} \left(\frac{2Zr}{na_0}\right)^l F\left(-n+l+1, 2l+2, \frac{2Zr}{na_0}\right) , \tag{10.84}$$

$f_{n\varkappa} \to 0$.

The function $F\left(-n+l+1, 2l+2, \frac{2Zr}{na_0}\right)$ can be expressed in terms of the generalized Laguerre polynomial

$$F\left(-n+l+1, 2l+2, \frac{2Zr}{na_0}\right) = (-1)^{2l+1}\frac{(2l+1)!(n-l-1)!}{[(n+l)!]^2}L_{n+l}^{2l+1}\left(\frac{2Zr}{na_0}\right) .$$

Using this relation it is easy to see that the radial function R_{nl} in (10.84) coincides with the non-relativistic Schrödinger radial function from (1.11).

Note that the constant C_0 in (10.82) differs from the usually given formula (see for example [2]) by the factor $\frac{\varkappa}{|\varkappa|}$, which ensures the same non-relativistic limit $g_{n\varkappa} \to R_{nl}$ for both positive and negative values of $\varkappa$. If this factor is omitted, $g_{n\varkappa} \to R_{nl}$ for $\varkappa > 0$ and $g_{n\varkappa} \to -R_{nl}$ for $\varkappa < 0$. Consider, further, the behavior of the functions $g_{n\varkappa}$ and $f_{n\varkappa}$ at large and small values of r in the case of light nuclei,

$$Z \ll 137 , \quad \alpha Z \ll 1 , \quad N \approx n - \frac{n-K}{2nK}\alpha^2 Z^2 .$$

Comparison of (10.79) and (10.83) shows that in the whole range $\varrho \simeq 1$ and $\varrho > 1$ the difference between the function g_{nx} and R_{nl} is extremely small. The ratio $(g_{nx} - R_{nl})/R_{nl}$ does not exceed $\alpha^2 Z^2$ in order of magnitude. The ratio $\left(\dfrac{f_{nx}}{g_{nx}}\right)^2$ also has the order of $\alpha^2 Z^2$ in this range. For small values of ϱ we have

$$g_{nx} \sim \left(\frac{2Zr}{Na_0}\right)^{\gamma-1} ; \quad f_{nx} \sim \left(\frac{2Zr}{Na_0}\right)^{\gamma-1} ; \quad R_{nl} \sim \left(\frac{2Zr}{na_0}\right)^{l} .$$

It is not difficult to show that for the states $j = l - \tfrac{1}{2}$, $\gamma \simeq |\varkappa| = l$,

$$\frac{g_{nx} - R_{nl}}{R_{nl}} \sim \alpha^2 Z^2 \left(\frac{2Zr}{Na_0}\right)^{-1} .$$

In the range $\alpha Z \ll \dfrac{2Zr}{na_0} \ll 1$ the difference $g_{nx} - R_{nl}$ does not exceed $\alpha Z R_{nl}$. For smaller values of r this difference rapidly increases.

For the states $j = l + \tfrac{1}{2}$, $\gamma \simeq |\varkappa| = |l-1|$ the difference $g_{nx} - R_{nl}$ also increases with decreasing r but more slowly than in the case $j = l - \tfrac{1}{2}$.

When $j = \tfrac{1}{2}$ in both cases $l = 0$, $\varkappa = -1$ ($s_{1/2}$ state) and $l = 1$, $\varkappa = 1$ ($p_{1/2}$ state) the functions g_{nx} and f_{nx} have a singularity at the origin, since

$$\gamma \simeq |\varkappa| \cdot \left(1 - \frac{\alpha^2 Z^2}{2\varkappa^2}\right) = \left(1 - \frac{\alpha^2 Z^2}{2}\right) ; \quad \gamma - 1 \simeq -\frac{\alpha^2 Z^2}{2} .$$

Thus for light nuclei, $Z \ll 137$, the functions g_{nx}^2 and $(g_{nx}^2 + f_{nx}^2)$ practically coincide with R_{nl}^2 everywhere with the exception of the range of very small values of r. For large values of $Z (\alpha Z \gtrsim 0.5)$ the difference becomes more marked.

For many applications it is useful to consider the range of small values of r in more detail. At small r, when in (10.70)

$$\frac{Ze^2}{r} \gg E - E_0 , \quad \frac{Ze^2}{r} \gg E + E_0 - 2mc^2$$

holds, it is possible to neglect the term $E - E_0$ and assume $E + E_0 = 2mc^2$. After that it is easy to obtain

$$f'' + \frac{3}{r} f' + \left[\frac{2Z}{a_0 r} + (1 - \gamma^2)\frac{1}{r^2}\right] f = 0 , \qquad (10.85)$$

$$g = -\frac{1}{\alpha Z}(rf)' + \frac{\varkappa}{\alpha Z} f . \qquad (10.86)$$

The solution of (10.85) which satisfies the boundary conditions (10.68) is

$$f(r) = \text{Const } r^{-1} J_{2\gamma}\left(\sqrt{\frac{8Zr}{a_0}}\right), \qquad (10.87)$$

where $J_{2\gamma}$ is a Bessel function of the first kind. Using the well-known formula for differentiation of Bessel functions

$$\frac{dJ_p}{dx} = \frac{p}{x} J_p - J_{p+1} \qquad (10.88)$$

and denoting the constant in (10.86) by $C\alpha Z$, we obtain

$$rg(r) = C\left[-(\gamma+\varkappa)J_{2\gamma}\left(\sqrt{\frac{8Zr}{a_0}}\right) + \frac{1}{2}\sqrt{\frac{8Zr}{a_0}} J_{2\gamma+1}\left(\sqrt{\frac{8Zr}{a_0}}\right)\right] \qquad (10.89)$$

$$rf(r) = C\alpha Z J_{2\gamma}\sqrt{\frac{8Zr}{a_0}}. \qquad (10.90)$$

The constant C is defined in the Sect. 10.3.1.

10.3 Relativistic Corrections

10.3.1 Calculation of Some Radial Integrals

In various applications, for example in the calculation of hyperfine splitting constants, the integrals

$$\int (g^2 + f^2) r^{-p} r^2 dr$$

$$\int g f r^{-p} r^2 dr \qquad (10.91)$$

are encountered. When $p \geq 2$ the principal contribution to these integrals comes from the region of low values of r. This allows, when calculating (10.91), the use of the approximate expressions (10.89) and (10.90) for the functions g and f.

By means of these functions the integrals (10.91) can be calculated in explicit form [35]:

$$\int (g^2 + f^2) r^{-p} r^2 dr = \int (g^2 + f^2) r^{-q+1} dr$$

$$= C^2 \left(\frac{2Z}{a_0}\right)^q \frac{(2q-2)!}{q!(q-1)!} \cdot \frac{[q(-2\varkappa+q)(-2\varkappa+q-1) + 4\alpha^2 Z^2(q-3)]}{(2\gamma+q)(2\gamma+q-1)\ldots(2\gamma-q)},$$

$$\qquad (10.92)$$

$$\int gfr^{-p}r^2\,dr = \int gfr^{-q+1}\,dr$$

$$= C^2 \frac{1}{2}\alpha a_0 \left(\frac{2Z}{a_0}\right)^{q+1} \frac{(2q-1)!}{q!(q-1)!} \cdot \frac{(2\varkappa - q)}{(2\gamma+q)(2\gamma+q-1)\ldots(2\gamma-q)} \,. \quad (10.93)$$

Formulas (10.92, 93) apply when $p \geq 1$ ($q \geq 2$).
In the non-relativistic approximation $\alpha Z = 0$, $\varkappa = l$, $\gamma = l$,

$$\langle r^{-p}\rangle \to C^2 \left(\frac{2Z}{a_0}\right)^{p-1} \frac{(2p-4)!}{(p-1)!(p-2)!} \cdot \frac{(p-1)}{(2l+p-1)(2l+p-2)\ldots(2l-p+3)} \,. \quad (10.94)$$

For $p = 3$,

$$\langle r^{-3}\rangle = C^2 \left(\frac{Z}{a_0}\right)^2 \frac{1}{(l+1)(l+1/2)l} \,.$$

Comparing this expression with $\langle r^{-3}\rangle$ from (1.14) we have

$$C^2 = \frac{Z}{a_0 n^3} \,. \quad (10.95)$$

The constant C^2 is best evaluated in terms of the doublet splitting of the levels $j = l + \frac{1}{2}$ and $j = l - \frac{1}{2}$ (10.80) or doublet splitting constant ζ_{nl} (5.149)

$$C^2 = \frac{l(l+1)}{\alpha^2 Z^3 H_r(lZ) a_0} \left(\frac{\Delta E_{j,j-1}}{\text{Ry}}\right) = \frac{l(l+1)(l+1/2)}{\alpha^2 Z^3 H_r(lZ) a_0} \left(\frac{\zeta_{nl}}{\text{Ry}}\right) \,. \quad (10.96)$$

This definition can also be used for nonhydrogenlike atoms because in the region of small r, which gives the main contribution to the $\Delta E_{j,j-1}$, the atomic field can be approximated by a Coulomb potential. Determination of the constant C^2 from the experimental value of the doublet splitting $\Delta E_{j,j-1}$ gives good results, especially for atoms and ions with one electron outside closed shells.

10.3.2 Hyperfine Splitting Constant A

It follows from (10.1) that the interaction of an electron with a magnetic field is defined by the expression

$$H' = e\alpha A = eA \begin{pmatrix} 0 & \sigma \\ \sigma & 0 \end{pmatrix}$$

where A is the vector-potential of the field. If the field is produced by a magnetic dipole moment μ, then

$$A = -ir^{-2}L \sum_{q=-1}^{1} C_{1q}(\theta,\varphi)\mu_q^* \, , \tag{10.97}$$

where $C_{1q} = \sqrt{\dfrac{4\pi}{3}} Y_{1q}(\theta,\varphi)$, $L = -i[r,\nabla]$ is the angular momentum operator, and μ_q are the spherical components of the vector μ. We shall introduce the notation

$$-ier^{-2}aLC_{1q}(\theta,\varphi) = T_q \, .$$

Then

$$H' = \sum_q T_q \mu_q^* \, . \tag{10.98}$$

Expression (10.98) is the scalar product of irreducible tensor operators of rank one; thus one can use the general formulas of Chap. 4 to calculate the matrix elements H'.

In the representation $\gamma j IFM$ (I is the spin of the nucleus and F the total angular momentum of the atom) the matrix element H' has the form

$$\langle \gamma j IFM | H' | \gamma j IFM \rangle = (-1)^{j+I+F} (\gamma j || T || \gamma j) \cdot (\gamma I || \mu || \gamma I) \begin{Bmatrix} j & I & F \\ I & j & 1 \end{Bmatrix} \tag{10.99}$$

$$(\gamma I || \mu || \gamma I) = g_I \sqrt{I(I+1)(2I+1)} \left(\frac{m}{m_p}\right) \mu_0 \; \langle \gamma j IFM | H' | \gamma j IFM \rangle$$

$$= -g_I \left(\frac{m}{m_p}\right) \mu_0 \frac{(\gamma j || T || \gamma j)}{\sqrt{2j(2j+1)(2j+2)}} [j(j+1) + I(I+1) - F(F+1)] \, . \tag{10.100}$$

Comparing this expression with (6.23), we find

$$A = g_I \left(\frac{m}{m_p}\right) \mu_0 \frac{(\gamma j || T || \gamma j)}{\sqrt{j(j+1)(2j+1)}} \, . \tag{10.101}$$

To determine the reduced matrix element of T it is sufficient to calculate the matrix element $\langle \gamma jm | T_0 | \gamma jm \rangle$ when $m = j$, since

$$\langle \gamma jj | T_0 | \gamma jj \rangle = \frac{(\gamma j || T || \gamma j)}{\sqrt{j(j+1)(2j+1)}} \, .$$

From (10.98) it follows that

$$\langle \gamma jm|T_0|\gamma jm\rangle = -ie\int U_{jm}^* r^{-2}(LC_{10})\begin{pmatrix}0 & \sigma \\ \sigma & 0\end{pmatrix} U_{jm}\,d\tau$$

$$= -ie[\int \psi_{jm}^* r^{-2}(LC_{10})\sigma\chi_{jm}\,d\tau + \int \chi_{jm}^* r^{-2}(LC_{10})\sigma\psi_{jm}\,d\tau]\,,\tag{10.102}$$

where the wave functions ψ_{jm} and χ_{jm} are determined by (10.64).

By taking into account that

$$r^{-2}(LC_{10})\sigma\chi = (L\sigma r^{-2}C_{10})\chi = (L\sigma r^{-2}C_{10}\chi) - r^{-2}C_{10}(L\sigma\chi)\,,$$

and also the Hermitian character of the operators

$$\int \psi_{jm}^*(L\sigma r^{-2}C_{10}\chi_{jm})\,d\tau = \int (L\sigma\psi_{jm})^* r^{-2}C_{10}\chi_{jm}\,d\tau\,,$$

instead of the first of the integrals on the right-hand side of (10.102) we have

$$\int (L\sigma\psi_{jm})^* r^{-2}C_{10}\chi_{jm}\,d\tau - \int \psi_{jm}^* r^{-2}C_{10}(L\sigma\chi_{jm})\,d\tau\,.$$

The second integral in (10.102) can be transformed in a similar way. Further, from the definition of the functions ψ_{jm} and χ_{jm} in (10.64), it follows that

$$\begin{aligned}\sigma L\psi_{jm} &= 2sL\psi_{jm} = [j(j+1)-l(l+1)-\tfrac{3}{4}]\psi_{jm} = -(\varkappa+1)\psi_{jm}\\ \sigma L\chi_{jm} &= 2sL\chi_{jm} = [j(j+1)-\tilde{l}(\tilde{l}+1)-\tfrac{3}{4}]\chi_{jm} = -(\varkappa-1)\chi_{jm}\,.\end{aligned}\tag{10.103}$$

Thus,

$$\langle \gamma jm|T_0|\gamma jm\rangle = ie2\varkappa(\int \psi_{jm}^* r^{-2}C_{10}\chi_{jm}\,d\tau + \int \chi_{jm}^* r^{-2}C_{10}\psi_{jm}\,d\tau)\,.\tag{10.104}$$

The functions ψ_{jm} and χ_{jm} are eigenfunctions of the operators j^2, s^2 and l^2; these operators have the eigenvalues $j(j+1)$, $\tfrac{3}{4}$, and $l(l+1)$ in the state ψ_{jm} and $j(j+1)$, $\tfrac{3}{4}$, and $\tilde{l}(\tilde{l}+1)$ in the state χ_{jm}. Therefore, by performing the integration over angular variables we obtain

$$\langle \gamma jm|T_0|\gamma jm\rangle = -4e\varkappa\langle sljm|C_{10}|s\tilde{l}jm\rangle \int gfr^{-2}r^2\,dr\,.\tag{10.105}$$

Thus,

$$A = -4eg_I\left(\frac{m}{m_p}\right)\mu_0\frac{\varkappa(slj\|C_1\|slj)}{\sqrt{j(j+1)(2j+1)}}\int gfr^{-2}r^2\,dr\,.\tag{10.106}$$

For the reduced matrix element of C_1 we have

$$\left(\tfrac{1}{2}lj\|C_1\|\tfrac{1}{2}\tilde{l}j\right) = -\frac{1}{2}\sqrt{\frac{2j+1}{j(j+1)}}\,.$$

10.3 Relativistic Corrections

Therefore

$$A = eg_I\left(\frac{m}{m_p}\right)\mu_0\frac{2\varkappa}{j(j+1)}\cdot\int gfr^{-2}r^2dr \ . \tag{10.107}$$

We use for the integral in (10.107) the approximate formula (10.93). Then

$$A = C^2a_0\alpha^2Z^2g_I\left(\frac{m}{m_p}\right)\frac{2\varkappa(2\varkappa-1)}{j(j+1)(4\gamma^2-1)\gamma}\text{Ry} \ . \tag{10.108}$$

Substituting $\varkappa = l$, $\gamma = l$, and also $C^2 = \dfrac{Z}{n^3 a_0}$, [see (10.94)] we obtain

$$A = \alpha^2 g_I\left(\frac{m}{m_p}\right)\frac{Z^3}{n^3 j(j+1)(l+1/2)}\text{Ry} \ , \tag{10.109}$$

which coincides with (6.53).
If the notation

$$F_r(jZ) = \frac{2j(j+1)(2j+1)}{\gamma(4\gamma^2-1)} \tag{10.110}$$

is introduced, (10.108) can be rewritten as

$$\int gfr^{-2}r^2 dr = C^2\frac{\hbar}{2mc}\left(\frac{2Z}{a_0}\right)^2\frac{(2\varkappa-1)F_r(jZ)}{4j(j+1)(2j+1)} \ , \tag{10.111}$$

and

$$A = \alpha^2 g_I\left(\frac{m}{m_p}\right)\frac{Z^3 F_r(jZ)}{n^3 j(j+1)(l+1/2)}\text{Ry} \ . \tag{10.112}$$

The factor $F_r(jZ)$ is called the relativistic correction to the hyperfine structure constant A. When $\alpha Z = 0$, $F_r = 1$. Values of $F_r(jZ)$ for the values $j = \frac{1}{2}, \frac{3}{2}$ are given in Table 6.3 (see also Fig. 6.1).

If (10.95) is used to determine the constant C, then when $l \neq 0$

$$A = g_I\left(\frac{m}{m_p}\right)\frac{l(l+1)}{j(j+1)(l+1/2)Z}\cdot\frac{F_r}{H_r}\delta E_{j,j-1} \ , \tag{10.113}$$

or

$$A = g_I\frac{1}{Z}\left(\frac{m}{m_p}\right)\frac{(l+1)F_r}{j(j+1)H_r}\zeta_l \ , \tag{10.114}$$

where ζ_l is the doublet splitting constant from (10.80).

10.3.3 Hyperfine Splitting Constant B

The interaction of an electron with an electric field $H' = -e\varphi$ does not contain the Dirac matrices α and β, so in calculating the constant B it is possible to proceed directly from (6.29, 30). The only difference is that now

$$\langle \gamma ljm|\eta_{20}|\gamma ljm\rangle = \int U_{jm}^* \eta_{20} \begin{pmatrix} I & 0 \\ 0 & I \end{pmatrix} U_{jm} d\tau = \int (\psi_{jm}^* \eta_{20} \psi_{jm} + \chi_{jm}^* \eta_{20} \chi_{jm}) d\tau$$

$$= \langle ljm|C_{20}|ljm\rangle \int g^2 r^{-3} r^2 dr + \langle \tilde{l}jm|C_{20}|\tilde{l}jm\rangle \int f^2 r^{-3} r^2 dr,$$
(10.115)

$$(\gamma j||\eta_2||\gamma j) = (slj||C_2||slj) \int g^2 r^{-3} r^2 dr + (s\tilde{l}j||C_2||s\tilde{l}j) \int f^2 r^{-3} r^2 dr .$$

Since the reduced matrix element $(slj||C_2||slj)$ does not depend on l and

$$(slj||C_2||slj) = (s\tilde{l}j||C_2||s\tilde{l}j) = -\frac{1}{4}\sqrt{\frac{(2j+3)(2j+1)(2j-1)}{j(j+1)}} ,$$

from (1.115) and (6.35) we find

$$B = \frac{3e^2 Q}{16 I(2I-1)j(j+1)} \int (g^2 + f^2) r^{-3} r^2 dr . \tag{10.116}$$

We use for the radial integral in (10.116) the approximate expression (10.92),

$$\int (g^2 + f^2) r^{-3} r^2 dr = C^2 \left(\frac{2Z}{a_0}\right)^2 \frac{2(2\varkappa-2)(2\varkappa-1) + 4\alpha^2 Z^2}{(2\gamma+2)(2\gamma+1)2\gamma(2\gamma-1)(2\gamma-2)} . \tag{10.117}$$

It is convenient to rewrite this formula in the following form:

$$\int (g^2 + f^2) r^{-3} r^2 dr = C^2 \left(\frac{2Z}{a_0}\right)^2 \frac{R_r}{l(2l+1)(2l+2)} , \tag{10.118}$$

$$R_r = \frac{l(2l+1)(l+1)}{\gamma(\gamma^2-1)(4\gamma^2-1)} [3\varkappa(\varkappa-1) - \gamma^2 + 1] . \tag{10.119}$$

In the non-relativistic approximation $\varkappa = l$, $\gamma = l$, $\alpha Z = 0$, $C^2 = \dfrac{Z}{n^3 a_0}$, $R_r = 1$, and (10.116) goes into (6.73). The factor R_r is called the relativistic correction to the hyperfine splitting constant B. Values of R_r for the states $l = 1$ are given in Table 6.3. Substitution of (10.118, 119) into (10.116) gives

$$B = \frac{3Qa_0^{-2}Z^3 R_r}{8n^3 I(2I-1)j(j+1)(l+1)(l+1/2)l} \text{Ry} . \tag{10.120}$$

If the constant C in (10.118) is expressed in terms of the doublet splitting constant ζ_l, according to (10.96), then

$$B = \frac{3Qa_0^{-2} R_r}{8a^2 Z I(2I-1)j(j+1)H_r} \zeta_l . \tag{10.121}$$

10.3.4 Nucleus Finite-Size Correction

Equations (10.71, 75) for the electron energy in a Coulomb field (Dirac energy) are obtained under the assumption of a point nucleus. In the case of a finite-size nucleus with radius r_0 the short-distance behavior of the Coulomb potential is significantly modified:

$$\begin{aligned} V(r) &= -\frac{Ze^2}{r}, \quad r \geq r_0 , \\ V(r) &\to V_0 \quad r \to 0 . \end{aligned} \tag{10.122}$$

The nuclear charge distribution usually is represented by that of a homogeneously charged sphere. Then

$$V(r) = -\frac{Ze^2}{2r_0} \left[3 - \left(\frac{r}{r_0}\right)^2 \right], \quad r \leq r_0 . \tag{10.123}$$

To determine the correction due to the finite size of the nucleus to the Dirac energy ΔE_{FS} it is necessary to solve numerically the radial Dirac equations in the potential (10.122, 123). The corresponding correction ΔE_{FS} can be defined by

$$\Delta E_{FS} = \frac{2a^3 Z^4}{\pi n^3} F_{FS} \cdot \text{Ry} . \tag{10.124}$$

For $s_{1/2}$ states a relatively simple result can be obtained by considering the difference between the potential $V(r)$ for a finite-size nucleus and a point nucleus as a perturbation [see (6.87)]:

$$F_{FS}(ns_{1/2}) = \frac{\pi}{\alpha^3 Z^2} \cdot \frac{3(\gamma+1)}{\gamma(2\gamma+1)(2\gamma+3)[\Gamma(2\gamma+1)]^2} \left(\frac{2Zr_0}{a_0}\right)^{2\gamma} , \tag{10.125}$$

where $\gamma = \sqrt{1-\alpha^2 Z^2}$, $\alpha = \dfrac{e^2}{\hbar c}$.

332   10. Relativistic Corrections in the Spectroscopy of Multicharged Ions

The nuclear radius r_0 can be expressed in terms of the root-mean-square radius. For a homogeneous charge distribution, $\langle r^2 \rangle^{1/2} = \sqrt{3/5}\, r_0$. The ΔE_{FS} in (10.124) is presented in terms of a dimensionless function F_{FS}. This expression is useful to compare ΔE_{FS} with radiative corrections to the Dirac energy discussed below in Sect. 10.3.5.

The dependence of the $\Delta E_{FS}(ns_{1/2})$ in (10.124) on Z is given by

$$\Delta E_{FS} \propto Z^{2+2\gamma} \left(\frac{r_0}{a_0}\right)^{2\gamma} = Z^{2+2\sqrt{1-\alpha^2 Z^2}} \left(\frac{e_0}{a_0}\right)^{2\sqrt{1-\alpha^2 Z^2}}.$$

In the range $1 \leqslant Z \leqslant 100$, $1 \geqslant \gamma \geqslant 0.7$, $10^{-5} \leqslant \dfrac{r_0}{a_0} \leqslant 10^{-4}$, ΔE_{FS} increases very rapidly, approximately be eleven orders of magnitude.

The numerical calculations of the corrections ΔE_{FS} to the energy levels $1s_{1/2}$, $2s_{1/2}$, $2p_{1/2}$, $2p_{3/2}$ of the hydrogenlike ions have been carried out in [36] under the assumption of a homogeneous nuclear charge distribution. For nuclear radii, either available experimental values or the following expression

$$\langle r^2 \rangle^{1/2} \simeq (0.836 A^{1/3} + 0.570) \cdot 10^{-13} \text{ cm}, \quad A \geqslant 10, \tag{10.126}$$

where A is the nucleon number, have been used. The obtained values of F_{FS} for $1s_{1/2}$, $2s_{1/2}$ and $2p_{1/2}$ levels are given in Table 10.1 for $Z = 1, 2, \ldots, 10$, $20 \ldots 100$. For the whole range of $1 \leqslant Z \leqslant 100$ they can be approximated by

$$F_{FS} = \frac{4}{3} 10^{-3} \cdot \frac{C_0 + C_1 Z^{3/2} + C_2 Z^3 + C_3 Z^{9/2}}{1 + C_1' Z^{3/2}} \tag{10.127}$$

where

	C_0	C_1	C_2	C_3	C_1'
$1s_{1/2}$	9.5680	0.11062	$-6.5123 \cdot 10^{-5}$	$2.7072 \cdot 10^{-7}$	$-6.7586 \cdot 10^{-4}$
$2s_{1/2}$	7.8413	0.14447	$-1.8386 \cdot 10^{-4}$	$4.9616 \cdot 10^{-7}$	$-7.1974 \cdot 10^{-4}$
$2p_{1/2}$	0.11523	$-1.0182 \cdot 10^{-3}$	$-1.3228 \cdot 10^{-5}$	$5.7860 \cdot 10^{-8}$	$-8.2446 \cdot 10^{-4}$

The most influenced by the nucleus finite-size correction are the s levels, $\Delta E_{FS}(2p_{1/2}) \ll \Delta E_{FS}(2s_{1/2})$. For the $2p_{3/2}$ level the correction ΔE_{FS} is negligible.

Comparing the values of $F_{FS}(1s_{1/2})$ from Table 10.1 or (10.127) with (10.125) we see that for small Z, the approximation of (10.125) gives quite reasonable results. The difference between (10.125) and (10.127) increases with Z. For $Z = 92$ and $A = 238$, (10.125) gives 1.54 instead of 0.80. For a hydrogen atom $Z = 1$, and (10.125) and (10.127) give practically the same value $F_{FS} = 0.0013$.

As was said above the nucleus finite-size correction is especially important for ions with large Z.

Table 10.1. Nucleus finite-size correction functions F_{FS}

Z	$1S_{1/2}$	$2S_{1/2}$	$2P_{1/2}$
1	0.0013	0.0013	0
2	0.0055	0.0055	0
3	0.0011	0.0011	0
4	0.012	0.012	0
5	0.011	0.011	0
6	0.012	0.012	0
7	0.013	0.013	0
8	0.015	0.015	0
9	0.017	0.017	0
10	0.018	0.018	0
20	0.027	0.028	0
30	0.041	0.043	0
40	0.060	0.065	0.001
50	0.093	0.105	0.003
60	0.144	0.171	0.007
70	0.236	0.300	0.017
80	0.398	0.544	0.044
90	0.711	1.060	0.117
100	1.343	2.214	0.332

10.3.5 Radiative Corrections. Lamb Shift

The Dirac equation (10.1) does not contain the interaction of an electron with the quantized radiation field, described by quantum electrodynamics (QED). This interaction leads to the so-called radiative corrections to the Dirac electron energy ΔE_{RC}. The corresponding shift of energy levels removes the degeneracy with respect to l in the Dirac formula (10.71) and provides the splitting of hydrogen atom levels $2s_{1/2}$, $2p_{1/2}$ discovered by Lamb and Retherford. The correction ΔE_{RC} is usually expressed in terms of a dimensionless, slowly varying functions F_{RC} defined as

$$\Delta E_{RC} = \frac{2\alpha^3 Z^4}{\pi n^3} F_{RC} \cdot \text{Ry} . \tag{10.128}$$

For an electron in a Coulomb field of a point nucleus the main radiative corrections, the electron self-energy and vacuum polarization, which can be relatively simply evaluated, are [2]

$$F_{RC}^{(0)} = \frac{4}{3} \left\{ \left[\ln\frac{1}{(\alpha Z)^2} + \frac{19}{30} + L_{ns} \right] \delta_{l0} \right.$$
$$\left. + \left[\frac{3}{8} \cdot \frac{j(j+1)-l(l+1)-3/4}{l(l+1)(2l+1)} + L_{nl} \right] (1-\delta_{l0}) \right\} , \tag{10.129}$$

where L_{ns} and L_{nl} are so-called Bethe logarithms, which do not depend on j, and for the lower s, p and d levels have the values

$$L_{1s} = -2.9841, \quad L_{2s} = -2.8118$$

$$L_{2p} = 0.0300, \quad L_{3p} = 0.0382$$

$$L_{3d} = 0.052.$$

For higher levels the following approximate expressions can be used in estimations

$$L_{ns} = -2.9841 + 0.2666(1 - n^{-3/2})$$

$$L_{nl} = \frac{1.36}{(2l+1)^3}\left[1 - 0.87\left(\frac{l+1/2}{n+1/2}\right)^{3/2}\right]. \tag{10.130}$$

According to (10.129) the radiative corrections are especially important for $S_{1/2}$ states. With increase of l, $F_{RC}^{(0)}$ rapidly decrases. For the $2S_{1/2}$ state the function $F_{RC}^{(0)}$ is approximately 10^2 times larger than for the $2P_{1/2}$ state. In the case of hydrogen atom the approximation (10.129) gives the shift of $S_{1/2}$ states and Lamb splitting of $2S_{1/2} - 2P_{1/2}$ levels with an accuracy of $\sim 0.5\%$. The accuracy of experiments is much higher. Moreover, the error of the approximation of (10.129) rapidly increase with increase of ion charge Z. To improve approximation (10.129) the higher order radiative corrections are treated using the standard QED perturbation expansion. For high ion charges, $Z \gtrsim 50$, the perturbation expansion in terms of αZ becomes meaningless. Because of that and in light of recently developed techniques for producing multicharged hydrogenlike ions (like uranium) non-perturbative numerical calculations have been carried out. We use below the results of Johnson and Soff [36]. In these calculations the nuclear size and reduced mass effects on radiative corrections and relativistic recoil corrections have also been included.

The values of F_{RC} obtained in the numerical calculations [36] for the energy levels $1S_{1/2}$, $2S_{1/2}$, $2P_{1/2}$, $2P_{3/2}$ of hydrogenlike ions are given in Table 10.2.

Summarizing the results of above-mentioned perturbation expansion and non-perturbative numerical calculations [36] we define the function F_{RC} in (10.128) by

$$F_{RC} = F_{RC}^{(0)} + \frac{4}{3}\left\{\alpha Z A_{50} + (\alpha Z)^2\left[A_{61}\ln\frac{1}{(\alpha Z)^2} + A_{62}\ln^2\frac{1}{(\alpha Z)^2}\right]\right.$$

$$\left. + (\alpha Z)^2 G(\alpha Z)\right\}. \tag{10.131}$$

This expression is constructed in such way that it contains "confirmed analytical terms" and also the term $(\alpha Z)^2 G(\alpha Z)$ which approximates the difference between the results of numerical calculations [36] and analytical terms. Thus (10.131) ap-

Table 10.2. Radiative correction functions F_{RC}

Z	$1S_{1/2}$	$2S_{1/2}$	$2P_{1/2}$	$2P_{3/2}$
1	10.041	10.271	−1.126	0.123
2	8.266	8.496	−0.126	0.124
3	7.245	7.476	−0.125	0.124
4	6.536	6.767	−0.124	0.125
5	5.997	6.229	−0.123	0.125
6	5.565	5.798	−0.121	0.126
7	5.206	5.441	−0.119	0.127
8	4.902	5.138	−0.118	0.128
9	4.638	4.875	−0.116	0.129
10	4.407	4.646	−0.115	0.130
20	3.008	3.008	−0.093	0.143
30	2.317	2.596	−0.067	0.160
40	1.898	2.203	−0.037	0.178
50	1.619	1.954	−0.002	0.198
60	1.424	1.796	+0.037	0.220
70	1.284	1.699	0.083	0.242
80	1.182	1.650	0.139	0.264
90	1.109	1.641	0.208	0.287
100	1.060	1.671	0.298	0.310

proximates the results of numerical calculations in the whole range $1 \leqslant Z \leqslant 100$. The error of approximation does not exceed 1%. The exceptions are $1S_{1/2}$ states of ions with $Z \gtrsim 80$. In some such cases the approximation (10.131) is less accurate, with an error of $(1-2\%)$.

The approximation (10.131) is not valid for the $2P_{1/2}$ states of a few ions with charges near $Z = 50$, where $F_{RC}(2P_{1/2})$ changes sign.

The first term $F_{RC}^{(0)}$ in (10.131) is determined by (10.129, 130). The constants A_{50}, A_{61} and A_{62} are

$$A_{50} = 7.214 \delta_{l0}$$

$$A_{62} = -0.750 \delta_{l0}$$

$$A_{60} = \begin{cases} 3.965 & 1S_{1/2} \\ 4.348 & 2S_{1/2} \\ 0.43 & 2P_{1/2} \\ 0.24 & 2P_{3/2} \end{cases}$$

(10.132)

The last empirical term in (10.131) is determined as

$$G(\alpha Z) = A_{60} + A_{70} \alpha Z + A_{71}(\alpha Z) \ln \frac{1}{(\alpha Z)^2}, \tag{10.133}$$

where

$$A_{60} = \begin{array}{rl} -23.05 & 1S_{1/2} \\ -24.70 & 2S_{1/2} \\ -0.66 & 2P_{1/2} \\ -0.23 & 2P_{3/2} \end{array}$$

$$A_{70} = \begin{array}{rl} 19.64 & 1S_{1/2} \\ 20.93 & 2S_{1/2} \\ 1.35 & 2P_{1/2} \\ 0.47 & 2P_{3/2} \end{array}$$

(10.134)

$$A_{71} = 5.10\delta_{l0} \ .$$

Comparing the values of F_{RC} in Table 10.2 with (10.131) we see that the first term $F_{RC}^{(0)}$ gives reasonable estimation of ΔE_{RC} only for a limited number of ions with small charges $Z \leqslant 3$. For larger Z the higher order corrections are very important. For example, $F_{RC}^{(0)}$ gives at $Z > 42$ a wrong sign for $\Delta E_{RC}(1S_{1/2})$. Without the empirical term $(\alpha Z)^2 G(\alpha Z)$, (10.131) can be used in estimation of ΔE_{RC} only up to $Z \simeq 20-30$. For $Z \geqslant 30$ the empirical term should be included in estimations. According to (10.128, 131), ΔE_{RC} is of the order of $\alpha^3 Z^4$ Ry and rapidly increases with Z and decreases with principle quantum number n. The dependence on Z is not so strong as in (10.124) for the nucleus finite-size correction ΔE_{FS}. For small $Z \leqslant 10$ ΔE_{RC} exceeds ΔE_{FS} at least by three orders of magnitude but for the larger Z both corrections ΔE_{RC} and ΔE_{FS} are of the same order.

In the general case of a hydrogenlike ion, taking into consideration all corrections discussed above, we have for the electron energy the expression

$$E(nlj) = E_{nk} - \frac{m}{m+M}\left[1 + \frac{(\alpha Z)^2}{4n^2}\right] E_{nk} + \Delta E_{FS} + \Delta E_{RC} \ , \qquad (10.135)$$

where E_{nk} is the Dirac energy for a nucleus finite-size Coulomb field (10.75); the second term gives the reduced mass correction to the Dirac energy, including the relativistic reduced mass correction. The last two terms ΔE_{FS} and ΔE_{RC} are the nucleus finite-size and radiative corrections. They remove the l degeneracy in the Dirac energy E_{nk}.

The sum of the corrections ΔE_{FS} and ΔE_{RC} is usually defined as the total Lamb shift and is expressed as

$$\Delta E_L = \Delta E_{FS} + \Delta E_{RC} = \frac{2\alpha^3 Z^4}{\pi n^3} F_L \cdot \text{Ry} \ , \qquad (10.136)$$

$$F_L = F_{FS} + F_{RC} \ . \qquad (10.137)$$

Table 10.3. Lamb shift ΔE_L [cm^{-1}]

Z	$(1S_{1/2})$		$(2S_{1/2})-(2P_{1/2})$	
1	272624	E-6	352868	E-7
2	35926	E-4	46840	E-5
3	15956	E-3	20922	E-4
4	45507	E-3	59972	E-4
5	10193	E-2	13495	E-3
6	19620	E-2	26084	E-3
7	3402	E-1	4541	E-2
8	5467	E-1	7325	E-2
9	8291	E-1	11151	E-2
10	12013	E-1	1621	E-1
20	13184	E0	1838	E0
30	5186	E1	7438	E0
40	1361	E2	2001	E1
50	2905	E2	4367	E1
60	5514	E2	8454	E1
70	9907	E2	1546	E2
80	1757	E3	2795	E2
90	3241	E3	5290	E2
100	6525	E3	1105	E3

The Lamb shift of the level $1S_{1/2}$ and the Lamb splitting of the levels $2S_{1/2}$, $2P_{1/2}$ in cm^{-1} are given in Table 10.3 [36].

The function F_L (LAMB) and also the largest contributions to this function, electron self-energy (SELF), vacuum polarization (V.P.) and nucleus finite-size correction (F.S.) are shown in Fig. 10.1.

The sign of the Lamb shift is the same as for the repulsive potential. The estimated errors in calculated values of ΔE_L in Table 10.3 are of the order of $\sim 0.01\%$ for $Z \leq 10$, $\sim 0.1\%$ for medium Z and $\sim 1\%$ for $Z \simeq 90-100$. For the heaviest elements the uncertainty in the nuclear radius dominates the total error in the Lamb shift.

The accuracy of the approximations given by (10.124, 127) and (10.128, 131) is satisfactory for the majority of cases in the spectroscopy of multicharged ions.

The calculations of radiative and nucleus finite-size corrections for hydrogenlike ions form the basis for theoretical studies of heliumlike ions and approximate estimations in the case of other multicharged ions. The energy levels of heliumlike ions are tabulated in [37].

A very special problem is the Lamb splitting of the $2S_{1/2}-2P_{1/2}$ levels of hydrogen atom. The theoretical explanation and description of this effect was one of the first successes of renormalized QED. To provide the test of QED comparing the theoretical and experimental values of Lamb shift in hydrogen it is necessary to perform a very accurate and refined treatment of the different types of higher order corrections. But at present the Lamb shift provides one of the less stringent tests of the theory. The main reason for this is that the correction ΔE_{FS} also con-

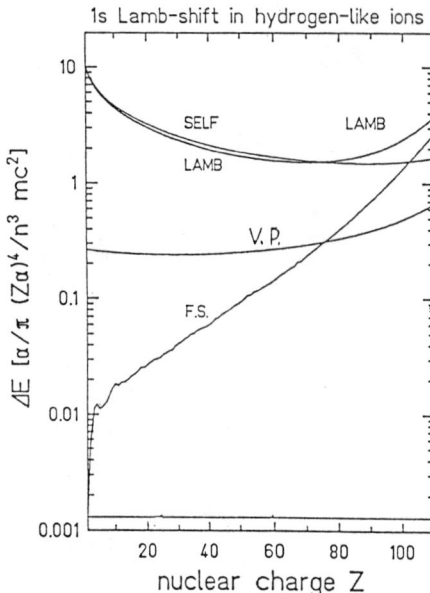

Fig. 10.1. $1s$ Lamb shift in hydrogenlike ions [36]. LAMB – Lamb shift, SELF – Electron self-energy, V.P. – Vacuum polarization, F.S. – Nucleus finite-size correction

tributes to the Lamb shift, and the proton's root-mean-square electromagnetic radius is not known to the needed accuracy. The last point can be thought of as a barrier to the test of QED by Lamb shift of hydrogen atom levels. For tests of strong-field QED the measurements of the energy levels of high Z hydrogenlike ions are of interest.

Chapter 11   Spectra of Multicharged Ions

The spectra of multicharged ions are of special interest to many modern fields of research, such as X-ray space astronomy and astrophysics, controlled thermonuclear fusion, extreme ultraviolet and soft X-ray lasers. Different methods of creating high-temperature laboratory plasmas, low-inductance sparks and other discharges, tokomak and laser-produced plasmas, have been used to study the spectra of ions with ion charges of $Z_i \lesssim 50$. Increasing activity in investigating the structure of highly charged heavy ions is stimulated by unique possibilities provided by modern accelerators and beam-foil spectroscopy. Fast heavy few-electron ions can be produced by stripping electrons from sufficiently accelerated ions shot through a thin foil. The higher the ion energy is, the higher the charge state that can be created behind the foil. A very impressive example of this technique is the production of hydrogenlike uranium.

In a brief discussion of the spectra of multicharged ions given below, our main focus is on the specific features of these spectra, which distinguish them from the spectra of neutral atoms and low-charged ions.

In this chapter an H-like ion with nuclear charge Z is denoted by [H]$_Z$-ion, an He-like ion as [He]$_Z$-ion and so on.

11.1 Energy Levels

The energy levels of an [H]$_Z$-ion are given by (10.134). In the first approximation

$$E_{nlj} = -\left[1 + \frac{(\alpha Z)^2}{n}\left(\frac{1}{j+1/2} - \frac{3}{4n}\right)\right]\frac{Z^2}{n^2}\text{Ry} . \tag{11.1}$$

The spectral series, such as Layman, Balmer and others, are similar to that in the neutral hydrogen spectrum, but scaled according to $E_{nl} \propto Z^2$, $\omega_{nn'} \propto Z^2$, $\lambda \propto Z^{-2}$ so that the corresponding spectral lines are shifted to extreme UV and X-ray spectral regions. Fine splitting is scaled as $\Delta E_{j,j-1} \propto Z^4$, hyperfine splitting as Z^3, see (10.80) and (10.112, 120). Thus the relative value of the fine splitting $\Delta E_{j,j-1}/E_{nlj}$ increases as Z^2. The resonance lines of neutral atoms H and He are in different regions of the spectrum, $\lambda = 1216$ Å and $\lambda = 593$ Å, respectively. With increase of Z the influence of the additional $1s$ electron in a [He]$_Z$-ion on the resonance transition $1s2p - 1s^2$ decreases. As a result, the resonance lines of [H]$_Z$- and [He]$_Z$-ions are in the same spectral region. For example, the resonance line in hydrogenlike Fe XXVI is $\lambda = 1.78$ Å and in heliumlike Fe XXV it is $\lambda = 1.85$ Å.

The plasma temperature needed to produce an [H]$_Z$-ion with nuclear charge Z is approximately proportional to Z^2. Because of this, fine splitting increases with Z more rapidly than the relative Doppler width $\Delta\omega_D/\omega = v/c \propto Z$, and the fine structure components of resonance lines are usually well resolved. For example, the resonance doublet $\lambda = 3.018$ Å and $\lambda = 3.024$ Å of the [H]$_Z$-ion Ca^{19+} is observed in spectra of laboratory and astrophysical plasmas as two non-overlapping lines. In beam-foil spectroscopy the conditions for resolving a resonance doublet are usually better than in plasma spectroscopy.

For high-charged ions, with increase of Z, QED corrections rapidly increase and also the nucleus finite-size correction to the energy levels. The sum of all corrections which remove the l degeneracy of energy levels in the Coulomb field of a point nucleus is usually called the Lamb shift. The Lamb shift ΔE_L increases approximately as Z^4 (Sect. 10.3). Different methods of experimental determination of the Lamb shift of the high-charged ions are discussed in [38].

In the extreme case of very large Z the Lamb shift of the levels $n = 1s$ and $2s$ has the order of magnitude of ~ 100 eV and $\Delta E_L/E_{nl} \sim 10^{-3}$. For example, in H-like uranium $Z = 92$

$$\Delta E_L(1s) \simeq 459 \text{ eV} , \quad \Delta E_L(2s) \simeq 75 \text{ eV} .$$

In ions with more than one electron, with increase of nuclear charge Z a more or less continuous transition from LS coupling to jj coupling takes place. The level structure of multicharged ions usually corresponds to an intermediate type of coupling.

With increase of Z the l degeneracy of the nl and nl' levels typical for a pure Coulomb field is restored. In [Li]$_Z$-ions at low Z, the structure of the $2s$ and $2p$ levels is similar to that in neutral Li (Fig. 3.1). The level spacing $E_{2p} - E_{2s}$ is almost of the same order as E_{2p}. The fine splitting $\Delta E(p_{3/2}, p_{1/2})$ is negligible compared to E_{2p}. With increase of Z the fine splitting increases as Z^4 and the $2P_{1/2}$ level moves close to the $2S_{1/2}$ level restoring the l degeneracy. In [He]$_Z$-ions, with increase of Z the Landé rule for triplet $1s2p$ $^3P_0$, $^3P_1$, $^3P_2$ is violated. The $^3P_2$ level moves up to the $1s2p\,^1P_1$ level and the levels $^3P_0$, $^3P_1$ move to the $1s2s\,^3S_1$, $^1S_0$ levels. In the limit of large Z two groups of close levels arise — two levels $2\,^1P_2$, $2\,^3P_2$ and four levels $2\,^1S_0$, $2\,^3S_1$, $2\,^3P_0$ and $2\,^3P_1$. The diagrams of the $1s2p$ levels in the [He]$_Z$-ions neon ($Z = 10$) and tungsten ($Z = 74$) are shown in Fig. 11.1.

In the case of high-charged ions the electron-nucleus interaction significantly exceeds the interaction between electrons. As a result, electron configurations with the same set of principal quantum numbers correspond to relatively close energy levels. The relative values $(E_{nl} - E_{nl'})/E_{nl}$ decrease as Z^{-1}. Whereas the frequencies of transitions $nl - n'l'$ for $n' \neq n$ scale as Z^2 the frequencies of transitions $\Delta n = 0$, $nl - nl'$ scale as Z. For example, in Be-like ions there are electron configurations $1s^22s^2$, $1s^22s2p$, $1s^22p^2$ and $1s^22s3s$, $1s^22s3p$, $1s^22s3d$ with the same sets of principal quantum numbers. For the transitions $\Delta n \neq 0$, such as $1s^22s3p - 1s^22s^2$, with increase of Z the corresponding wavelength λ decreases as Z^{-2}, for the transition $\Delta n = 0$, $1s^22s3p - 1s^22s3s$, λ decreases only as Z^{-1}.

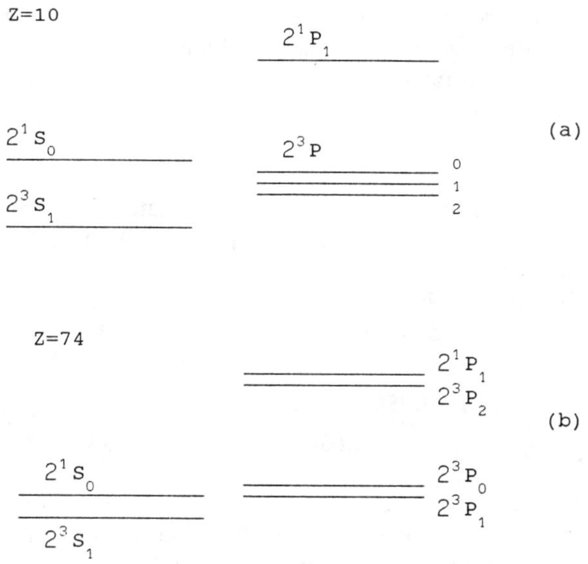

```
Z=10                    2¹P₁
                        ─────

2¹S₀                                    (a)
─────           2³P
                ═══════   0
2³S₁                      1
─────                     2


Z=74
                        ─────── 2¹P₁
                        ─────── 2³P₂
                                        (b)
2¹S₀                    ─────── 2³P₀
─────                   ─────── 2³P₁
─────
2³S₁
```

Fig. 11.1. Multiplet structure of the $n = 2$ states of the He-like ions a) neon $Z = 10$ and b) tungsten $Z = 74$

Configurations with the same set of principal quantum numbers and the same parity are often called the complex.

11.2 Forbidden Transitions

In the spectra of multicharged ions some forbidden transitions are observed which are not present in the spectra of neutral atoms and low-charged ions. Relativistic effects lead to the violation of the usual selection rules. The main difference from the spectra of neutral atoms is that the spectra of multicharged ions in very hot astrophysical and laboratory plasmas are formed, as a rule, under so-called coronal conditions.

In the case of thermodynamic equilibrium the intensity of a spectral line I_{ik} is proportional to the probability of the radiative transition A_{ik}. In the coronal limit the intensity I_{ik} is determined by the rate of upper level excitation q_i and does not depend on A_{ik}, or depends only on the branching ration A_{ik}/W_i:

$$I_{ik} = \hbar \omega_{ik} q_i A_{ik}/W_i , \tag{11.2}$$

where W_i is the total probability of the decay of state i. In the absence of branching, when only one radiative transition $i \to k$ is possible or when the radiative channel $i - k$ dominates, $W_i = A_{ik}$ and $I_{ik} = \hbar \omega_{ik} q_i$. The independence of the intensities

I_{ik} from A_{ik} is the important feature of the coronal limit. As a result, lines with very different probabilities A_{ik} are observed in the spectra with intensities proportional to the corresponding excitation rates.

11.2.1 H-like Ions

The state $2s$ of an $[H]_Z$-ion has the same parity as the ground state $1s$ and can decay to the ground state either by a two-photon electric dipole transition $2E1$ or by a magnetic dipole transition $m1$.

The probability of a $2E1$ transition in $[H]_Z$-ions has been calculated in [39]. The results for the whole region of $1 \leq Z \leq 100$ are expressed in the following simple form

$$2E1: A(2S_{1/2} - 1S_{1/2}) = 8.230 \, Z^6 \frac{1 + 3.95 \,(\alpha Z)^2 - 2.04 \,(\alpha Z)^4}{1 + 4.60 \,(\alpha Z)^2} \, [\text{sec}^{-1}] \, . \quad (11.3)$$

The selection rules for magnetic dipole radiation, given in Sect. 9.3.5, forbid $M1$ transitions between levels of different electron configurations. For multicharged ions this restriction is violated by relativistic effects.

Calculations taking into consideration the relativistic corrections of the order of α^2 give [40]

$$M1: A(2S_{1/2} - 1S_{1/2}) = \frac{\alpha^9 Z^{10}}{972} \frac{me^4}{\hbar^3} \approx 2.46 \cdot 10^{-6} Z^{10} \, [\text{sec}^{-1}] \, . \quad (11.4)$$

The dependence of $A(M1)$ on Z is more strong than $A(2E1)$. For $Z \leq 30$, $A(2E1)$ exceeds $A(M1)$, but for $Z > 40$ the magnetic dipole transition gives the main contribution to radiative decay of the state $2S$. Note that both probabilities $A(2E1)$ and $A(M1)$ increase with Z more rapidly than the probability of allowed electric dipole transitions $2P_{3/2} - 1S_{1/2}$; $2P_{1/2} - 1S_{1/2}$ $A(E1) \propto Z^4$. For small Z, $A(2E1)$, and $A(M1)$ are much less than $A(1E)$. With increase of Z the ratio $A(M1)/A(E1)$ rapidly increases, and in the extreme case of large Z it is $\sim 10^{-2}$. For H-like uranium the radiative transition probabilities are (in sec^{-1})

$A(M1) = 2 \cdot 10^{14}$

$A(2E1) = 0.04 \cdot 10^{14}$

$A(E1) = 4.7 \cdot 10^{16}$

$A(E1) = 3.5 \cdot 10^{16}$.

11.2.2 He-like Ions

The ground state of $[He]_Z$-ion is $1s^2 \, ^1S_0$. All excited states below the first limit of ionization belong to electron configurations $1snl$. Each configuration $1snl$ gives

11.2 Forbidden Transitions

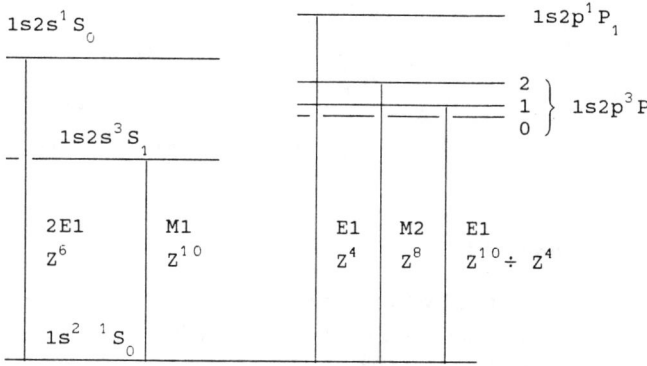

Fig. 11.2. Level diagram and radiative decay modes of the $n = 2$ states of a He-like ion

singlet and triplet terms $1snl\,^1L$ and $1snl\,^3L$. In the spectrum of He atom two independent systems of lines are observed, corresponding to singlet and triplet terms. Intercombination transitions between triplet and singlet terms are spin-forbidden (Sect. 3.3.1). Between the excited states $1s2s$, $1s2p$ and the ground state $1s^2$ there is only one allowed electric dipole transition $1s2p\,^1P_1 - 1s^2\,^1S_0$. All other transitions are forbidden. In the spectra of multicharged $[\text{He}]_Z$-ions, the restrictions on almost all such transitions are violated, but the dependence on Z is much more complicated than in the case of $[\text{H}]_Z$-ions.

The radiative decay of $n = 2$ states of $[\text{He}]_Z$-ions is shown in Fig. 11.2. In Fig. 11.1 we showed the energy level diagrams of the excited states $1s2s$ and $1s2p$ for two $[\text{He}]_Z$-ions, He-like neon ($Z = 10$) and He-like tungsten ($Z = 74$). The diagrams differ in the reordering of the levels $^3P_0$, $^3P_1$, $^3P_2$ with increase of Z. For small Z the diagram is typical of LS coupling. As Z increases along the $[\text{He}]_Z$ sequence, the $^3P_2$ level moves up close to the $^1P_1$ level, which is typical of the terms $(\frac{1}{2}, \frac{1}{2})$, $(\frac{1}{2}, \frac{3}{2})$ of jj coupling (Sect. 2.3). Three levels, $^3P_0$, $^3P_1$, $^1S_0$, on the other hand, stay close together for all Z. The detailed treatment of the radiative decay of $n = 2$ states of $[\text{He}]_Z$-ions, taking into consideration the evolution of $^3P_{0,1,2}$ levels along the $[\text{He}]_Z$ sequence, has been carried out in [41]. We give below the main results of this work.

The $1s2s\,^3S_1$-$1s^2\,^1S_0$ Transition

The $^3S_1$ term is lowest in energy among the $n = 2$ terms. This state can decay to the ground state by an $M1$ transition or by a two-photon $2E1$ transition. The decay is dominated by the $M1$ mode for all Z. The probability of a $^3S_1 - {}^1S_0\ M1$ transition is tabulated in [41]. For estimations the following simple expression can be used [40]:

$$M1: A(1s2s\,^3S_1 - 1s^2\,^1S_0) = \frac{2^5}{3^9} \alpha^9 Z^{10} \left(\frac{\Delta E}{Z^2 \text{Ry}}\right)^3 \frac{me^4}{\hbar^3}$$

$$\simeq 0.39\, Z^{10} \left(\frac{\Delta E}{Z^2 \text{Ry}}\right)^3 [\text{sec}^{-1}], \quad Z \gtrsim 10. \quad (11.5)$$

The $1s\,2s\,^1S_0 - 1s^2\,^1S_0$ Transition

The $^1S_0$ state decays to the ground state by a two-photon transition $2E1$. The one photon $M1$ transition is forbidden as a $0-0$ transition. The probability of a $2E1$ transition can be estimated by using (11.3). The references to calculations of the lifetime of the $2\,^1S_0$ state are given in [40].

The $2\,^1S_0$ state can also decay to the excited $2\,^3P_1$ state by a spin-forbidden $E1$ transition. Since the $2\,^1S_0$ and $2\,^3P_1$ states are close in energy for all Z this decay mode never competes with the $2E1$ mode. It is necessary to note that there is some range of Z in which the level $2\,^1S_0$ lies lower than the level $^3P_1$.

The $1s\,2p\,^1P_1 - 1s^2\,^1S_0$ Transition

As has been said before, only this transition between the states $n = 2$ and $n = 1$ is the spin-allowed electric dipole transition $E1$ with probability increasing as Z^4. The transitions to the excited states $2\,^1S_0$ and $2\,^3S_1$ states are insignificant compared to the resonance transition $2\,^1P_1 - 1\,^1S_0$.

The $1s\,2p\,^3P_0 - 1s^2\,^1S_0$ Transition

The three fine structure components $^3P_0$, $^3P_1$, $^3P_2$ decay to the ground state by entirely different modes. The transition $^3P_0 - {}^1S_0$ is an odd-even, $J = 0 - J = 0$ transition. So the $^3P_0$ state can decay to the ground state only by a two-photon $E1 + M1$ mode or by a three-photon $3E1$ mode. Both these modes are ineffective in the whole range of Z. The state $^3P_0$ can decay to the excited state $^3S_1$ by an allowed $E1$ transition. The probability of a $2\,^3P_0 - 2\,^3S_1$ transition is practically the same as for a $2\,^3P_1 - 2\,^3S_1$ transition, increasing approximately as Z. For $Z = 2$, $A(2\,^3P_0 - 2\,^3S_1) \simeq 1.15 \cdot 10^7 \text{ sec}^{-1}$ and for $Z = 50$, $A(2\,^3P_0 - 2\,^3S_1) \simeq 1.7 \cdot 10^9 \text{ sec}^{-1}$. The transition probability is influenced by the decrease of the spacing between the levels $^3P_0$, $^3P_1$ and $^3S_1$ with Z as angular momenta coupling tends to jj-coupling.

The $1s\,2p\,^3P_1 - 1s^2\,^1S_0$ Transition

The $^3P_1$ state decays to the ground state by intercombination (spin-forbidden) transition. The numerical calculations [41] show that the probability of this intercombination transition strongly depends on Z but this dependence is different for different ranges of Z. For relatively low $Z < 20$ the probability grows as Z^{10}, then

Table 11.1. Radiative transition probabilities for $1s2p\,^3P_1 - 1s^2\,^1S_0$ in $[\text{He}]_Z$-ions

Z	$A(^3P_1 - ^1S_0)$ [sec^{-1}]	Z	$A(^3P_1 - ^1S_0)$ [sec^{-1}]
2	2.33 E 2	16	5.87 E 11
4	4.21 E 5	18	1.82 E 12
6	2.89 E 7	20	4.85 E 12
8	5.56 E 8	25	3.12 E 13
10	5.40 E 9	30	1.22 E 14
12	3.40 E 10	50	2.12 E 15
14	1.58 E 11	100	4.27 E 16

dependence on Z becomes more weak, approximately proportional to Z^8 in the range $20 < Z < 40$, and proportional to Z^4 for $Z > 50$. Such dependence on Z is determined by the reordering of the ^3P_J levels with increase of Z discussed above. The probabilities $A(1s2p\,^3P_1 - 1s^2\,^1S_0)$ from [41] for different values of Z are given in Table 11.1. The $^3P_1$ state can decay also to the excited $^3S_0$ by spin-allowed $E1$ transitions but for $Z \geq 10$ the decay to the ground state dominates, because the spacing of the levels $2\,^3P_1$, $2\,^3S_0$, is small.

The $1s\,2p\,^3P_2 - 1s^2\,^1S_0$ Transition

The state $1s2p\,^3P_2$ can decay to the ground state $1s^2\,^1S_0$ by a magnetic quadrupole transition $M2$. The probability of this transition is given by [40]

$$M2: A(1s\,2p\,^3P_2 - 1s^2\,^1S_0) = \frac{\alpha^7 Z^8}{1215}[1 + 0.28(\alpha Z)^2]\frac{me^4}{\hbar^3}\left(\frac{\Delta E}{Z^2 \text{Ry}}\right)^5$$

$$\simeq 0.037\, Z^8 [1 + 0.28(\alpha Z)^2]\left(\frac{\Delta E}{Z^2 \text{Ry}}\right)^5 \;[\text{sec}^{-1}]\;. \tag{11.6}$$

The comparison of (11.6) with numerical calculations tabulated in [41] show that for $Z > 30$ (11.6) is accurate to within $10-15\%$.

The decay mode to the excited state $^3S_1$ for $Z < 18$ is more effective than the $M2$ decay mode to the ground state. But for larger Z the $M2$ decay $2\,^3P_2 - 1\,^1S_0$ dominates. For $Z > 16$ the level $2\,^3P_2$ lies higher than the level $2\,^1S_0$. But the decay $2\,^3P_2 - 2\,^1S_0$ is ineffective for all Z. For He-like uranium the probabilities of the radiative transitions $n = 2 - n = 1$ are (in sec^{-1})

$2\,^1S_0 - 1\,^1S_0 \quad A(2E1) \quad = 8 \cdot 10^{12}$

$2\,^3S_1 - 1\,^1S_0 \quad A(M1) \quad = 1 \cdot 10^{14}$

$2\,^1P_1 - 1\,^1S_0 \quad A(E1) \quad = 5 \cdot 10^{16}$

$2^3P_2 - 1\,^1S_0 \quad A(M2) \quad\quad = 2 \cdot 10^{14}$

$2^3P_1 - 1\,^1S_0 \quad A(E1) \quad\quad = 3 \cdot 10^{16}$

$2^3P_0 - 1\,^1S_0 \quad A(E1+M1) = 6 \cdot 10^9 \ .$

The Z dependence of the transitions $n = 2 - n = 1$ is shown in Fig. 11.2.

The examples of [H]$_Z$- and [He]$_Z$-ions discussed above show that in the spectra of high-charged ions different forbidden transitions like magnetic multidipole transitions and intercombination transitions are much more important than in the spectra of neutral atoms or low-charged ions. The two-photon decay modes also have to be taken into considerations. The Z dependence of the forbidden transitions is very sensitive to the angular momenta coupling scheme.

Different spectral lines were identified in solar spectra as $M1$ magnetic dipole transitions in [He]$_Z$-ions C V, O VII, Ne IX, Si XIII, S XV, Fe XXV. Similar lines were also observed in spectra of tokomak plasmas. The magnetic quadrupole lines $1s\,2p\,^3P_2 - 1s^2\,^1S_0$ were observed in Ca XIX and Fe XXV solar spectra. Recently the $M1$, $M2$ and $2E1$ decay modes were determined for [He]$_Z$-ion Kr^{34+}. There are many other examples of observations of forbidden lines in spectra of high-charged ions.

11.3 Satellite Structure

The main difference between spectra of neutral atoms and spectra of high-charged ions in hot plasma under conditions of coronal limit is the appearance of additional spectral lines called satellites [Ref. 42]. The origin of satellites of the Ly_α resonance line in an [H]$_Z$-ion spectrum is illustrated by Fig. 11.3, in which the energy levels of [H]$_Z$- and [He]$_Z$-ions are shown. Besides the ground and the first excited states of [He]$_Z$-ion the doubly excited states $2s^2$, $2s2p$, $2p^2$ are also shown which are usually populated in the process of dielectronic recombination. The energy levels of these states are higher than the first ionization limit. Because of this, in the general case, they can decay via two channels — radiative transition and autoionization. For example, for the $2p^2$ state we have

$$\text{radiative transition} \quad [\text{He}(2p^2)]_Z \rightarrow [\text{He}(1s2p)]_Z + \hbar\omega \ , \quad\quad (11.7)$$

$$\text{autoionization} \quad [\text{He}(2p^2)]_Z \rightarrow [\text{H}]_Z + e^- \ . \quad\quad (11.8)$$

The probability of autoionization W_a has a weak dependence on Z [38, 42], whereas the radiative probability A increases as Z^4. Therefore for small Z the autoionization channel (11.8) dominates. For $Z \simeq 10-20$ the radiative and autoionization probabilities are of the same order of magnitude. For $Z > 20$ the

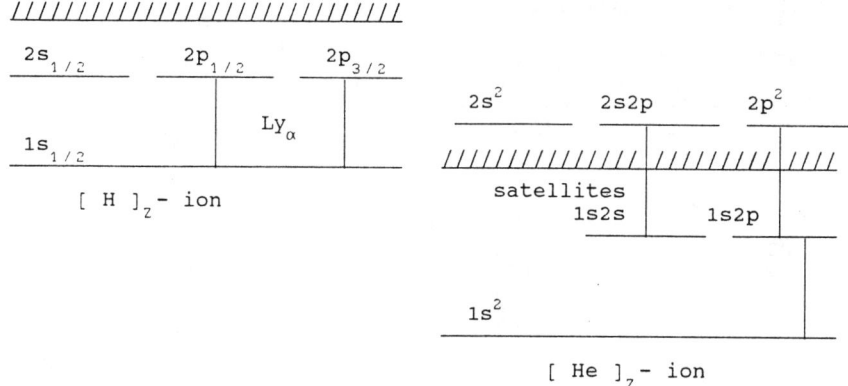

Fig. 11.3. The origin of the satellites in spectra of H-like ions

radiative channel (11.7) dominates and the intensities of the corresponding lines are completely determined by the $2p^2$ states excitation. At large Z the frequency of a $2p-1s$ electron transition in a strong Coulomb field of nuclear charge Z is not changed much by the presence of the additional $2p$ electron. Therefore, the wavelengths of the $2p-1s$ transitions in $[H]_Z$-ion and in $[He]_Z$-ion are almost the same.

The $[H]_Z$- and $[He]_Z$-ions are usually present together in hot plasma and the resonance doublet $2P_{1/2}-1S_{1/2}$, $2P_{3/2}-1S_{1/2}$ of $[H]_Z$-ions has a group of satellites arising from the transitions $2s2p-1s2s$, $2p^2-1s2p$ of $[He]_Z$-ions.

Satellite structure can be observed near some other transitions in $[H]_Z$-ions. The satellites to the $n_0 l_0 - n_1 l_1$ parent line of $[H]_Z$-ion correspond to the $n_0 l_0, nl - n_1 l_1, nl$ transitions of $[He]_Z$-ion, where n, l are the quantum numbers of the additional electron. Such satellites are called n satellites. In a spectrum of an $[H]_Z$-ion the $n=2$ satellites are well separated from the parent line and are shifted to the red, the $n=3$ satellites are much closer to the parent line and may be located

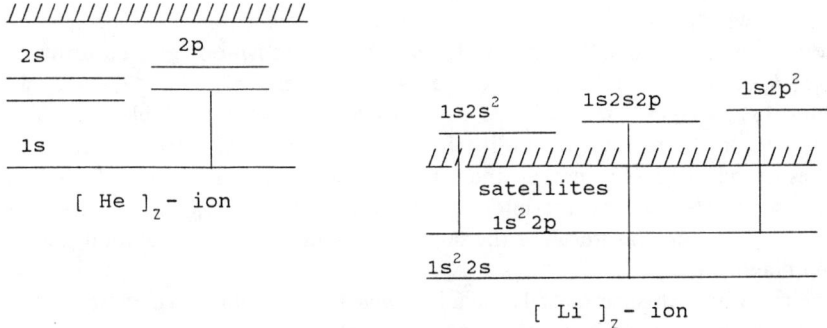

Fig. 11.4. The origin of the satellites in spectra of He-like ions

Table 11.2. The system of notation in the spectra of He-like ions [44]

Array	Multiplet	$J_0 - J_1$	Notation
$1s2p - 1s^2$	$^1P - {^1S}$	1–0	w
	$^3P - {^1S}$	2–0	x
		1–0	y
$1s2s - 1s^2$	$^3S - {^1S}$	1–0	z
$1s2p^2 - 1s^2 2p$	$^2P - {^2P}$	3/2–3/2	a
		3/2–1/2	b
		1/2–3/2	c
		1/2–1/2	d
	$^4P - {^2P}$	5/2–3/2	e
		3/2–3/2	f
		3/2–1/2	g
		1/2–3/2	h
		1/2–1/2	i
	$^2D - {^2P}$	5/2–3/2	j
		3/2–1/2	k
		3/2–3/2	l
		1/2–3/2	m
		1/2–1/2	n
$1s2s^2 - 1s^2 2p$	$^2S - {^2P}$	1/2–3/2	o
		1/2–1/2	p
$1s2p2s - 1s^2 2s$	$[^1P]^2P - {^2S}$	3/2–1/2	q
		1/2–1/2	r
	$[^3P]^2P - {^2S}$	3/2–1/2	s
		1/2–1/2	t
	$^4P - {^2S}$	3/2–1/2	u
		1/2–1/2	v

on both the red and the blue sides of it. The satellites corresponding to $n = 4$ are usually located within the Doppler width of the parent line.

The satellite structure in [He]$_Z$-ion spectra is more rich than in the spectra of [H]$_Z$-ion. The origin of the satellite structure in [He]$_Z$-ion spectra is illustrated in Fig. 11.4 in which [He]$_Z$-ion and [Li]$_Z$-ion level diagrams are shown. The satellite structure is shown schematically. The $1s2p^2 - 1s^2 2p$ transitions give an array of 14 $S_0 L_0 J_0 - S_1 L_1 J_1$ satellites. There are two satellites $1s2s^2 - 1s^2 2p$ and six satellites $1s2p2s - 1s^2 2s$. All these satellites are included in Table 11.2 The $1s2pnl - 1s^2 nl$ transitions of Li-like ions were among the first identified as satellites to the $1s2p - 1s^2$ parent lines of He-like ions [43]. The ratio of satellite intensities $I(\gamma_0 nl - \gamma_1 nl)$ to the intensity of the parent line $I(\gamma_0 - \gamma_1)$ is very sensitive to plasma temperature. Thus the satellite structure is a very important tool in plasma diagnostics.

The main lines in spectra of He-like ions and their satellites, which are of the most interest in plasma diagnostics, were designated in [44] by latin letters. This notation is given in Table 11.2. The first four lines, w, x, y, z, correspond to the

11.3 Satellite Structure   349

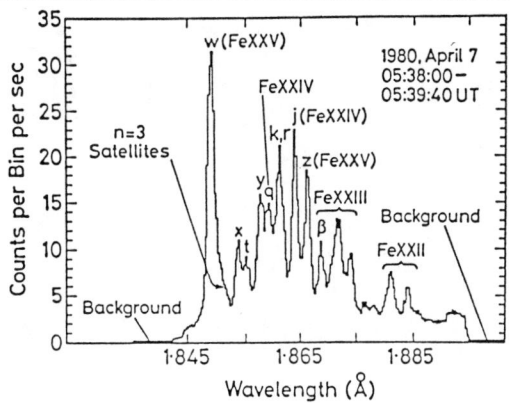

Fig. 11.5. Solar spectrum of He-like Fe XXV. Satellites are labeled according to the notation of Table 11.1. From [45]

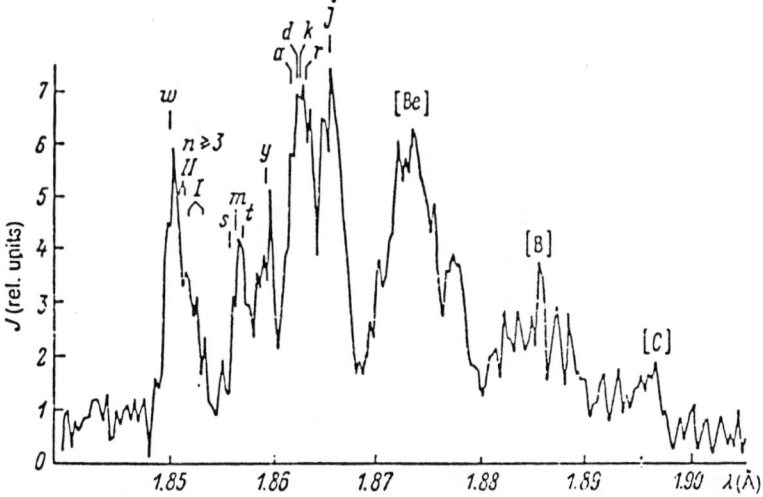

Fig. 11.6. Fe spectrum from laser produced plasma in the vicinity of the resonance line of the He-like ion (w). The notation of satellites corresponds to Table 11.1. From [46]

transitions of [He]$_Z$-ions. All others are satellites in the [He]$_Z$-ion spectrum arising from transitions of [Li]$_Z$-ions.

With the increase of the number of electrons in an ion, i.e. in the spectra of Li-, Be-, B-like ions and so on, the number of satellites significantly increases. Not all of them are resolved and the groups of overlapped satellites are often observed as broad lines. Examples of spectra with satellite structure are given in Fig. 11.5 [45] and Fig. 11.6 [46].

The detailed review of the spectra of highly ionised atoms with numerous references to experimental results is published by I. Martinson [47].

References

1. E. Condon, G. Shortley: *The Theorey of Atomic Spectra* (Cambridge University Press, Cambridge 1951)
2. G. Bethe, E. Salpeter: *Quantum Mechanics of Atoms with One and Two Electrons* (Springer, Berlin, Göttingen, Heidelberg 1957)
3. L. Landau, E. Lifshitz: *Quantum Mechanics* (Addison-Wesley, Reading, Mass. 1958)
4* M. Elyashevich: *Atomic and Molecular Spectroscopy* (Fizmatgiz, Moscow 1962)
5* S. Frish: *Optical Spectra of Atoms* (Fizmatgiz, Moscow 1963)
6. G. Racah: Phys. Rev. *62*, 438 (1942)
7. G. Racah: Phys. Rev. *63*, 367 (1943)
8. A. Edmonds: *Angular Momentum in Quantum Mechanics* (Princeton University Press, Princeton 1957)
9* A. Yutsis, I. Levinson, V. Vanagas: *Mathematical Methods of Angular Momentum Theory* (Giz Lieturos TSR, Vilnius 1960)
10* D. Varshalovich, A. Moskalev, V. Hersonsky: *Quantum Theory of Angular Momentum* (Nauka, Moscow 1975)
11. A. Simon: *Table of Numerical Values of Clebsch-Gordan Coefficients* (ORNL-1718, 1954)
12. G. Racah: "Group Theory and Spectroscopy", in Springer Tracts in Modern Physics, Vol. 37 (Springer, Berlin, Heidelberg, New York 1965) p. 28
13. G. Racah: Phys. Rev. *76*, 1352 (1949)
14. R. Rosenzweig: Phys. Rev. *88*, 580 (1952)
15* G. Bukat, A. Dolginov, R. Zhitnikov: Opt. Spektrosk. *8*, 285 (1960)
16. R.G. Barnes, W.V. Smith: Phys. Rev. *93*, 95 (1954)
17. H. Matsunobu, H. Takebe: Prog. Theor. Phys. *14*, 589 (1955)
18. H.A. Robinson, G.H. Shortley: Phys. Rev. *52*, 713 (1937)
19. L.L. Foldy: Phys. Rev. *111*, 1093 (1958)
20. A. Sommerfeld: *Atombau und Spektrallinien* (Vieweg, Braunschweig 1951)
21* V. Lisitsa: Opt. Spektrosk. *31*, 862 (1971)
22. W.L. Wiese, M.W. Smith, B.M. Glennon: *Atomic Transition Probabilities* (NBS: I, 1966; II, 1969)
23. D.R. Bates, A. Damgaard: Phil. Trans. R. Soc. London *242*, 101 (1949)
24. G. Drake, A. Dalgarno: Astrophys. J *157*, 459 (1969)
25. R.H. Garstang, L.S. Shamey: Astrophys. J. *148*, 665 (1967)
26* A. Akhieser, V. Berestetsky: *Quantum Electrodynamics* (Fizmatgiz, Moscow 1959)
27* D.A. Frank-Kamenetsky: *Physical Processes inside Stars* (Fizmatgiz, Moscow 1959)
28* V.I. Kogan: "Recombination Radiation of the Hydrogen Plasma" in *Collected Papers: Papers Physics and the Problem of Controlled Thermonuclear Reactions,* Vol. 3 (Izd. AN SSSR, Moscow 1958)
29. A. Burgess, M. Seaton: Rev. Mod. Phys. *30*. 992 (1958)
30. G. Peach: Mem. R. Astron. Soc. *71*, 13 (1967)
31. L. Landau, E. Lifshitz: *Classical Theory of Fields* (Addison-Wesley, Reading, Mass. 1962)
32* V. Gervids, V.I. Kogan: JETP Lett *22*, 308 (1975)
33. A. Burgess: Monthly Notice Roy. Astzon. Soc. *118*, 477 (1959)
34. D.H. Mezzel: Nature *218*, 756 (1968)
35. C. Schwartz: Phys. Rev. *97*, 380 (1955)

---

[1] An asterisk indicates a work in Russian.

36 W.R. Johnson, G. Soff: Atomic Data Nucl. Data Tables *33*, 405 (1985)
37 G.W.F. Drake: Canad. J. Phys. *66*, 586 (1988)
38 P.H. Mokler: *High Energy Ion-Atom Collisions*, Springer Lect. Notes Phys, v.294 (Springer, Berlin, Heidelberg 1988) p. 463
39 S.P. Goldman, G.W.F. Drake: Phys. Rev. *24A*, 183 (1981)
40 E. Marrus, P.J. Mohr: in *Advances in Atomic and Molecular Physics,* vol. 14 (Academic, New York 1978) p. 181
41 C.D. Lin, W.R. Johnson, A. Dalgarno: Phys. Rev. *15A*, 154 (1977)
42 R.K. Janev, L.P. Presnyakov, V.P. Shevelko: *Physics of Highly Charged Ions*, Springer Ser. Electrophys. v. 13 (Springer, Berlin Heidelberg 1985)
43 B. Edlen, F. Tyren: Nature *143*, 940 (1939)
44 A.H. Gabriel: Mon. Not. R. Astron Soc. *160*, 99 (1972)
45 J.L. Culhane: in *Advances in Space Research* v. 8 (1988) p. 67
46 M.A. Mazing: Opt. Spectrosc. (USSR) *66*, 303 (1989)
47 I. Martinson: Rep. Prog. Phys. *52*, 157 (1989)

List of Symbols

Constants

$a_0 = \hbar^2/me^2$ Bohr radius
c Velocity of light
e Elementary charge
m Mass of electron
m_p Mass of proton
$me^4/\hbar^2$ Atomic unit of energy
$R = me^4/4\pi c\hbar^3$ Rydberg constant
$Ry = me^4/2\hbar^2$ Rydberg unit of energy

Quantum Numbers

j Electron angular momentum
J Atomic angular momentum
l Orbital angular momentum
L Atomic orbital momentum
m, M Magnetic orbital momentum
n Principal orbital momentum
n_1, n_2 Parabolic orbital momentum
s, S Spin orbital momentum

Basic Notations

$A_{\gamma\gamma'}$ Einstein coefficient for spontaneous emission — Eq. (9.24)
$B_{\gamma\gamma'}, B_{\gamma'\gamma}$ Einstein coefficients for stimulated emission and absorption — Eq. (9.24)
$C_{lm} = C_m^l = \sqrt{\dfrac{4\pi}{2l+1}}\, Y_{lm}(\theta,\varphi)$ Spherical functions — Eq. (4.7)
$C_{m_1 m_2}^{j} = (j_1 j_2 m_1 m_2 | j_1 j_2 j m)$ Clebsch-Gordan coefficients — Eq. (4.19), Sect. 4.1
f Oscillator strength — Eq. (9.48), Sect. 9.2.2
f_κ Eq. (9.136)
$F_1(nl;n'l')$ Eqs. (9.77)–(9.79)
$F_\kappa(nl;n'l')$ Eq. (9.139)
g Gaunt factor — Eq. (9.273), Landé factor — Eq. (8.7), statistical weight

$G_{S'L'}^{SL} = (l^{n-1}[S'L']lSL\}l^n SL)$ Coefficients of fractional parentage — Sect. 5.1.5
$k = \omega/c$ Wave number
k_ω Absorption coefficient
n_*, n^* Effective principal quantum number — Eq. (3.3)
$P_\gamma(r) = r R_\gamma(r)$ Radial function
$P_l^m, P_l^m(\cos\theta)$ Associated Legendre polynomials
$P_l, P_l(\cos\theta)$ Legendre polynomials
Q Quadrupole moment, Q-factor — Sect. 9.6
$R, R(r)$ Radial function
S Line strength — Eq. (9.54)
$T_q^k = T_{kq}, T^k$ Spherical (irreducible) tensor operator — Sect. 4.3
$[T^k \times U^r]^s$ Tensor product of operators T^k, U^r — Sect. 4.3.4
$(T^k U^k)$ Scalar product of operators T^k, U^k — Sect. 4.3.4
v Seniority number — Sect. 5.1.6
W Transition probability
$Y_{lm}(\theta,\varphi)$ Spherical functions — Sect. 1.1.2
$\Gamma(x)$ Gamma function
Δ_l Quantum defect — Eq. (3.4)
ν Effective quantum number ($\nu \equiv n_*$ in Eq. (3.3))
ζ Charge of atomic core — Eq. (9.345)
σ Effective cross sections of radiative processes
$\langle F \rangle$ Mean value, diagonal matrix element
$\begin{pmatrix} j_1 & j_2 & j_3 \\ m_1 & m_2 & m_3 \end{pmatrix}$ $3j$ Symbol – Sect. 4.2.1
$\begin{Bmatrix} a & b & c \\ d & e & f \end{Bmatrix}$ $6j$ Symbol – Sect. 4.2.3
$\begin{Bmatrix} a & b & c \\ e & d & f \\ g & h & k \end{Bmatrix}$ $9j$ Symbol – Sect. 4.2.5
$(\gamma \| T \| \gamma')$ Reduced matrix element — Eq. (4.121)

Subject Index

Angular momenta 53, 54
  addition of three 57, 58, 66, 67
  addition of two 55, 56
Angular momentum operator 54, 55

Balmer series 8, 9, 14
Bates-Damgaard approximation 236, 237, 256, 283
Bohr magneton 11
Breit interaction 133
Bremsstrahlung 245, 247, 257

Clebsch-Gordan coefficients 60 – 62, 64
Configuration mixing (interaction) 118, 125
Correspondence principle 203
Coulomb integral 117
Coupling 22, 23, 27, 154, 155
  intermediate 144, 147
  jj- 27, 28, 29, 30, 141
  jl- 152, 153
  LS- 23, 27
  non-homogeneous 115
Cross sect
  absorption 205, 249
  stimulated emission 205

Defect, quantum 36, 256
Diffuse series 37, 39, 41
Dipole radiation 203
  electric 203, 205
  magnetic 225 – 227
Dirak equation 303
Displaced terms 42

Eckart-Wigner theorem 77
Einstein's coefficients 204
Electron configuration 19
Energy unit
  atomic 4
  Rydberg 4
Equivalent electrons 20, 25
  states 20
Exchange integral 117

Fermi-Segré formula 165
Filled shells 20

Fine splitting 13, 21, 126 – 140, 313
Fine structure 12, 21, 37, 126 – 140
  of alkali spectra 37
  of helium 39, 40, 133 – 139
  of hydrogen 13, 14, 15
  constant 22, 127
Forbidden transition 27, 237, 341
Fractional parentage coefficients 96 – 98, 100 – 105
Fundamental series 37, 39, 41

Gaunt factor 253, 257, 271, 272
g-factor 156, 189 – 191

Hund's rule 21
Hyperfine splitting 159, 227, 229, 326, 330

Interaction
  electrostatic 20, 21, 115
  exchange 117
  spin-orbit 11, 21, 127, 132
  spin-other orbit 134, 139
  spin-own orbit 134
  spin-spin 134, 139
Intercombination lines (transitions) 27, 39, 40, 42, 237
Inverted multiplets 22
Isotope shift 170 – 172

$3j$-symbols 60 – 66
$6j$-symbols 66 – 72
$9j$-symbols 72 – 74, 145, 146

Kramers approximation 252 – 254, 272, 273, 282, 283

Lamb shift 15, 333
Landé factor 190
Landé interval rule 22, 39, 126, 127, 160
Line strength 209
Lyman series 8, 9

Magnetic moment, nuclear 154
Metastable level 42
Metastable term 39, 41, 44
Multiplet 22, 27

Multiplicity 21
Multipole radiation 216
Multipole moments 218, 219

Nucleus finite size correction 331

Operator irreducible
  direct product 86–88
  scalar product 81
  spherical 74, 75, 76
  symmetrical 106
  tensor product 80, 81
Oscillator strength 208, 214, 281
  sum rule 209, 232, 234

Parabolic coordinates 179
Parabolic quantum numbers 179
Parity 18, 19, 22, 27
Paschen-Back effect 194–195
Pauli equation 308
Pauli principle 20, 25, 26, 30, 32
Photoionization 239, 251, 255
Polyad 25, 27
Potential
  ionization 4, 5, 34
  resonance 5
Principal series 37, 39, 41

Quadrupole moment, electric 156, 157, 158, 182
Quadrupole radiation (transition) 225, 228, 283
Quantum numbers
  effective principal 36
  magnetic 4
  parabolic 179
  principal 4, 19
  seniority 98, 99

Racah V coefficients 60, 61
Racah W coefficients 66, 68
Racah, Rosenthal, and Breit formula 172
Recombination, radiative (photorecombination) 240, 245, 251, 255, 284, 301, 302
Reduced matrix elements 77
Relativistic corrections 165, 166, 168, 169, 325

Satellite structure 346
Selection rules 7, 26, 170, 180, 206, 211, 222, 225, 228
Self-consistent field 16
Seniority number 98, 99
Sharp series 37, 39, 41
Spherical harmonics 54
  theorem of addition 54
Stark effect 173
  linear 173–177
  quadratic 177–181
  quadrupole 181
  relative intensities in 177, 215, 216
Statistical weight 25, 26, 278
Supermultiplet 27
Wigner-Eckart theorem 77
Zeeman effect 189
  anomalous 192
  normal 192
  relative intensities in 215, 216

I. I. Sobelman, L. A. Vainshtein, E. A. Yukov

Excitation of Atoms and Broadening of Spectral Lines

1981. X, 315 pp. 34 figs. 40 tabs. (Springer Series in Chemical Physics, Vol. 7) Hardcover ISBN 3-540-09890-9

Contents: Elementary Processes Giving Rise to Spectra Excitation. – Theory of Atomic Collisions. – Approximate Methods for Calculation Cross-Sections. – Collisions between Heavy Particles. – Some Problems of Excitation Kinetics. – Tables and Formulas for the Estimation of Effective Cross-Sections. – Broadening of Spectral Lines. – References. – Tables. – Captions.

H. Haken, H. C. Wolf

Atomic and Quantum Physics

An Introduction to the Fundamentals of Experiment and Theory

Translated from the German by W. D. Brewer

2nd enl. ed. 1987. XVI, 454 pp. 265 figs. Softcover ISBN 3-540-17702-7

Contents: Introduction. – The Mass and Size of the Atom. – Isotopes. – The Nucleus of the Atom. – The Photon. – The Electron. – Some Basic Properties of Matter Waves. – Bohr's Model of the Hydrogen Atom. – The Mathematical Framework of Quantum Theory. – Quantum Mechanics of the Hydrogen Atom. – Lifting of the Orbital Degeneracy in the Spectra of Alkali Atoms. – Orbital and Spin Magnetism. Fine Structure. – Atoms in a Magnetic Field: Experiments and their Semiclassical Description. – Atoms in a Magnetic Field: Quantum Mechanical Treatment. – Atoms in an Electric Field. – General Laws of Optical Transitions. – Many-Electron Atoms. – X-Ray Spectra, Internal Shells. – Structure of the Periodic System. Ground States of the Elements. – Nuclear Spin, Hyperfine Structure. – The Laser. – Modern Methods of Optical Spectroscopy. – Fundamentals of the Quantum Theory of Chemical Bonding. – Appendix. – Solutions to the Problems. – Bibliography. – Subject Index. – Fundamental Constants of Atomic Physics. – Energy Conversion Table.

German edition:

H. Haken, H. C. Wolf,
Atom- und Quantenphysik.
4., erw. Aufl. 1990. ISBN 3-540-52198-4

W. Demtröder

Laserspektroskopie

Grundlagen und Techniken

2., überarb. u. erw. Aufl. 1991. XIII, 632 S. 489 Abb. Geb. ISBN 3-540-52601-3

Die 1. Auflage erschien in der Reihe Hochschultext

Die Spektroskopie gewinnt immer größere Bedeutung bei der Untersuchung von Atomen und Molekülen.

W. Demtröder stellt jetzt die Neuauflage seines Lehrbuchs vor, das die Brücke schlägt zwischen den „klassischen" Werken über Optik und Spektroskopie und den „modernen" Beiträgen zur Laserspektroskopie. Er erläutert die verschiedenen Techniken und die instrumentelle Ausrüstung der Laserspektroskopie und illustriert sie anhand konkreter Beispiele. Ein ausführliches Literaturverzeichnis weist den Weg zur Originalliteratur.

Die zweite Auflage wurde um neue Methoden der Laserspektroskopie und der für sie relevanten Meßtechniken erweitert. Das Buch richtet sich an Studenten und Wissenschaftler, die sich in das Gebiet einarbeiten wollen.

English edition in preparation

A. Unsöld, B. Baschek

The New Cosmos

Translated from the German by W. D. Brewer

4th, completely rev. ed. 1991. XVI, 438 pp. 242 figs.
Hardcover ISBN 3-540-52593-9

Astronomy, astrophysics and space research have developed extensively and rapidly in the last few decades. The new opportunities for observation afforded by space travel, the development of high-sensitivity light detectors and the use of powerful computers have revealed new aspects of the fascinating world of galaxies and quasars, stars and planets. The fourth, completely revised edition of **The New Cosmos** bears witness to this explosive development. It provides a comprehensive but concise introduction to all of astronomy and astrophysics.

It stresses observations and theoretical principles equally, requiring of the reader only basic mathematical and scientific background knowledge. Like its predecessors, this edition of **The New Cosmos** will be welcomed by students and researchers in the fields of astronomy, physics and earth sciences, as well as by serious amateur astronomers.

German edition:

A. Unsöld, B. Baschek, **Der neue Kosmos.**
5., überarb. u. erw. Aufl. 1991.
ISBN 3-540-53757-0

Springer-Verlag
Berlin
Heidelberg
New York
London
Paris
Tokyo
Hong Kong
Barcelona
Budapest